AF531442

CHEMICAL KINETICS

By
Dr. Praveen Tyagi
Lecture
D.P.B.S., P.G. College
Anoopshahr, Bulandshahr
(U.P.)

DISCOVERY PUBLISHING HOUSE PVT. LTD.
NEW DELHI-110 002

First Published – 2004

Reprinted – 2024

ISBN: 978-81-8356-159-4

Chemical Kinetics

Published by:

DISCOVERY PUBLISHING HOUSE PVT. LTD.

4383/4B, Ansari Road, Darya Ganj
New Delhi-110 002 (India)
Phone: +91-11-23279245; 23253475; 43596065
Mobile: +91 9811179893 / +91 9871656464
E-mail: discoverypublishinghouse@gmail.com
orderdphbooks@gmail.com
web: www.discoverypublishinggroup.com

Printed at:
Infinity Imaging Systems
Delhi

Preface

This book has been written for the students of under-graduate and post-graduate level of the various universities in India. A special feature of the book is that the text has been illustrated with a large number of line diagrams and the data presented in the form of numerous tables for reference and comparison. In the preparation of text standard works and review by renowned author have been freely consulted and the reference given chapter wise. At the end of the book will be found useful by those who wish to make a more detailed study of the topics discussed.

We are extremely grateful to our respected Managing Director Shri Tilak Wasan, Discovery Publishing House, for valuable and active cooperation. We humble request our students and chemistry teachers to send us their constructive criticism and suggestions which we shall be using in the publication of the next edition.

Author

Contents

1

CHEMICAL KINETICS

INTRODUCTION

Chemical kinetics is that branch of chemistry which deals with the study of the rate of chemical reaction and the mechanism by which they occur. In other words, it deals with how fast and through what mechanism a particular chemical reaction occurs.

The chemical reaction can be classified in to the following categories on the basis of their speeds.

(a) Instantaneous or fast reaction which proceed a very fast speed and it is particularly impossible to measure the speed of such reactions. Typical examples of fast reactions include :

  (i) several ionic reactions, *i.e.,* neutralisation of acids and bases,

  (ii) organic substituted reaction,

  (iii) reactions of biological importance, and

  (iv) explosive reactions of oxygen with hydrogen and hydrocarbons. The rates of such reactions can be measured by using special methods.

(b) Extremely slow reactions-which proceed at a very slow speed and the speed is so slow that it is again not possible to measure the speed of such reactions.

(c) Reactions which proceed at a measurable speed.. Reactions involving organic substance belongs to this category, *e.g.*, inversion of cane sugar, saponification of ethyl acetate etc.

Reactions which proceed at a measurable speed. Reactions involving organic substance belong to this category.

Reactions belongings to the above third category, viz, (c) are utilised in the study of *chemical kinetic*.

REACTION VELOCITY

Reaction velocity is different at different intervals of time and hence it can not be determined by dividing the quantity of the substance transformed by time. It is, therefore, defined as,

"the rate of which the concentration of a reactant changes with time.

It is represented by dc/dt, where dc is the concentration of the reactant left behind after a short intervals of time dt. The negative sign implies that the concentration that the reactions decreases with time or the reaction velocity decreases with time.

The velocity of the reaction is also represented by dx/dt, where dx is the amount of the reactants change during a small interval of time dt.

RELATION BETWEEN REACTION VELOCITY AND TIME

Guldberg and Waage (1867) emphasised the dynamic nature of chemical equilibrium. According to them, "the rate of chemical reaction is directly proportional to the product of the active masses or molecular concentrations of the reactants. "As the chemical reaction proceeds, the reactants are consumed where by their molecular concentration decrease.

Therefore, the rate of the chemical reaction will also decrease with them. If a graph is plotted between reaction velocity and time, it is seen that the reaction velocity falls rapidly in the initial stage of the reaction and after some time the curve becomes asymptotic to the time axis (Fig 1.1). It can therefore, be easily concluded that the reaction velocity is maximum to start with, which then falls gradually with time. In the later stage the reaction velocity becomes so low that it takes a very long time for it to become zero.

i.e., for the reaction to be completed. If can be safely assumed that in the later stages, the reaction is nearly completed. Consider the reaction,

$$aA + bB \rightleftharpoons cC + dD.$$

In it, a moles of A react with b moles of B to form C moles of C and d moles of D. The rate of such reaction can be expressed either in terms of decreases in concentration of a reactant per mole or increase in concentration of a product per mole.

Thus, we can write as follows.

$$r = \frac{-dC_A}{dt} \times \frac{1}{a} = K\ k\ C_A^a\ C_B^b \quad ...(1)$$

$$= \frac{-dC_B}{dt} \times \frac{1}{b} = k\ C_A^a\ C_B^b \quad ..(2)$$

$$r = \frac{+dC_C}{dt} \times \frac{1}{c} = k\ C_A^a\ C_B^b \quad ...(3)$$

$$r = \frac{+dC_D}{dt} \times \frac{1}{d} = k\ C_A^a\ C_B^b \quad ...(4)$$

From (1), (2), (3) and (4), are get,

$$r = \frac{-dC_A}{dt} \times \frac{1}{a} \times \frac{-dC_B}{dt} \times \frac{1}{b}$$

$$= \frac{+dC_C}{dt} \times \frac{1}{C} = \frac{+dC_D}{dt} \times \frac{1}{d}.$$

The unit of reaction rate is expressed in mole per unit time, *i.e.*, mol^{-1} $time^{-1}$.

MACROSCOPIC AND MICROSCOPIC KINETICS

Most rate measurements are made on systems that are in thermal equilibrium, which implies that energy is distributed among the molecules according to the Boltzmann distribution. Experiments of this kind can be referred to as "bulk" or "bulb" experiments, the latter expression relating to the glass bulbs commonly used for kinetic studies on gases. The term *macroscopic kinetics* describes the branch of kinetics because the results relate to the behaviour of a very large group of molecules in thermal equilibrium. Much of this book is concerned with macroscopic kinetics, because bulk experiments are by far the most common.

The macroscopic kinetic investigations that have been carried out have led to considerable insight into what is taking place at the molecular level during the course of reaction. Information about this, however must be arrived at by inference, hypotheses are proposed and then tested with respect to the behaviour of bulk systems. There is a limit to what can be learned in this way. During recent years, further insight into chemical reactions has been provided by kinetic studies in which the reacting molecules are in well-defined state. Such studies are termed as *microscopic.* A very important technique in this field is that of *crossed molecular beams.* It a bimolecular reaction $A + B \rightarrow Y + Z$ is to be studied, each reactant A and B gets formed into a beam, characterized by well-defined velocities and internal energy states. The beams are made to cross one another so that some reaction takes place. The directions of motion of the reaction products and of the unreached A and B molecules are then determined, together with their velocities and internal energy states. Investigations of this kind give valuable information about the dynamics of both reactive and unreactive collisions, because the results of well-defined collisions are being observed.

SLOW AND FAST REACTIONS

(a) **Slow Reactions :** These are the reactions which proceed slowly and their rates can be measured by conventional methods. An example is the inversion of cane-sugar in aqueous solution.

$$\underset{\text{Cane-sugar}}{C_{12}H_{22}O_{11}} + H_2O \longrightarrow \underset{\text{Glucose}}{C_6H_{12}O_6} + \underset{\text{Fructose}}{C_6H_{12}O_6}$$

In slow reactions a large number of bonds have to be broken in reactant molecules and a large number of new bonds have to be formed in the product molecules.

(b) **Fast Reactions :** The rates of many reactions are too fast to be measured by the conventional methods. These methods cannot deal with reactions whose half-lives are less than a second or so. Such reactions are classified as *fast*. Typical examples of fast reactions include :

(i) many ionic reactions such as the neutralisation of acids by bases,

(ii) reactions of biological significance,

(iii) organic substitution reactions, and

(iv) the explosive reactions of oxygen with hydrogen and hydrocarbons.

The rates of fast reactions can be measured by employing special methods. It may be noted that in case of simple ionic reactions, no bonds are to be broken. The reactions, therefore, are very fast.

RATE OF REACTION

Definition : A simple definition of the term *rate of reaction* is as follows.

"The rate of reaction is defined as the amount of product (in gm mols) formed in a given time under stated conditions."

There are two objections to the above definition.

(i) The concentration of reactants changes as the reaction proceed, and hence constant conditions cannot be maintained. This can be seen from a typical curve in which the percentage composition is plotted against time as shown in Fig. (1.1). From the curve it is evident that the amount of reactants in a given time interval decreases as the reaction proceeds. The amounts of reactant for two equal time intervals are shown on the graph. The values of AB, which corresponds to the earlier time, is much greater than CD. It means that the rate of reaction clearly decreases with increasing time.

From the graph is also evident that there is no definite instant of time at which the reaction is completed, as the curve approaches 100 per cent asymptotically at the point E.

(ii) The amount of product formed depends on the initial quantity of reactants as well as on their concentrations and chemical reactivities.

The above mentioned difficulties (i) and (ii) are overcome by defining the rate of reaction as *the rate of change of concentration of either reactant of product per unit time.*

Expressing Reactions Rates : Let us consider a simple hypothetical reaction of the type.

$$A \rightarrow L.$$

We know that the rate of reaction at any given instant depends upon the concentration of A, at the time. As the reaction proceeds in the forward direction, the concentration of A goes on decreasing with time. The rate of reaction r (= dx/dt) at any given instant, will be represented by the following expression.

$$r = dx/dt = -\, dC_A/dt = k\ C_A \qquad ...(1)$$

where dX_A denotes the infinitesimally small decreases in the concentration of A in an infinite similarly small interval of time dt, C_A denotes the concentration of reactant A at the given instant and k is known as the rate constant of a reaction.

The rate of the reaction can also be expressed in terms of a increase in concentration of the product L. But the concentration of the product goes on increasing with time. Therefore, the rate of reaction may be represented as follows.

$$r = dx/dt = + dC_L/dt = kC_A \qquad ...(2)$$

where dC_L denotes an infinitesimally small increase in the concentration of the product L during an infinitesimally small interval of time. From equations (1) and (2), we get

$$r = dx/dt = - dC_A/dt = + dC_L/dt \qquad ...(3)$$

From equation (3), it follows that the rate of a reaction can be expressed either in terms of decrease in the concentration of a reactant or increase in the concentration of a product during an infinitesimally small interval of time dt.

In equation (1), the minus sign indicates that the concentration of reactant is decreasing with time. In equation (2), the positive sign indicates that the concentration of product is increasing with time..

Suppose we consider the following reaction involving more than one molecular species.

$$A + B \rightarrow L + M.$$

As the rate of the reaction may be expressed either in terms of decrease in concentration of any of the reactants (A or B) or in terms of increase in concentration of any of the products (L or M) one can write as follows.

$$r = - dC_A/dt = kC_AC_B$$

or $$r = - dC_B/dt = kC_AC_B$$

or $$r = + dC_L/dt = kC_AC_B$$

or $$r = + dC_M/dt = kC_AC_B.$$

As all the four expressions in the above equations measure the same quantity, *i.e.*, the rate of the reaction at any given instant, one can write as follows.

$$r = - dC_A/dt = - dC_B/dt$$
$$= + dC_L/dt = + dC_M/dt.$$

Let us consider a reaction of the following type.

$$aA + bB \rightarrow lL + mM.$$

In the above reaction, a moles of A react with b moles of B to yield l moles of L and m moles of M. But the rate of such a reactant per mole or increase in the concentration of a product per mole. Therefore, we can write as follows.

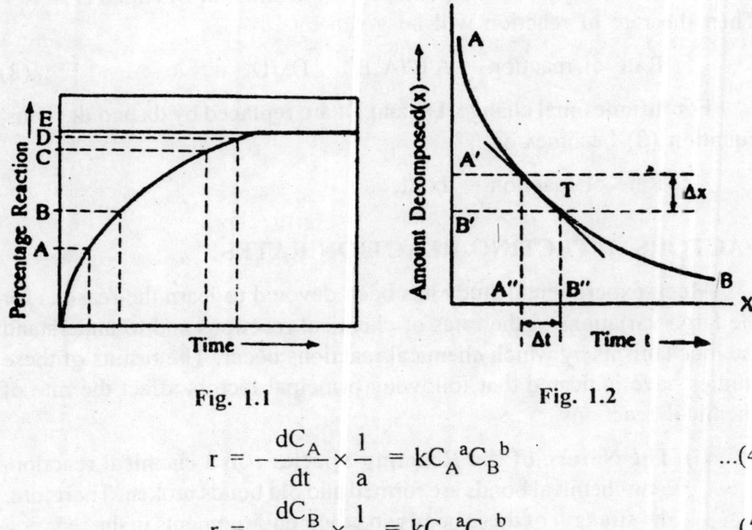

Fig. 1.1

Fig. 1.2

$$r = -\frac{dC_A}{dt} \times \frac{1}{a} = kC_A{}^aC_B{}^b \quad ...(4)$$

$$r = -\frac{dC_B}{dt} \times \frac{1}{b} = kC_A{}^aC_B{}^b \quad ...(5)$$

$$r = +\frac{dC_L}{dt} \times \frac{1}{l} = kC_A{}^aC_B{}^b \quad ...(6)$$

$$r = +\frac{dC_M}{dt} \times \frac{1}{m} = kC_A{}^aC_B{}^b \quad ...(7)$$

From Eqs. (4), (5), (6) and (7), we get

$$r = -\frac{dC_A}{dt} \times \frac{1}{a} = -\frac{dC_B}{dt} \times \frac{1}{b}$$

$$= +\frac{dC_L}{dt} \times \frac{1}{l} = +\frac{dC_M}{dt} \times \frac{1}{m}.$$

Unit : The unit of rate of reaction is expressed in gm moles per litre per unit time (written as mole^{-1} litre^{-1} time^{-1}). The time is an minutes or seconds.

Determining Reaction Rates : In order to determine the reaction rate, a curve AB of they type shown in Fig. (1.2) is considered in which the amount decomposed (x) is plotted against time t. A tangent is drawn at the point T on the curve AB. Then, two points are selected on the tangent at which the rate remains the same. After this, the lines, paralleled to X-axis and Y-axis the drawn corresponding to the two points. Then, the value of Dx is A'B' corresponding to time interval Dt which is A"B" Then the rate of reaction will be given by

Rate of reaction = A'B'/A"B" = Dx/Dt ...(8)

For infinitesimal change, Dx and Dt are replaced by dx and dt. Thus, equation (8) becomes as

Rate of reaction = dx/dt.

FACTORS AFFECTING REACTION RATES

Much experimental study has been devoted to learn the reasons for the large variations in the rates of chemical reactions and to understand the mechanisms by which chemical reactions occur. The results of these studies have indicated that following principal factors affect the rate of chemical reactions.

(i) **The Nature of the Reacting Species :** In a chemical reaction, new chemical bonds are formed and old bonds broken. Therefore, the strength of these bonds and their environments in the reacting molecules will affect the rate of the reaction. Some reactions, as for example, the reaction of acidic with basic solutions, proceed so rapidly as to be instantaneous. Others such as, for example, the reaction of N_2 with H_2 molecules to form NH_3 molecules at room temperature in the absence of a catalyst takes place so slowly as to be incomplete even after millions of years.

$$NaOH + HCl \rightarrow NaCl + H_2O$$

$$N_2 + 3H_2 \rightleftharpoons 2NH_3$$

(ii) **Concentration of Reactants :** Since according to law of mass action, the rate of reaction is proportional to molar concentration of reactants, the rate changes with time because the concentration of various reactants decreases as time passes. This has been experimentally shown to be so. That the rate of reaction depends on concentration of reactants is illustrated by rusting of iron which is enhanced during rainy season, when water vapour concentration in air is higher as compared to that of dry air in actually increased concentration increases collisions and hence increases reaction rates. This is illustrated in Fig. 1.3.

(iii) Effect of Temperature : Temperature has a striking effect on the rate of chemical reactions. Reaction rate negligibly slow at ordinary temperatures may becomes appreciable and even explosive at elevated temperatures. As a very rough but useful rule, the rate constants is doubled for a rise in temperature of 10°C. Thus, a temperature change of 100°C may alter the rate more or less by a factor of $2^{100/10}$ or 2^{10} or 10^{3}. An example is the effect of temperature on the rate of decomposition of hydrogen iodide, the rate constant increases by a factor of 1.7 for each 1°C rise in temperature.

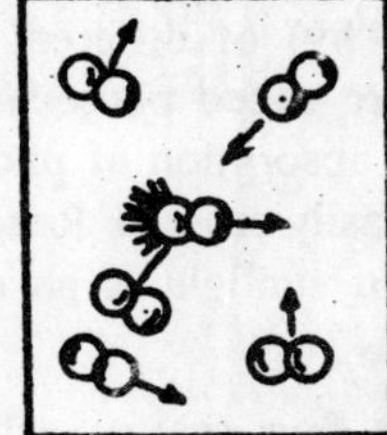

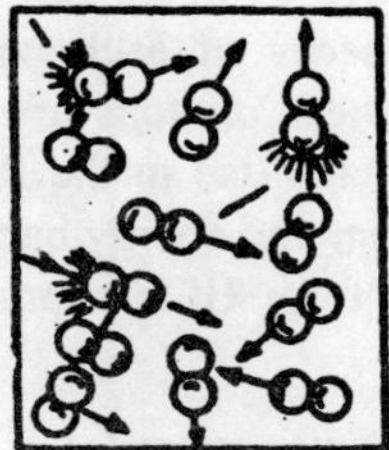

Fig. 1.3

When the logarithm of reaction velocity is plotted against the reciprocal of absolute temperature, the various points obtained have been found to fall on a straight line for most of the chemical reactions. This variation, therefore, can be represented by the following equation

$$\log k = a - \frac{b}{T}.$$

Here k is the specific reaction rate, T the absolute temperature and a and b positive imperial constants, the values of which are obtained from the intercept and slope respectively of the linear plot.

(iv) Effect of the Presence of a Catalyst : The rates of certain reactions may be increased by the presence of certain substances in the reacting system. These substances have been termed catalyst. For example, maganese dioxide speeds up the rate of decomposition of hydrogen peroxide to water and oxygen.

(v) **Particle Size in Heterogeneous Reactions or Exposed Area :** Since heterogeneous reaction occur only at the surface boundary between the reacting phases, the rate of such reaction is proportional to the surface area. When a given mass is subdivided into smaller particles, the surface area is increased, hence as the particle size is decreased, the rate of reaction increases. Lumps of coal or zinc, for example, are difficult to ignite in air, but when powdered and dispersed in air, they react explosively.

(vi) **Effect of Radiations :** Not only heat energy but radiant energy may also alter the rate of certain chemical reactions. The term radiant energy means the energy derived from X-rays, ultraviolet rays, visible and infrared rays, radio waves etc.

The rates of some reactions are enhanced due to absorption of photons (energy of each photon = hv) of different energies. These reactions where photons are used are called photochemical reactions. This is perhaps due to the fact that absorption of photon by reactants energies them and energy barrier is easily crossed. Reaction between H_2 and Cl_2 to form HCl in presence of sunlight is an example of such reactions.

Conclusive remark : The first four factors affect the rates of homogeneous reactions while the fifth factor along the first four factors affect the rates of heterogeneous reactions. The last factor (vi) affects the radiation reaction, like photochemical reactions, only.

VELOCITY CONSTANT OR RATE CONSTANT

The mathematical relation between the rate of the reaction and the concentration of the reaction components is known as the rate equation or *rate law expression for reaction.* According to the law of mass action, the rate of a chemical reaction varies directly as the active concentrations of the reactants. Consider a homogeneous reaction of the type

$$A + B \rightarrow \text{Products.}$$

At a particular temperature, the rate of this reaction is given by

$$R = dx/dt = kC_A\ C_B \qquad ...(1)$$

where k is a constant which is characteristic of the reaction and is known as the rate constant or velocity constant of the reaction at the given temperature, CA and BB are the concentrations of reactants A and B respectively and x denotes the amount of product at time t. If the concentration of each of the reactants A and B is unity. *i.e.*, CA = CB = 1, equation (1) becomes

$$R = k \qquad ...(2)$$

From equation (2) it follows that at a given temperature, *the rate constant or velocity constant of a reaction may be defined as equal to the rate of reaction when the concentration of each of the reactant is unity.*

Units of k : Let us consider the general form of the reaction of the type.

$$aA + bB + ... \rightarrow \text{Products}$$

Rate of this reaction is given by ...(3)

$$R = dx/dt = k\ CAa\ CBb... \qquad ...(4)$$

Dimensions of R = moles/litres/sec. ...(5)

CA = CB = ... = moles/'litres and

Substituting Eqs. (4) and (5) in (3), we get

$$\frac{\text{moles}}{\text{litre}} \times \sec^{-1} = k\left(\frac{\text{moles}}{\text{litre}}\right)^a \times \left(\frac{\text{moles}}{\text{litre}}\right)^b \times ...$$

or

$$k = \left[\left(\frac{\text{moles}}{\text{litre}}\right)^{((a+b+...)-1)}\right] \times [\sec^{-1}]$$

MOLECULARITY OF A REACTION

It is the sum of the umber of molecules of various reactants that take part in a chemical reaction as represented by a balanced chemical equation. It is sometimes convenient to classify reactions according to their molecularity.

(a) Unimolecular Reactions : *A unimolecular reaction is one in which only one molecule of reactant is involved.* Examples re :

(b) Bimolecular Reactions : *A bimolecular reaction is one in which two molecules are involved.* Examples are :

(i) *Inversion of cane sugar :*

$$C_{12}H_{22}O_{11} + H_2O \rightarrow \underset{\text{Glucose}}{C_6H_{12}O_6} + \underset{\text{Fructose}}{C_6H_{12}O_6}.$$

As two molecules ($C_{12}H_{12}O_{11}$ and H_2O) are taking part in this reaction, the reaction is bimolecular. Thus, the molecularity of this reaction is two.

(iii) Hydrolysis of an ester : It is also a bimolecular reaction.

$$CH_3COOC_2H_5 + H_2O \rightarrow CH_3COOH + C_2H_5OH.$$

It is also a bimolecular reaction.

(c) Termolecular Reaction : *A termolecular reaction is one in which three molecules are involved.* Examples are :

(i) $2NO + O_2 \rightarrow 2NO_2$

(ii) $2FeCl_3 + SnCl_2 \rightarrow 2F_2Cl_3 + SnCl_4$

(iii) $2CH_3COOAg + HCOONa$

$$\rightarrow 2Ag + CO_2 + CH_3COOH + CH_3COONa$$

ORDER OF A REACTION

It is not necessary that the concentrations of all the reactants taking part in a reaction can determine the rate of reaction. Therefore, a new term is being introduced, *i.e., the order of a reaction may be define a as follows.*

"The order of a reaction is given by the number of atoms or molecules whose concentrations alter during the chemical change."

or

" The order or reaction is given by the total number of molecules or atoms whose concentrations determine the velocity of the reaction."

It is convenient to classify reactions according to the order of a reaction.

(a) First Order Reaction : *It is defined as one in which the reaction rate is determined by the variation of one concentration term only.* Thus, a reaction is said to be of the first order if its rate (R) is expressed by the expression of the type.

$$R = kC_A$$

(b) Second Order Reaction : *It is defined as one in which the two concentration terms determine the rate of the reaction.* Thus, a reaction is said to be of the second order if the rate is given by the expression of the type

$$R = kC_A^2 \text{ [For one type of reactants]}$$

or $$R = kC_A\, C_B \text{ [For two types of reactants]}$$

(c) Third order Reaction : *A reaction is said to be of third order if three concentration terms determine its rate.* Thus, a reaction is of third order if its rate R is given by the expression of the type

$$R = kC_A^3 \text{ [For one type of reactants]}$$

or $$R = kC_A^2\, C_B \text{ [For two type of reactants]}$$

$$R = kC_A\, C_B^2 \text{ [For two type of reactants]}$$

or $$R = kC_A\, C_B \text{ [For three type of reactants].}$$

In all the above expressions, A,. B, and C stand for the various reactants.

(d) General Order Reactions : Consider a reactions of the type

$$aA + bB \rightarrow \text{Products}$$

The rate of such a reaction $\propto C_A^a \times C_B^b \times ....$

or the rate of such a reaction $= k\, C_A^a \times C_B^b \times ...$

Then, the order for such a reaction a + b +

Therefore, the order of the reaction may be defined as.

" *The sum of powers to which the concentration (or pressure) terms are raised in order to determine the rate of the reaction.*"

The order of a reaction is an *experimental quantity.* It is not necessary that the order of reaction may not always be a whole number. It can be fractional quantity also.

ORDER AND MOLECULARITY OF SIMPLE REACTIONS

Previously it was through that molecularity and order of reaction were identical. It has now been observed that certain examples are known in which these two are not identical. For example.

(i) *Inversion of cane sugar :* This reaction is

$$C_{12}H_{22}O_{11} + H_2O \rightarrow C_6H_{12}O_6 + C_6H_{12}O_6.$$

As this reaction involves two reactants ($C_{12}\ H_{22}\ O_{11}$ and H_2O), the molecularity of the reaction is two. When its rate is determined experimentally, it has been found to observe the following rate law

$$-\frac{d[C_{12}H_{22}O_{11}]}{dt} \propto C_{C12}\ H_{22}\ O_{11}.$$

As the rate of the reaction is determined by the variation of one concentration term only $[C_{12}H_{22}O_{11}]$, the order of the reaction is one. Thus, the molecularity of this reaction is two but order of the reaction is one.

This is due to the fact that water being present in such a large excess that its concentration does not *measurably change during the chemical reaction.*

(ii) Saponification of ethyl acetate : This reaction is

$$CH_3COOC_2H_5 + NaOH \rightleftharpoons CH_3COONa + C_2H_5OH.$$

The molecules ($CH_3COOC_2H_5$ and NaOH) are involved in this reaction and hence, the molecularity of the reaction is two. This rate of this reaction has been found to obey the following rate.

$$-\frac{d[CH_3COOC_2H_5]}{dt} \propto [CH_3COOC_2H_5][NaOH].$$

Hence the order to this reaction is two.

From the above examples it is evident that the order of reaction is the number of chemical species (molecules, atoms or ions) whose concentrations affect the rates of the reaction while the molecularity of are action is the number of chemical species taking part in a chemical reaction.

Molecularity is a theoretical concept which is applied once the mechanism of a reaction is known whereas the order is empirical and must be determined experimentally.

The order of a reaction has been never found to exceed three because the probability of more than three molecules meeting simultaneously with required activation energy to react is extremely small.

The various differences between *molecularity* and *order* of a reaction are given in the following Table 1.1:

Table 1.1

Order of a reaction	Molecularity of a reaction
1. It is an experimentally determined quantity which is obtained from the rate for the overall reaction.	1. It is a theoretical concept which depends on the rate determining step in the reaction mechanism.
2. It possesses all sorts of values including zero.	2. It is generally not exceeding three and never zero.
3. It may be whole number of fractional value.	3. It os always a whole number.
4. It cannot be obtained from a balanced chemical equation.	4. It is obtained from a single balanced chemical equation.
5. It does not reveal anything about the mechanism of the reaction.	5. It reveals some basic facts about reaction mechanism
6. It is equal to the sum of the exponents of the molar concentrations of the reactants in the rate equation	6. It is equal to the number of molecules of the reactants which are taking part in a single step chemical reaction.

ORDER AND MOLECULARITY OF COMPLEX REACTIONS

Many reactions are known which occur in two or more steps. These reactions are after termed as *complex-rations.* In such reactions each step will have its own *molecularity* which depends upon the number of molecules of this reactant of reactants taking part in that reaction. All these steps may proceed at the same or different rates. But the slowest step determines the *order of overall reaction.* In other words, the slowest step is the rate determining step of the reaction as a whole. In order to understand the meaning of order and molecularity of complex reactions, we will consider the following examples :

(a) The reduction of bromic acid to hydrobromic acid by hydriodic acid may be represented as

$$HBrO_3 + HI \rightarrow HBr + 3H_2O + 3I_2 \quad ...(1)$$

Although, this reaction involves seven molecules yet it is of the second order. This order of the reaction is justified by the following mechanism.

(i) $HBrO_3 + HI \rightarrow HBrO_2 + HIO$ [slow]

(ii) $HBrO_2 + 4HI \rightarrow HBr + 2H_2O + 2I_2$ [fast]

(iii) $HIO + HI \rightarrow H_2O + I_2$ [fast]

As the first step is slow, it means that the reaction is of the second order. The molecularity of complex reaction (1) as such has no significance.

In elementary step involved in the complex reaction each has its own molecularity. For example, the modularities os steps (i), (ii) and (iii) are 2, 5 and 2 respectively.

(b) The oxidation of hydriodic acid by hydrogen peroxide in aqueous solution can be represented as

$$H_2O_2 + 2H^+ + 2I^- \rightarrow 2H_2O + I_2.$$

Experimentally, it is of the second order. In order to account this order of the reaction, the following mechanism has been suggested.

(i) $H_2O_2 + I^- \rightarrow H_2O + IO^-$ [slow]

(ii) $H^+ + IO^- \rightarrow HIO$ [fast]

(iii) $HIO + H^+ + I^- \rightarrow H_2O + I_2$ [fast]

The first step is slow and determines the over-all order of the reaction. The molecularities of steps (i), (ii) and (iii) are 2, 2 and 3 respectively. From the above examples we conclude that.

(a) The order of a complex reaction is given by the order of the slowest step in the sequence of various steps which are involved in that reaction.

(b) The concept of molecularity of complex reaction has no significance. Each step involved in the complex reaction has its own molecularity.

KINETICS OF ZERO ORDER REACTIONS

A reaction is said to be of zero order, when its rate is independent of the concentration of the reactants. Alternatively, if the concentration of the reactants remains unaltered during the course of a reaction, it is said to be of zero order. As the velocity of a zero order reaction is constant, it can be written as

$$\text{rate} = \text{constant}$$

or,in mathematical symbols,

$$dx/dt = k_0 \quad \text{...(1)}$$

where x is the concentration of product formed and k_0 is the rate constant for zero order reaction. On integrating equation (1) with respect to t, we get

$$\int dx = k_0 \int dt$$

or $$x = k_0 t + \text{constant} \quad \text{...(2)}$$

Since x = 0, when t = 0 (*i.e.*, at the beginning of the reaction, no product has formed), the constant must be zero. Hence,

$$0 = k_0 \times 0 + \text{constant} \quad \text{...(3)}$$

or constant = 0.

On substituting equation (3) in (2), we get

$$x = k_0 t \quad \text{...(4)}$$

The dimensions of k_0 are concentration/time, *i.e.*, mole/litre/time. In order to calculate the half-life of this reaction the condition that t = t_1/t_2 when x = a/2 (a being the initial concentration of the reactant) is substituted into eq. (4) to give

$$a/2 = k_0 t_1/2 \text{ or } t_1/2 \;\; t_1/2 \propto a\,.$$

Thus in zero order reaction the half-life is proportional to the initial concentration of the reactant.

Examples of zero order reaction are.

(i) The reaction between acetone and bromine is said to be of zero order with respect to the reactant bromine as this reaction proceeds at the same rate irrespective of the concentration of the bromine.

$$CH_3COOCH_5 + Br_2 \rightarrow CH_3COCH_2Br + HBr.$$

(ii) The phosphine decomposition on the surface of molybdenum or tungsten at high pressure is of zero order.

$$2PH_3 \longrightarrow 2P + 3H_2.$$

The reaction becomes of the first order if the pressure is lowered.

(ii) The reaction between acetone and iodine is found to be zero order in respect of io‹'·ne.

$$CH_3COCH_3 + I_2 \xrightarrow{H^+} CH_3COCH_2I + HI.$$

The observed rate law for this reaction is

$$\frac{-d[I_3]}{dt} = k\,[H_3O+]\,[CH_3COCH_3],$$

which does not include $[I_2]$ factor and hence the reaction is of zero order in respect of iodine concentration.

The plot of rate, r, versus concentration for a zero order reaction is shown in Fig. 1.4.

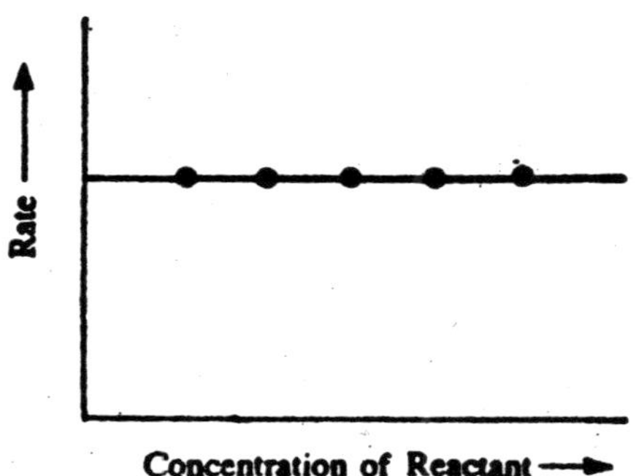

Fig. 1.4 : The plot rate, r, versus concentration for a zero-order reaction.

KINETICS OF FIRST ORDER REACTIONS

The first order reaction is that in which the rate depends upon the concentration of a single reactant only. Consider a general first order reaction,

$$A \rightarrow \text{Products}. \qquad ...(1)$$

Suppose a is the initial concentration of the reacting substance A in gm moles per litre. Let x gm moles/litre of A decompose in time t. The rate of the reaction at time t is, according to the law of mass action, proportional to the concentration of A at that instant, viz, (a – x). This can be expressed mathematically as

$$dx/dt \propto (a - x) \text{ or } dx/dt = k_1 (a - x) \quad ...(2)$$

where dx/dt is called the reaction velocity and k1 is usually referred to as the specific reaction rate or velocity coefficient. On rearranging Eq. (2) one gets the following equation.

$$\frac{dx}{a - x} = k_1 dt \quad ...(3)$$

On integrating, we get

$$-\ln (a - x) = k_1 t + I_1 \left[\because \int \frac{dx}{x} = \ln x + I_1 \right] \quad ...(4)$$

where I_1 is the constant of integration. When t = 0, x = 0 becomes as $- \ln a = I_1$. Substituting this in Eq. (4), we have

$$- \ln (a - x) = k_1 t - \ln a \text{ or } a - \ln (a - x) = k_1 t$$

$$\text{or} \quad \ln \frac{a}{a - x} = k_1 t \quad \left(\ln \frac{A}{B} = \ln A - \ln B \right)$$

$$\text{or} \quad k_1 = \frac{1}{t} \ln \frac{a}{a - x} \quad ...(6)$$

$$\text{or} \quad k_1 = \frac{2.303}{t} \log \frac{a}{a - x} \quad [\because \ln x = 2.303 \log_{10} x \quad ...(7)$$

Equation (7) is known as *kinetic equation for the first order reactions.*

Second From : In certain reactions the initial concentration is not known definitely because the time when the reaction starts is uncertain.. The value of a is eliminated by integrating Eq. (3) within limits as

$$\int_{x_1}^{x_2} \frac{dx}{a - x} = \int_{t_1}^{t_2} k_1 . dt \quad ...(8)$$

where x_1 and x_2 represented gram moles of A which have decomposed in time t_1 and t_2 respectively, Eq. (8). is simplified as

$$\ln \frac{a - x_1}{a - x_2} = k_1 (t_2 - t_1) \text{ or } k_1 = \frac{2.303}{t_2 - t_1} \log \frac{a - x_1}{a - x_2} \quad ...(9)$$

$$\text{or} \quad k_1 = \frac{2.303}{t_2 - t_1} \log \frac{c_1}{c_2} \quad ...(10)$$

where c_1 (= $a - x_1$) and c_2 (= $a - x_2$) are the molar concentrations of the reactant 'A' at time t_1 and t_2 respectively.

Characteristics of First Order Reactions : These are given below.

(i) All reactions of the first order obey equation (7) *i.e.,* the concentration of the reactants is determined at different intervals of time and values substituted in Eq. (7), the values of k obtained should always be same.

(ii) Suppose the new concentration unit be m times the original values. Then, we have

$$k_1 = \frac{2.303}{t} \log \frac{ma}{ma - mx} = \frac{2.303}{t} \log \frac{a}{a - x}$$

which is the same is the original first order kinetic Eq. (7) Thus, *the values of k1 is not altered by the change in concentration units.*

(iii) *The time taken for the completion of a definite fraction of the first order reaction is independent of the initial concentration.* Consider the case when half of the reaction species A decomposes in $t_1/2$, *i.e.,* $x/_2$. Eq. (7) becomes as

$$k_1 = \frac{2.303}{t_{1/2}} \log \frac{a}{a - a/2}$$

or $$t_1/2 = \frac{2.303}{k_1} \log 2 = \frac{0.693}{k_1} \qquad ...(11)$$

Eq. (11) is independent of the initial concentration 'a' of the reaction species A.

Units of First Order Rate Constant (k_1) : From Eq. (7), We have

$$k_1 = \frac{2.303}{t} \log \frac{a}{a - x}$$

$$= \text{time}^{-1} \log \frac{\text{moles/litre}}{\text{moles/litre}} = \text{time}^{-1}$$

Hence k_1 has the dimensions of reciprocal time or time^{-1}. If time is expressed in seconds then the unit of k_1 is sec^{-1}.

Physical Significance of k_1 : Suppose the value of k_1 for the decomposition of N_2O_5 is .00863 sec^{-1}. It follows that .00863 × 100 = 0.863% of N_2O_5 would dissociated per sec, provided the initial concentration is had constant in that one second interval. Thus, the physical significance of k_1 is that *it represents the fraction of the reactant decomposed per unit time of constant concentration.*

Evaluation of k_1 : The value of k_1 can be evaluated by using the rate equation (7), knowing the concentration x of the reactant at different time intervals t provided the value of initial concentration a is known. Equation (7) can be expressed as

$$t = \frac{2.303}{k_1} \log a - \frac{2.303}{k_1} \log (a - x).$$

If log (a – x) is plotted against time t, a straight line should be obtained whose slope is equal to – 2.303/k_1 from which k_1 can be evaluated (Fig. 2).

(iv) From Eq. (7) it follows that the concentration of a first order reaction decreases exponentially with time. This is shown in Fig. 1.5. The plot of log (a – x) versus t is shown in Fig. 1.6.

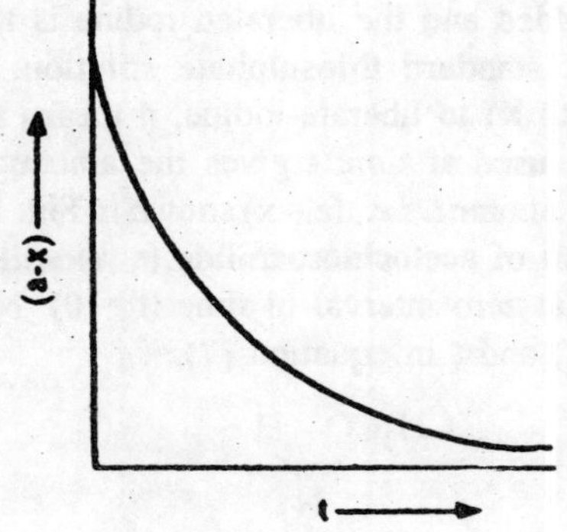

Fig. 1.5 : The plot of (a–x) versus time fora first-order reaction.

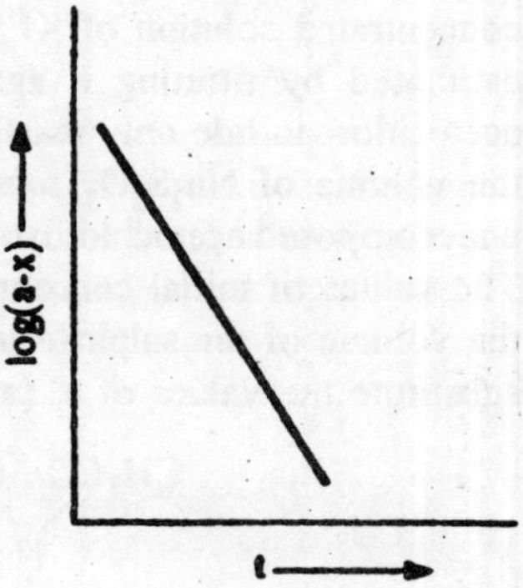

Fig. 1.6 : The plot of log (a–x) versus t for a first-order reaction.

Examples of First Order Reaction : There are many reactions which are unimolecular and also follows the first order equation (7). Examples are.

(a) Decomposition of H_2O_2 : This reaction is supposed to be taking place in two steps :

Step I $H_2O_2 \rightarrow H_2 + O$ *Slow*

Step II $O + O \rightarrow O_3$ *Fast.*

As the combination of oxygen atoms is very rapid, the observed rate of the reaction is due to the first step only. Therefore, the decomposition of H_2O_2 is an examples of the true first order reaction. The usual way to record the progress of the reaction is as follows.

(i) Withdraw equal volumes of H_2O at different time intervals and then titrate with standard $KMnO_4$ solution. The values of $KMnO_4$ used corresponds to the concentration of unchanged hydrogen peroxide (a – x) at that instant.

(ii) Measure the volume of oxygen evolved at different time intervals. The former method is preferred.

(b) *Conversion of N-accetochloroanilide into p-chloroacetanilide.* This reaction takes place in the presence of hydrochloric acid which acts as a catalyst.

The progress of the reaction can be measured by removing equal volumes of the reaction mixture at different intervals of time. A concentrated solution of KI is added and the liberated iodine is then estimated by titrating it against standard thiosulphate solution. As acetochloroanilide only reacts with KI to liberate iodine, it means that the volume of $Na_2S_2O_3$ solution used at time t gives the amount of undecomposed acetochloroanilide at time t. *i.e.*, (a – x) shown in Fig. 1.7. The values of initial concentration of acetochloroanilide (= a) will be the volume of thiosulphate used at zero interval of time (t = 0). Now substitute the values of a, (a – x) and t in equation (7).

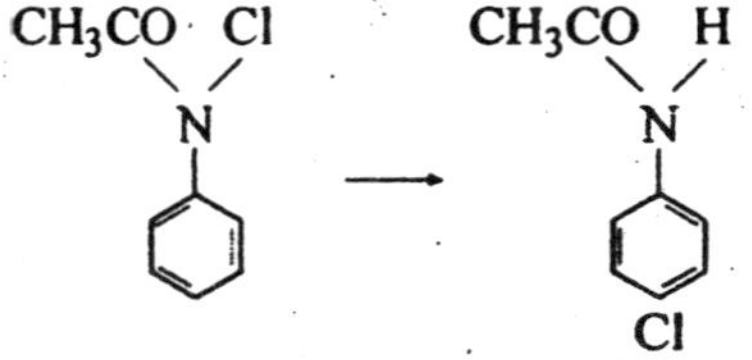

CH_3CO Cl, N — C_6H_5 $\longrightarrow$ CH_3CO H, N — C_6H_4 — Cl

Fig. 1.7

(c) *Decomposition of ammonium nitrite.* When an aqueous solution of ammonium nitrite is warmed, the following reaction occurs

$$NH_4\,NO_2 \rightarrow 2H_2O + N_2.$$

The progress of the reaction can be determined by collecting the N_2 gas in the a gas burette at different intervals of time. The volume of nitrogen at any time t will correspond to the amount of ammonium nitrite decomposed at that time (*i.e.*, x). The total volume of the nitrogen collected at the end of the reaction will correspond to total quantity of the ammonium nitrite initially (*i.e*.,. a).

If V_∞ is the final volume of the nitrogen gas collected when the reaction is compete and Vt the volume of the nitrogen collected at any time, t the value of k_1 is given by

$$k_1 = \frac{2.303}{k_1} \log_{10} \frac{V_\infty}{V_\infty - V_t}$$

(d) *Decomposition of* N2O5. The reaction is of first order as shown in the following two stages.

(i) $N_2O_5 \rightarrow 2NO_2 + O$ (slow rate)

(ii) $O + O \rightarrow O_2$ (fast).

Hence the rate is determined by the state (i). The progress of the reaction may be determined by measuring the volume of pressure of oxygen evolved at regular intervals of time. If V_t is the volume at any time t, V_∞ is the final volume when the reaction is complete. Thus V_∞ is the original concentration of N_2O_5, *i.e.*, a and (V_∞ – Vt) will be (a – x). Hence,

$$k = \frac{2.303}{t} \log \frac{a}{a-x} = \frac{2.303}{t} \log \frac{V_\infty}{V_\infty - V_t}.$$

Pseudo-Unimolecular Reactions : There are other reactions which are not unimolecular but follows the first order equation (7). It means that in such reactions ore than one molecule is involved. Examples are

(a) *Hydrolysis of methyl acetate.* The reaction is catalysed by the presence of mineral acids.

$$CH_3COOCH_3 + H_2O \overset{H^+}{\rightleftharpoons} CH_3COOH + CH_3OH.$$

Again, it is an example of pseudo-unimolecular reaction as it involves two molecules but the rate is determined by methyl acetate only and not by water because its active mass does not change appreciably due to its presence in large excess.

As acetic acid is produced during the reaction, the progress of the reaction can be measured by withdrawing equal volumes of reaction mixture and titrating with standard NaOH solution at different intervals of time. The final reading is usually taken after 24 hours or more. The amount of alkali used at time t will give the quantity of ester decomposed upto that time. The values of k_1 can be calculated from the expression,

$$k_1 = \frac{2.303}{t} \log_{10} \frac{V_\infty - V_0}{V_\infty - V_t}$$

where $V_\infty - V_0 = a$ and $V_\infty - V_t = a - x$.

(b) *Inversion of cane sugar.*

The reaction is

$$\underset{\text{Cane sugar}}{C_{12}H_{22}O_{11}} + H_2O \xrightarrow{H^+} \underset{\text{Glucose}}{C_6H_{12}O_6} + \underset{\text{Fructose}}{C_6H_{12}O_6}$$

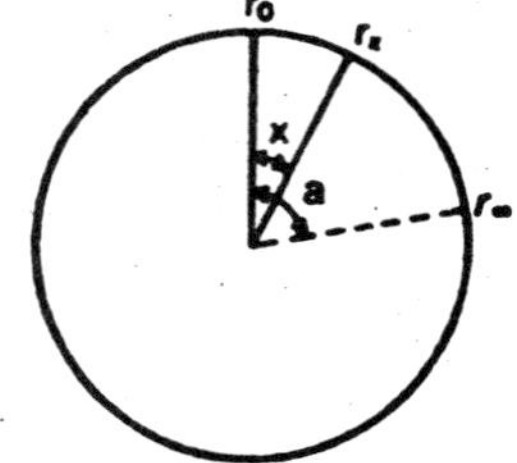

Fig. 1.8

As it involves more than one molecule but follows the first order equation, it shows that this is an example of pseudo-unimolecular reaction. As the active mass of water is not changing appreciably] in dilute solution, the rate of reaction is determined by the concentration of sucrose only. That is why it follows the first order equation (7).

The progress of the reaction can be observed by measuring the change in angle of rotation by means of a polarimeter. As all the three substances can sugar, glucose and fructose are optically active, this change in rotation is proportional to the amount of the sugar decomposed. The change produced in rotation in the time 't' gives the value of x and that produced at the completion of the reaction gives 'a', *e.g.*, r_0 is the rotation at the beginning of the time and r_t after time 't' ant r_∞ at the end of the reaction Fig. 1.8. Then

$$a = r_0 - r_\infty \text{ and } x = r_0 - r_t$$

$$a - x\ (r_0 - r_\infty - r_0 + r_t) = r_t - r_\infty.$$

Substitute the various values in eq. (7) and calculate the value of velocity constant.

RADIOACTIVE DECAY

The radioactive decays have been found to follows first order kinetics. These reactions have been studied by recording the number of counts per seconds because this number is proportional to the number of atoms disintegrating per second. The results obtained on a radioactive substance such as protoactinium –234 have been shown in Fig. 1.9. It is evident at the half-life is independent of initial concentration and is approximately 60 –70 sec., revealing first order kinetics for the decay. This has been verified by plotting $\log_{10}$ (counts sec^{-1}) against time Fig. 1.10. These results have been interpreted statistically in the following way.

Suppose each atom of the radioactive element is having a chance l that it decays in the next second. Thus, for a sample having N radioactive atoms, the number of disintegrations in the next second be λ N. The rate of decay is thus as follows.

$$\frac{9.2\times10^{-6}}{(0.02)^2\times1.00} = \lambda\, N_t \qquad ...(1)$$

The constant λ is generally known as the *decay constant.* Its units are s^{-1}.

On integrating Eq. (1) we have $-\ln N_t = \lambda t +$ constant

When $t = 0$, $N_t = N_0$. Then constant $= -\ln N_0$

Hence $\ln N_t = l_t - \ln N_0$

or $\lambda_t = \ln\left|\frac{N_0}{N_t}\right|$ or $t = \frac{2.303}{\lambda}\log\frac{N_0}{N_t}$.

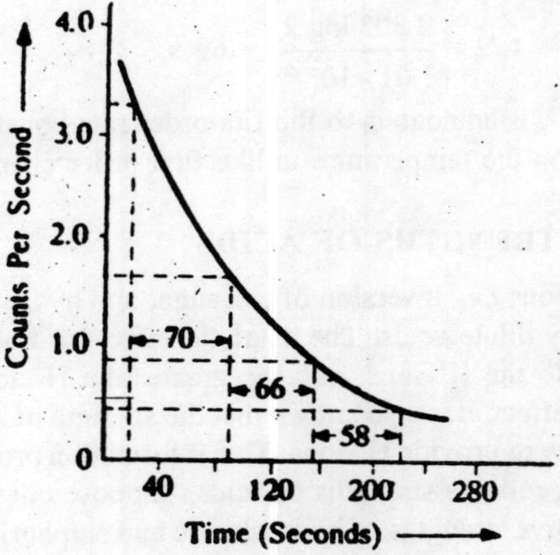

Fig. 1.9 : Radioactive decay of Protoactinium-234

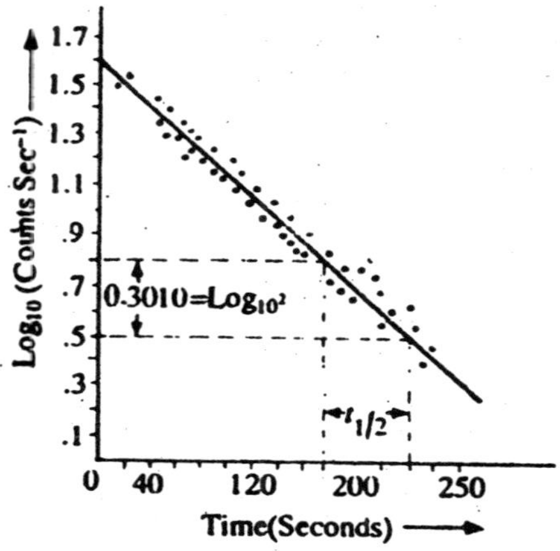

Fig. 1.10 : (Log_{10} counts/sec.) vs/me plot for the decay of protoactinium-34.

The time required for this change $N_1 = \frac{N_0}{2}$ is called *half-life* period, $t_1/2$, and is given as follows.

$$t_{1/2} = \frac{2.303}{\lambda} \log 2.$$

From the data plotted in Fig. 1.10, slope of log (counts s^{-1}) – time plot is 4.52×10^{-3} giving l = 1.01×10^{-2} s^{-1}. This gives

$$t_1/2. = \frac{2.303 \log 2}{1.01 \times 10^{-2}} = 69 \text{ s}.$$

Although, λ is analogous to the firs order rate constant, yet it does not depend upon the temperature unlike first order chemical reactions.

RELATIVE STRENGTHS OF ACIDS

Both reactions *i.e.*, inversion of can sugar and hydrolysis of an ester are catalysed by dilute acids. The catalytic effect for both the reactions is largely due to the H^+ ions, *i.e.*, the greater the H^+ ions, the greater is the catalytic effect. It is also known that the strength of an acid depends upon its capacity to provide H^+ ions. This information provides a method for determining relative strengths of acids. Suppose one is interested in comparing relative strengths of hydrochloric and sulphuric acid. One has to determine the velocity constant, k_1, fir the inversion of cane sugar or hydrolysis of ethyl acetate, first in the presence of hydrochloric acid and then in the presence of sulphuric acid provided the normality of both the acids is same. Then

$$\frac{\text{Strength HCl}}{\text{Strength of } H_2SO_4} = \frac{k_1 HCl}{k_1 H_2SO_4}$$

where k_1 HCl and $k_1 H_2SO_4$ are the values of velocity constants of the reaction in the presence of acids A and B respectively.

KINETICS OF SECOND ORDER REACTIONS

If the rate of a reaction is determined by the change in concentration of two reactants or the square of the concentration of a single reactant, it said to be of the second order. Such a reaction can be represented in the general way as

(I) A + B → Products

or (II) 2A → Products

Equation of Second Order Reactions : There are two cases.

Case I. Let us consider the case I, A + B → Products.

Suppose a and b are the initial concentrations in gm, moles/litre of A and B respectively. Let x gm. moles/litre of A and B decompose in time t.

Then the concentrations of A and B will be (a – x) and (b – x) gm. moles/litre respectively. According to the law of mass action, the rate of such a second order reaction is represented as

$$dx/dt \propto (a - x)(b - x) \text{ or } dx/dt = k_2 (a - x)(b - x) \qquad ...(1)$$

where k_2 is the rate constant for the second order reaction.

On separating the variables, equation (1) may be put as

$$\text{or} \quad \frac{dx}{(b-x)} = k_2 \, dt \qquad ...(2)$$

By taking partial fractions, equation (2) may be written as

$$\frac{1}{(a-b)}\left[\frac{1}{(b-x)} - \frac{1}{(a-x)}\right] dx = k_2$$

$$\text{or} \quad \frac{1}{(a-b)} [-\ln(b-x) + \ln(a-x)] = k_2 t + I_2 \qquad ...(3)$$

where I_2 is known as constant of integration. When t = 0, x = 0, equation (3) takes the form $\frac{1}{(a-b)}$ [– ln b + ln a] I_2

or $$\frac{1}{(a-b)} \ln \frac{a}{b} = I_2 \qquad ...(4)$$

Substituting the value of I2 in equation (3), we obtain

$$\frac{1}{(a-b)} [-\ln (b-x) + \ln (a-x)] = k_2 t + \frac{1}{(a-b)} \ln \frac{a}{b}$$

or $$\frac{1}{(a-b)} \ln \frac{(a-x)}{(b-x)} = k_2 t + \frac{1}{(a-b)} \ln \frac{a}{b}$$

or $$k_2 t = \frac{1}{a-b} \ln \frac{a-x}{b-x} - \frac{1}{a-b} \ln \frac{a}{b} \left[\ln \frac{a-x}{b-x} \ln \frac{a}{b} \right]$$

or $$k_2 = \frac{1}{(a-b)t} \left[\ln \frac{(a-x)b}{(b-x)a} \right]$$

$$= \frac{2.303}{(a-b)t} \left[\log \frac{(a-x)b}{(b-x)a} \right] \qquad ...(5)$$

Equation (5) is knows as the *kinetic equation for the second order reaction.*

The plot of $\frac{2.303}{(a-b)t} \log \frac{(a-x)b}{(b-x)a}$ versus t gives a straight line (Fig. 1.11) whose slope is equal to the rate constant k2.

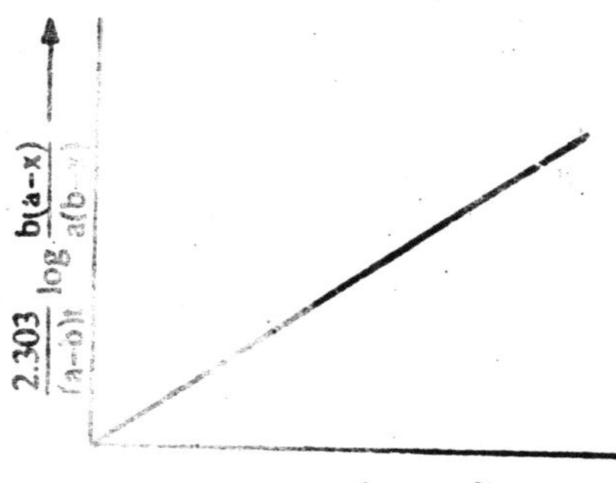

Fig. 1.11 : Evaluation of rate constant for a second-order reaction.

Case II. Let us consider the second case, 2A → Products.

Suppose a is the initial concentration of A in gm. moles/litre. Let x gm. moles/litre decompose in time t. According to the law of mass action, the rate of such reaction is given by

$$\frac{dx}{dt} \propto (a-x)(a-x) \text{ or } \frac{dx}{dt} = k_2 (a-x)^2 \quad ...(6)$$

where k_2 is the rate constant for the second order reaction. On separating the variables, equation (6) may be written as

$$\frac{dx}{(a-x)^2} = k_2 \, dt. \quad ...(7)$$

On integrating it, we obtain $\frac{1}{(a-x)} = k_2 t + I_2$ ...(8)

where I_2 is the integrating constant. When t = 0, x = 0, eq. (8) becomes as $I_2 = 1/a$. Substituting this values of I_2 in eq. (8) we obtain

$$\frac{1}{(a-x)} = k_2 t + \frac{1}{a} \text{ or } k_2 t = \frac{1}{(a-x)} - \frac{1}{a}$$

or $$k_2 t = \frac{x}{a(a-x)} \text{ or } k_2 = \frac{1}{t} \frac{x}{a(a-x)} \quad ...(9)$$

Characteristics of Second Order Reactions

(i) When we insert the values of a, x and t in equation (5) or (9) we will obtain a constant value of k_2 if the reaction is of the second order.

(ii) Consider the equation (5)

$$k_2 = \frac{1}{t} \frac{x}{a(a-x)}$$

or $$k_2 = \frac{1}{\text{sec}} \frac{\text{moles/litre}}{(\text{moles/litre}) \times (\text{moles/litre})}$$

$$= \frac{1}{\text{sec}} \times \frac{1}{(\text{moles/litre})}$$

$$= \text{litre/moles/sec} = \text{litre, moles}^{-1}, \text{sec}^{-1}.$$

Thus, the units of k_2 for a second order reaction is expressed in litre moles^{-1} sec^{-1}

(iii) Let us consider the half change, *i.e.*, $x = a/2$ in time $t_1/2$ Therefore, equation (9) becomes as

$$k_2 = \frac{1}{t_{1/2}} \frac{a/2}{a(a - a/2)} = \frac{1}{t_{1/2}\, a} \quad ...(10)$$

In general, we can say that the *time required for a certain fraction of a second order reaction is inversely proportional to the initial concentration of the reactant.*

(iv) The second order velocity constant k_2 depends on the concentration units in which it is expressed. Suppose the new unit be m times the first one. Then, equation (9) takes the form

$$k_2' = \frac{1}{t} \frac{mx}{ma(ma - mx)} = \frac{1}{mt} \frac{x}{a(a-x)}$$

$$= \frac{1}{m} \times k_2 \text{ [From equation (2)]}$$

Thus, the new value of k_2', becomes 1/m times the original value.

(v) When one of the reactants is present in large excess, the second order reaction gives first order results. Let us consider

$$k_2 t = \frac{2.303}{(a-b)} \left[\log \frac{(a-x)\, b}{(b-x)\, a} \right].$$

Suppose the reactant A of initial concentration 'a' is present in large excess. Then x and be can be neglected as compared to a. Thus, the equation (5) is modified, to

$$k_2 t = \frac{2.303}{a} \log \frac{b}{a(b-x)} \text{ or } k_2 a = \frac{2.303}{t} \log \frac{b}{b-x}$$

or $$k_1 = \frac{2.303}{t} \log \frac{b}{b-x}$$

where k_1 (= $l_2 a$) is the velocity constant of the first order reaction. Equation (11) is identical with the first order equation (7).

Examples of Second Order Reaction : These are

(a) *Reaction between per sulphate and iodide :* This reaction is

$$K_2S_2O_8 + 2KI \rightarrow 2K_2SO_4 + I_2.$$

It seems that the reaction is of the third order. In actual practice, it is of the second order. This can be explained by considering the mechanism of this reaction which is as follows.

First stage	$S_2O_8^{-2} \rightarrow 2SO_4^-$	fast
Second stag	$SO_4^- + I^- \rightarrow I + SO_4^{2-}$	fast
Third stage	$I + I \rightarrow I_2$	slow.

As the first and third stage are fast, the rate of the reaction is determined by the second stage which is slow. Hence, the order cf the reaction is two and also conforms to the experimental rate of the reaction which is given by

$$\frac{dx}{dt} = k_2 [I^-] [S_2O_8^{2-}]$$

(b) Saponification of ester : The reaction between ethyl acetate and sodium hydroxide is of the second order. This reaction takes place as follows.

$$CH_3COOC_2H_5 + NaOH \rightarrow CH_3COONa + C_2H_5OH$$

Prepare M/100 ethyl acetate and M/50 sodium hydroxide solution. Take 50 ml of M/100 ethyl acetate in one flask and 50 ml of M/50 NaOH in another flask. Keep both these flasks in the same thermostat to attain the same temperature.

When we observe that both the flasks have attained the same temperature, the alkali solution from this flask in put into the flask which is having ethyl acetate solution. Shake the flask. Then, pipette out 5 ml of the reaction mixture in a flask which is already having 10 ml of M/ 25 HCl to check the reaction. The excess of the acid is then titrated back by means of standard alkali solution.

Pipette out 5 ml of the reaction mixture after every 10, 20, 30, 40 minutes and follows the same procedure as above. Let the titre value at any time t be Vt. The infinite reading is then taken after 24 hours. Let this volume be V_∞. Then, substitute these values in equation (5), we obtain

$$k_2 = \frac{1}{t(a-b)} \log_e \frac{(V_\infty \, 0 \, V_0)(V - V_t)}{(V - V_0)(V_\infty - V_t)} \quad ..(12)$$

The value of V_0 can be obtained by taking 5 ml of the alkali at time t = 0 and adding it to 10 ml of HCl. Titrate the excess of the acid with 0.02 M NaOH. Let the titre value be V_0.

The value of V can be calculated by titrating 10 ml of HCl acid against 0.02 M NaOH. Let the titre value be V.

(c) Conversion of ammonium cyanate into urea : This reaction is

$$NH_4CNO \longrightarrow O{=}C\langle^{NH_2}_{NH_2}$$

The mechanism of this reaction is

(I) $NH_4CNO \rightleftharpoons NH_4NCO \rightleftharpoons H-N{=}C{=}O + NH_3$ [fast]

(II) $H{-}N{=}C{-}O + NH_2 \longrightarrow O{=}C\langle^{NH_2}_{NH_2}$

As the first step is very fast, the rate of the reaction is determined by step II which is slow.

The reaction II involves two concentration terms and hence this reaction is of the second order.

(d) (i) Decomposition of Ozone into oxygen

$$2O_3 \xrightarrow{100°C} 3O_2$$

(ii) Thermal decomposition of N_2O

$$2N_2O \longrightarrow 2N_2 + O_2$$

(iii) Thermal decomposition of NO_2

$$2NO_2 \longrightarrow 2\dot{N}O + O_2$$

KINETICS OF THIRD ORDER REACTIONS

A chemical reaction is said to be of third order if its rate depends on three concentration terms. In general, a third order reaction may be represented as.

I $3A \rightarrow$ Products

II $2A + B \rightarrow$ Products

III $A + B + C \rightarrow$ Products

Type I : Now consider the type I in which initial concentration of all three reacting molecules is the same

$$3A \rightarrow \text{Products} \quad ...(1)$$

Suppose a is the initial concentration of each reactant in gm mole/litre. Suppose x is the concentration change after time t. Then, the concentration of A at time t is (a – x) gm moles/litre.

According to the law of mass action, the rate of such a reaction at time t is given by

$$\frac{dx}{dt} \propto (a - x)\,(a - x)\,(a - x) \text{ or } \frac{dx}{dt} = k_2\,(a - x)^2 \qquad ...(2)$$

where k_3 is the real constant or velocity constant for the third order reaction. On separating the variables, equation (2) becomes as

$$\frac{dx}{(a-x)^3} = k_2\,dt \qquad ...(3)$$

Integrating equation (3), we obtain

$$\frac{1}{2\,(a-x)^2} = k_3\,t + I_3 \qquad ...(4)$$

where I_3 is known as constant of integration. Its value is determined from the initial conditions of the experiment, *i.e.,* when t = 0, x = 0 equation (4) takes the form

$$I_3 = \frac{1}{2a^2} \qquad ...(5)$$

Substituting the value of I_3 in equation (4), we obtain

$$\frac{1}{2\,(a-x)^2} = k_3\,t + \frac{1}{2a^2}$$

$$k_2 = \frac{1}{2t}\left[\frac{1}{(a-x)^2} - \frac{1}{a^2}\right] \qquad ...(6).$$

Equation (6) is known as the *kinetic equation of the third order reaction* when the concentration of three reactants is the same. Equation (6) can be simplified as

$$k_2 = \frac{1}{2t}\left[\frac{x\,(2a-x)}{a^2\,(a-x)^2}\right] \qquad ...(7)$$

Type II : We will now consider the type II in which two reactants are equal and third one different. 2 A + B → Products. Suppose a and b are initial concentrations of A and B respectively. Suppose x is the concentration change after time t from the commencement. Then the rate of the reaction is given by

$$dx/dt = k_3\,(a - 2x)^2\,(b - x) \qquad ...(8)$$

where the amount of A decomposed at any instant is 2x which is twice that of B *i.e.,* on separating the variables and integrating the equation (8), we obtain

$$k_3 = \frac{1}{t(a-2b)^2}\left(\frac{2x(2b-a)}{a(a-2x)} + \ln\frac{b(a-2x)}{a(b-x)}\right) \quad ...(9)$$

Again, equation (9) is known as the kinetic equation for a third order reaction.

Type III : The last case is A + B + C → Products.

Suppose a, b and are the initial concentrations of A, B and C respectively. Suppose x is the concentration change after time t from the commencement. Then the rate of the reaction is given by

$$dx/dt = k_3 (a-x)(b-x)(c-x) \quad ...(10)$$

or

$$\frac{dx}{(a-x)(b-x)(c-x)} = k_3\, dt \quad ...(11)$$

Integrating it, we obtain

$$k_3 = \frac{1}{t}\,\frac{(b-c)\ln\frac{a-x}{a} + (c-a)\ln\frac{b-x}{b} + (a-b)\ln\frac{c-x}{c}}{(b-c)(a-b)(c-a)} \quad ...(12)$$

Equation (12) also is known as the *kinetic equation for the third order reaction.*

CHARACTERISTICS OF THIRD ORDER REACTIONS

(i) All reactions of the third order must obey either of the equations (7), (9), (12), depenuing upon the concentrations of three reactants taking part in such reactions.

(ii) *Units for a third order rate constant* : Equation (7) is

$$k_3 = \frac{1}{2t}\frac{x(2a-x)}{a^2(a-x)^2}$$

$$= \frac{1}{sec} \times \frac{moles/litre \times moles/litre}{(moles/litre)^2 \times (moles/litre)^2}$$

$$= (moles/litre)–2\ sec^{-1}.$$

Thus, the units for k_3 will be $(moles/litre)^{-2}\ sec^{-1}$

(iii) *The time required to complete a definite fraction of the third order reaction is inversely proportional to the square of the initial concentration.* This may be seen by putting x = a/2 in time $t_1/2$ in equation (7) to give

$$k_3 = \frac{1}{2t_{1/2}} \frac{a/2(2a - a/2)}{a^2 (a - a/2)^2}$$

or $$t_1/2 = \frac{1}{2k_2} \times \frac{a/2 \times 3a/2}{a^2 \times a^2/4}$$

or $$t_1/2 = \frac{3}{2} \times \frac{1}{a^2} \text{ or } t_{1/2} \propto \frac{1}{a^2}.$$

Similarly, it can be proved that the time taken for the completion of any definite fraction of a reaction varies inversely as the square of the initial concentration.

EXAMPLES OF THIRD ORDER REACTIONS

(i) *Gaseous Reactions :* There are only five gaseous reactions of the third order. Each of these reactions involves the nitric oxide as one of the reactants. These reactions are.

(a) $2NO + Cl_2 \rightarrow 2NOCl$

(b) $2NO + Br_2 \rightarrow 2NOBr$

(c) $2NO + O_2 \rightarrow 2NO_2$

(d) $2NO + H_2 \rightarrow N_2O + H_2O$

(e) $2NO + D_2 \rightarrow N_2O + D_2O$

(ii) *Reactions in Solution :* Noyes and Cottle showed that the reduction of ferric chloride by stannous chloride was of the third order.

$$2F_2Cl_3 + SnCl_2 \rightarrow 2FeCl_2 + SnCl_4.$$

They studied this reaction by mixing equal quantities of stannous chloride and ferric chloride solution in a flask which was kept in a thermostat. They withdrew equal volumes of reaction mixture at regular intervals of time.

The excess of stannous chloride was removed by adding mercuric chloride to it and the ferrous chloride was estimated by titrating it against standard $K_2Cr_2O_7$ solution. It has now been proved that the reaction of the second order only.

(iii) The reaction between potassium oxalate and mercuric chloride is of third order.

$$2HgCl_2 + K_2C_2O_4 \rightarrow 2KCl + 2CO_2 + Hg_2Cl_2$$

KINETIC OF Nth ORDER REACTIONS

Consider a reaction of nth order which takes place as follows,

$$nA \rightarrow \text{Products.}$$

In this reaction, all the reactants are at the same concentration. Suppose a is the initial concentration of A in gm moles/litre. Let x gm moles/litre decompose in time t. Then the concentration of A at that time will be (a – x gm moles/litre. According to the law of mass action, the rate of nth order is given by

$$\frac{dx}{dt} \propto (a-x)^n \text{ or } \frac{dx}{dt} = k_n (a-x)^n \qquad ...(1)$$

where kn is the rate constant for the nth order reaction. Separating the variables, equation (1) may be put as

$$\frac{dx}{(a-x)^n} = k_n \, dt.$$

Integration of this leads to

$$\frac{1}{(n-1)(a-x)^{n-1}} = k_n t + I_n \qquad ...(2)$$

where In is the integration constant. When t = 0, x = 0 equation (2) becomes as $\frac{1}{(n-1)a^{n-1}} = I_n$ ...(3)

Substituting this in equation (2), we get

$$\frac{1}{(n-1)(a-x)^{n-1}} = k_n t + \frac{1}{(n-1)a^{n-1}}$$

or
$$k_n t = \frac{1}{(n-1)}\left[\frac{1}{(a-x)^{n-1}} - \frac{1}{a^{n-1}}\right]$$

or
$$k_n = \frac{1}{t(n-1)}\left[\frac{1}{(a-x)^{n-1}} - \frac{1}{a^{n-1}}\right] \qquad ...(4)$$

Suppose it is required to calculate the time required for the completion of half of the reaction, *i.e.*, x = a/2 at t = $t_{1/2}$.

Hence equation (4) becomes as

$$k_n = \frac{1}{t_{1/2}\,(n-1)}\left[\frac{1}{(a-a/2)^{n-1}} - \frac{1}{a^{n-1}}\right]$$

$$= \frac{1}{(n-1)\,t_{1/2}}\left[\frac{2^{n-1}}{a^{n-1}} - \frac{1}{a^{n-1}}\right]$$

$$k_n = \frac{1}{(n-1)\,k\,a^{n-1}}[2^{n-1}-1] \text{ or } t_{1/2} \propto \frac{1}{a^{n-1}}.$$

Similarly, it can be proved that $t_3/4$ $t_{3/4} \propto \frac{1}{a^{n-1}}$ ...(5)

Thus, the time required to complete a definite fraction of the reaction of nth order is inversely proportional to the initial concentration raised to the power which is one less than the order of reaction.

SOLVED EXAMPLES

Example 1:

The rate constant of a zero order reaction is 0.2 (moles/litre) hour^{-1}. What will be the initial concentration of the reactant if after half an hour its concentration is 0.05 moles/litre?

Solution:

We know

$$k_0 = \frac{C_0 - C}{t} \qquad \text{...(1)}$$

Here k_0 = 0.2 (moles/litre) hour^{-1}

$t = \frac{1}{2}$ hour

C = 0.05 moles/litre.

Substituting all these values in equation (1), we get

$$0.2 = \frac{(C_0 - 0.05)}{\frac{1}{2}}$$

C_0 = 0.15 moles/litre.

Example 2:

Find the order of the reaction and the rate constant of the decomposition of ammonia on a tungsten wire at 856°C. Data obtained in an experiment are given below.

Total pressure,	p (torr)	228	250	273	318
Time	t (sec.)	200	400	600	1000.

Solution:

The above results can be put as follows.

ΔP;	0	22	45	90
Δt;	0	200	400	800
$\frac{\Delta p}{\Delta t}$;	–	1.100	1.125	1.125.

It is seen from the data that the rate of change of pressure remains constant (almost). Hence

$$-\frac{dp}{dt} = \text{constant (k)}.$$

It indicates that reaction rate does not depend upon the concentration (or partial pressure) of the reactant.

Therefore, the reaction is a zero order reaction.

The specific reaction rate, $k = -\frac{dp}{dt} = \mathbf{0.1117\ torr^{-1}}$.

Example 3:

Following observations were made during hydrolysis of methyl acetate at 20°C, using 0.05 N HCl as catalyst.

t (Sec)	0	75	119	183	∞
Volume of the alkali used (ml.)	9.62	12.10	13.10	14.75	21.05

Show that the reaction is of the first order.

Solution:

$$V_\infty - V_0 = 21.05 - 9.62 = 11.43 \text{ ml.}$$

$$k_1 = \frac{2.303}{t} \log \frac{V_\infty - V_0}{V_\infty - V_t}.$$

Time (sec.) $V_\infty - V_t$.

$$75 \quad 21.05 - 12.10 = 8.95 \quad k_1 = \frac{2.303}{75} \log \frac{11.43}{8.95} = 0.003259$$

$$119 \quad 21.05 - 13.10 = 7.95 \quad k_1 = \frac{2.303}{119} \log \frac{11.43}{7.95} = 0.003264$$

$$183 \quad 21.05 - 14.75 = 6.30 \quad k_1 = \frac{2.303}{183} \log \frac{11.43}{6.93} = 0.003254.$$

A constant value of k_1 shows that the reaction is of the first order.

Example 4:

The decomposition of H_2O_2 *was studied by titrating it at different intervals of time with potassium permanganate. Calculate the velocity constant from the following data, if the reaction if of the first order.*

t (min)	0	10	30	40
$KMnO_4$ (ml.)	25.0	20.0	12.5	9.6

Solution:

Equation is $\quad k_1 = \frac{2.303}{t} \log \frac{a}{a-x}$

Here $\quad a = 25.0$ ml

t (min.) a – x ml. $k_1 = \frac{2.303}{t} \log \frac{a}{a-x}$

$$10 \quad 20.0 \quad k_1 = \frac{2.303}{10} \log \frac{25}{20.0} = 0.022287$$

$$30 \quad 12.5 \quad k_1 = \frac{2.303}{30} \log \frac{25}{12.5}$$

$$40 \quad 9.6 \quad k_1 = \frac{2.303}{40} \log \frac{25}{9.6} = 0.023897.$$

The value of k_1 is constant which shows that reaction is of the first order. The average value of k_1 is 0.0231 per min.

Example 5:

The optical rotations of cane sugar in 0.5 HCl at 35°C and at various intervals are given below. Show that the reaction is of the first order.

Time (min)	0	20	40	80	∞
Rotation (degrees)	+ 32.4	+ 25.5	+ 19.6	+ 10.3	– 14.1

Solution:

$$k = \frac{2.303}{t}\log\frac{a}{a-x}$$

Here $a \propto r_0 - r_\infty \propto (32.4) - (-14.1) = 46.5$

and $a - x \propto r_1 - r\infty$

t (min.)	$r_1 - r\infty$ (ml.)	$k_1 = \frac{2.303}{20}\log\frac{a}{a-x}$
20	$25.5 - (-14.1) = 39.6$	$k_1 = \frac{2.303}{20}\log\frac{46.5}{39.6} = 0.008039$/min.
40	$19.6 - (-14.1) = 33.7$	$k_1 = \frac{2.303}{40}\log\frac{46.5}{33.7} = 0.008050$/min.
80	$10.3 - (-14.1) = 24.4$	$k_1 = \frac{2.303}{80}\log\frac{46.5}{24.4} = 0.008035$/min.

The constant value of k_1 is obtained which shows that the reaction is of the first order. The average value of k_1 is 0.008053/min.

Example 6:

Use the following data to show that the reaction

$CH_3COOC_2H_5$ + NaOH → CH3COONa + C_2H_5OH *is a second order reaction.*

Initial concentration of ethyl acetate = caustic soda=10 moles/litre

Time (min.)	0	15	25	35	55
a – x	10	4.9	3.6	2.9	2.1.

Solution:

Here a = 10

t	a – x	x	$k_2 = \frac{1}{at}\cdot\frac{x}{a-x}$
0	10 (= a)	0	
15	4.9	5.1	$k_2 = \frac{1}{15\times 10}\cdot\frac{5.1}{4.9} = 0.0069$

25	3.6	6.4	$k_2 = \frac{1}{25 \times 10} \cdot \frac{6.4}{3.6} = 0.0071$
35	2.9	7.1	$k_2 = \frac{1}{35 \times 10} \cdot \frac{7.1}{2.9} = 0.0070$
55	2.1	7.9	$k_2 = \frac{1}{55 \times 10} \cdot \frac{7.9}{2.1} = 0.0068.$

The value of k_2 is constant and hence the reaction is of the second order.

Example 7:

Decomposition of a gas is of second order. When the initial concentration of the gas is 5 × 10^{-4} mole/litre it is 40% decomposed in 50 minutes. What is the value of velocity constant ?

Solution:

Here $a = 0.0005$ mole/litre

or $x = \frac{0.0005 \times 40}{100} = 0.0002$

$\therefore$ $k_2 = \frac{1}{at} \cdot \frac{x}{a - x}$

$$k_2 = \frac{1}{0.0005 \times 50} \times \frac{0.0002}{(0.0005 - 0.0002)}$$

$= $ **26.67 litre/mole/min.**

Example 8:

For the second order reaction

$CH_3COOC_2H_5 + OH^- \rightarrow CH_3COO^- + C_2H_5OH$ *at 25°C,* $k_2 = 6.21 \times 10^{-3}$ *litre/mole/sec. Calculate the time required for the hydrolysis of 90% ester if the initial concentrations of the reaction in the reaction mixture are*

(a) 0.05 M ester + 0.1 M NaOH

(b) 0.05 M ester + 0.5 M NaOH

Solution:

(a) When the initial concentrations of the reactants are not equal the second order equation can be written as or $$t = \frac{2.303}{k_2 (a-b)} \log \frac{b(a-x)}{a(b-x)}$$

Here $a = 0.1$ M; $b = -.05$ M;

$$x = 0.05 \times \frac{90}{100} = 0.045 \text{ M. Therefore,}$$

$$t = \frac{2.303}{0.00621 \times (0.1-0.05)} \log \frac{0.05(0.1-0.045)}{0.1(0.05-0.045)}$$

$$= 5.5 \times 103 \text{ sec.}$$

(b) When the concentration of the reactants are equal, equation (9) for the second order reaction is

$$t_2 = \frac{1}{ak_2} \frac{x}{a-x}$$

Here $a = 0.1$ M, $x = 0.1 \times \frac{90}{100} = 0.09$ M

$$\therefore \quad t = \frac{1}{0.1 \times 0.00621} \times \frac{0.09}{(0.1-0.09)} = \mathbf{14.49 \times 103 \text{ sec.}}$$

Example 9:

Equivalent amounts of $FeCl_3$ and $SnCl_2$ were taken and the progress of the reaction was followed by withdrawing and the titrating equal amounts of the solution against $K_2Cr_2O_7$. The results obtained were as follows.

Time (min)	1	3	7	11	∞
Vol. of $K_2Cr_2O_7$ used (ml)	7.17	13.32	18.06	20.50	31.25

Solution:

The equation for termolecular reactions involving equivalent concentrations of reactants is

$$k_3 = \frac{1}{2t} \frac{x(2a-x)}{a^2 (a-x)^2} \qquad \text{[equations (7)]}$$

The volume of K_2CrO_7 used at infinite time corresponds to the total amount of $FeCl_3$ initially present. Thus, $a = 31.25$ ml.

The volume of K_2Cr2O_7 used after time t corresponds to the amount of $FeCl_3$ decomposed, *i.e.*, it gives the values of x at time t.

$$a = 31.25$$

t	x	(a–x)	2a – x	$\frac{1}{2t}\frac{x(2a-x)}{a^2(a-x)^2} = k_3$
1	7.17	24.08	55.33	$\frac{1}{2}\times\frac{7.17\times 55.33}{(31.25)^2\times(24.08)^2} = 0.000.360$
3	113.32	17.93	49.18	$\frac{1}{6}\times\frac{13.32\times 49.18}{(31.25)^2\times(17.93)^2} = 0.000347$
7	18.06	13.10	44.44	$\frac{1}{14}\times\frac{18.06\times 44.44}{(31.25)^2\times(13.19)^2} = 0.000358$
11	20.50	10.75	42.00	$\frac{1}{22}\times\frac{20.50\times 42.00}{(31.25)^2\times(10.75)^2} = 0.000346$

A fairly constant value of k_3 shows that the reaction between ferric chloride and stanous chloride is of third order.

Example 10:

In an experiment in the reduction of ferric chloride with stannous chloride, the following results were obtained.

Initial concentration, a = 0.0625 M

t (min)	1	3	7	11
x	0.01434	0.02664	0.03612	0.04102

Show that the reaction is of third order.

Solution:

$$a = 0.0625 \text{ M}$$

t	x	(a = x)	$k_3 = \frac{1}{2t}\left(\frac{1}{(a-x)^2} - \frac{1}{a^2}\right)$
1	0.01434	0.04816	87.58
3	0.02664	0.03586	80.95
7	0.03612	0.02638	84.33
11	0.04102	0.02148	87.92.

Since the value k3 is constant, hence the reaction is of the third order.

Example 11:

For the third-order reaction 2A + B → Products, the differential rate equation is dx/dt = k_3 (a – 2x)2 (b – x). Integrate this rate equation.

Solution.

$$dx/dt = k_3 (a - 2x)^2 (b - x) \quad ...(1)$$

where a and b are the initial molar concentrations of A and B, respectively and x is the concentration of the product formed at time t so that at time t, the concentrations of A and B are (a – 2x) and (b – x), respectively. Separating the variables in Eq. (i) and integrating, we get,

$$\int \frac{dx}{(a-2x)^2 (b-x)} = k_3 \int dt = k_3 t + C \quad ...(2)$$

where C is the constant of integration.

Resolving into partial fractions, we have

$$\frac{1}{(a-2x)^2 (b-2x)}$$

$$= -\frac{2}{(2b-a)^2 (a-2x)} + \frac{2}{(2b-a)(a-2x)^2} + \frac{1}{(2b-a)^2 (b-x)} \quad ...(3)$$

Carrying out the integration of Eq. (iii), making use of the partial fractions, we obtain

$$\int \frac{dx}{(a-2x)^2 (b-x)} = -\frac{2}{(2b-a)^2} \int \frac{dx}{a-2x} + \frac{2}{(2b-a)} \int \frac{dx}{(a-2x)^2} + \frac{1}{(2b-a)^2} \int \frac{dx}{b-x}$$

$$= -\frac{2}{(2b-a)^2}\left(-\frac{1}{2}\right) \ln (a - 2x) - \frac{1}{(2b-a)^2} \ln (b - x) + \frac{2}{(2b-a)}\left(-\frac{1}{a-2x}\right)\left(-\frac{1}{2}\right)$$

$$= \frac{1}{(2b-a)^2} \ln (a-2x) - \frac{1}{(2b-a)^2} \ln (b-x) + \frac{1}{(2b-a)}\left(\frac{1}{a-2x}\right)$$

$$= \frac{1}{(2b-a)^2} \ln\left(\frac{a-2x}{b-x}\right) + \frac{1}{(2b-a)(a-2x)}$$

$$= k_2t + C \qquad ...(4)$$

At t = 0, so that from Eq. (4),

$$C = \frac{1}{(2b-a)^2} \ln(a/b) + \frac{1}{(2b-a)a} \qquad ...(5)$$

Substituting for C in Eq. (4), and transposing, we have

$$\frac{1}{(2b-a)^2}\left\{\ln\left[\frac{(a-2x)/(b-x)}{a/b}\right] + (2b-a)\left[\frac{a-(a-2x)}{a(a-x)}\right]\right\} = k_3t$$

$$\therefore \quad \frac{1}{(2b-a)^2}\left\{\ln\left[\frac{b(a-2x)}{a(b-x)}\right] + \frac{(2b-a)(2x)}{a(a-2x)}\right\} = k_3t \qquad ..(6)$$

This is the desired integrated rate equation.

2

DETERMINATION OF ORDER OF REACTION

Different methods of determining the order of reaction are discussed below.

(i) Isolation Method : This method was given by Ostwald in 1902. In this method, the concentration of all reactants except one is taken in excess and the order of reaction is then determined by any method with respect top that reactant (which is not taken in excess). Then in other reperate experiment, the concentration of any other reactant is not taken in excess, keeping that of all the others be isolated from other, reactants which are not taken in excess. The total order of reaction will be the sum of the order of all isolated reactions.

Consider the reaction

$$n_1\ A + n_2\ B + n_3\ C \longrightarrow \text{Products.}$$

The reaction velocity is given as follows.

$$\frac{dx}{dt} = k.\ C_A^{n_1} . C_B^{n_2} . C_C^{b_3} .$$

Let the order of reaction be n, when A is isolated, n_2 when B is isolated, and n_3 when C is isolated. The order of reaction as a whole will then be given by $n_1 + n_2 + n_3$.

The advantage of this method is that the mode of action of each components can be determined separately and disturbing effects can be traced to the origin.

(ii) Van't Hoff Differential Methods : In this method, the initial rate of reaction is measured when the concentration of one of the recatant is varied and that for all others are kept constant.

Initial reaction rates are determined by measuring the slopes of concentration time curves at zero time. These initial rates are related to the concentration of the reactants through the expression.

$$\left(\frac{dx}{dt}\right)_0 = k,\ [A]_0^x\ [B]_0^y.$$

When [B] is kept constant,

$$\left(\frac{dx}{dt}\right)_0 = k'\ [A]_0^x$$

where, $k' = k\ [B]_0^y$

or $\ln\left(\frac{dx}{dt}\right)_0 = \ln k' + x \ln [A]_0.$

So, if we measure the initial reaction rates while varying the initial concentration of the reactant, and keeping that of the others constant, a plot of ln (dx/dt) against in (A_0) gives a straight line whose slope gives the value of x, the order of reaction with respect to A.

Similarly, the order of reaction with respect to the others reactant may be determined by measuring the initial reaction rates with varying concentration of B, at constant [A].

(iii) Integration Method : In this method, the initial concentration of all the reactants taking part are determined. The concentration of the reacting substance is then determined at differents intervals of time. The different values of a and x are thus determined. These values are substitute in virions order rate expressions and the equation which gives the most constant values of velocity constant for a series of intervals of time gives the order of the reaction. The method, therefore, involves the trial of one equation after another till the correct one is found.

(iv) Fractional Change Methods of Half life Methods : As discussed before, the time taken to complete one half of the reaction is independent of initial concentration for a first order reaction, inversely proportional to the initial concentration for a second order reaction inversely proportional to the require of it for a third order reaction. This is so, provided the reactants are all the taken at the same initial concentration. Thus, in general, for a reaction of nth order.

$$t_{0.5} \propto \frac{1}{a_1^{n-1}}$$

Suppose we start with two independent reactions with initial concentrations a_1 and a_2. Let the corresponding times be t_1 and t_2 respectively, then

$$t_1 \propto \frac{1}{a_1^{n-1}}$$

and $$t_2 \propto \frac{1}{a_2^{n-1}}$$

or $$\frac{t_1}{t_2} = \left(\frac{a_2}{a_1}\right)^{n-1}$$

or $$\log = (n-1) \log \left(\frac{a_2}{a_1}\right)$$

or $$n = 1 + \frac{\log (t_1 / t_2)}{\log (a_2 / a_1)}$$

From this equation the order of reaction, n can be calculated.

In case of gaseous reactions the initial pressure (P) can be taken instead of initial concentration (a), so that,

$$n = 1 + \frac{\log (t_1 / t_2)}{\log (P_2 / P_1)}$$

(v) **Method of Ration Variation :** A complex reaction involving different substance may be studied by appreciably increasing the concentration of the reactions, one at a time and observing the reactions rates over a period in which the change in composition is not over 10% Consider the reaction.

$$A + B \longrightarrow \text{Products}$$

the rate of this reaction is given by,

$$\frac{dx}{dt} = k\,[A]_m\,[B]_n.$$

The index m can be determined by doubling [A], while keeping [B] constant and determining experimentally the change in the rate, $\Delta x/\Delta t$, since

$$\frac{(\Delta x/\Delta t)_2}{(\Delta x/\Delta t)_1} = \frac{k[A]^m . 2^m [B]^n}{k[A]^m [B]^n} = Z^m$$

where the subscripts Z and 1 refer to concentrations [2A] and [A], respectively. The same procedure is adopted for evaluating n.

The over all order of reaction = m + n.

(vi) Graphical Methods : B As discussed before, the reaction velocity in a first order reaction varies as one concentration term, while in a second order reaction, the reaction velocity is dependent on two concentration terms and so on. Mathematically, we can express the reaction velocity for a reaction of nth order as

$$\frac{dx}{dt} = k(a - x)^n$$

where a is the initial concentration and x is the amount of reactant in time t.

If a straight line is obtained by plotting dx/dt and $(a - x)^n$, then the reaction is of the nth order. In the case of zero order reaction, a straight line is obtained by plotting x against t.

The values of dx/dt at different intervals of time can be determined by plotting a curve between x (the amount of the substance de composed) and time t. The values of dx/dt at a particular time corresponding to a particular values of (a – x) is given by the slope of the curses at that point. On taking logarithms of equations (23), we get

$$\log \frac{dx}{dt} = \log k + n \log (a - x).$$

A curve when plotted between log dx/dt and log (a – x) will give a straight line, the slope of which will give the values of n, the order of reaction. The intercept on the axis of Y will give the values of log k.

METHODS OF DETERMINATION OF ORDER OF A REACTION

Essentially, the order of a reaction is determined by finding a rate expression which fits observed data over a wide range of conditions. This is simple enough statement, but carrying out the procedure is not always easy or obvious. Several special procedures used in determining such consistency are very much worth mentioning explicitly, as are some pitfalls of which to be aware.

The various methods of determining the order of a reaction are:

1. **Differential Method :** The differential method, was first suggested in 1884 by Van't Hoff, the procedure of this method is to determine by measuring the slopes of concentration time curves. One does this at various concentrations c of a reactant, and if the reaction has an order n with respect to this particular reactant, then

$$v = \frac{dc}{dt} = kcn \qquad ...(1)$$

A double logarithmic plot of in v versus ln c would yield a straight line of slope n, the intercept where ln c = 0 is then ln k. If a straight line plot is not obtained, the rate cannot be represented by Eq. (1), that is, the reaction is not having an order with respect to that particular reactant.

The procedure can be applied in two different ways. In one of them, shown schematically in (Fig. 2.1a) runs are performed at different initial concentrations and initial rates are determined by measuring initial slopes. A double-logarithmic plot then would give the order of reaction (Fig. 2.1b). The procedure, dealing with initial rates, avoids possible complications because of interference by products. Because of this, Letort referred to the order determined in this way as the *order with respect to concentration, or the true order*. The symbol n_c has been used to denote this order.

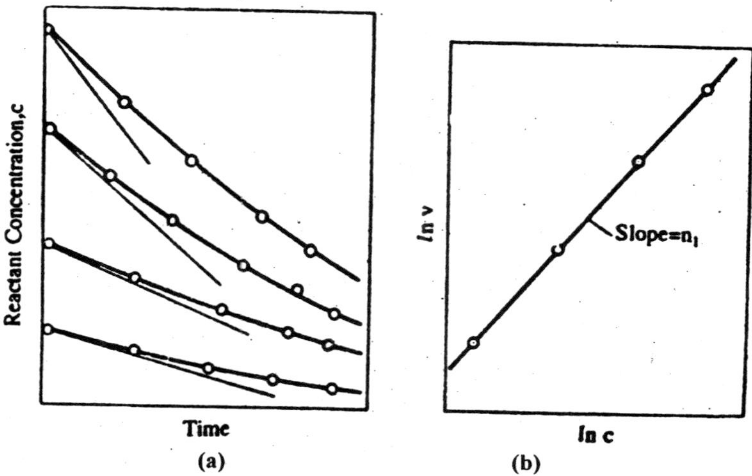

Fig. 2.1 : (a) Plot of reactant concentrations versus time for various initial concentrations. (b) A plot of $ln\ v_i$ versus $ln\ c_i$.

The second procedure involves considering a single run and measuring slopes at various times, corresponding to a number of values of the reactant concentration. This method is illustrated schematically in (Fig. 2.2a) and again the logarithms of the rates are plotted against the logarithms of the corresponding reactant concentration (Fig. 2.2b). The slope is the order, since time is now varying, letort referred to this order as the *order with respect to time* n_r

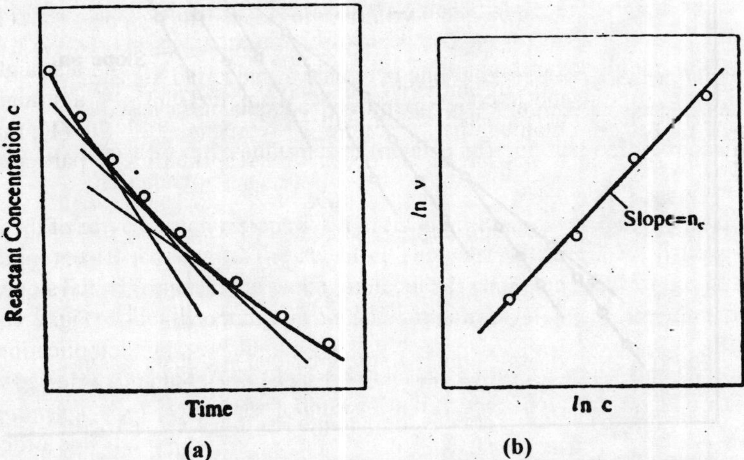

Fig. 2.2 : (a) A single concentration-time curve, with slopes measured at various reactant concentrations. (b) A plot of *l*n v versus *l*n c.

Fig. 2.3 shows schematic plots in which the two procedures have been combined. The points at the extremities of the n_t plots correspond to initial conditions and give rise to the n_c plot.

The two orders are not always the same for a given reactant. In the thermal decomposition of accetaldehyde, for example, as discussed further Letort found that the order with respect to concentration (the true order) is 3/2, and that the order with respect to times 2. T

he fact that the order with respect to time is greater than the order with respect to concentration means that as the reaction proceeds the rate falls off more rapidly than if the true order applied to the time course of the reaction. This abnormally large falling off can only imply that some substance produced in the reaction is acting as an inhibitor. Conversely, if n_t is less then n_c, the rate is falling off less rapidly with time than expected on the basis of the true order. Hence, some activation by the products of reaction exists, and the reaction is said to be *autocatalytic.*

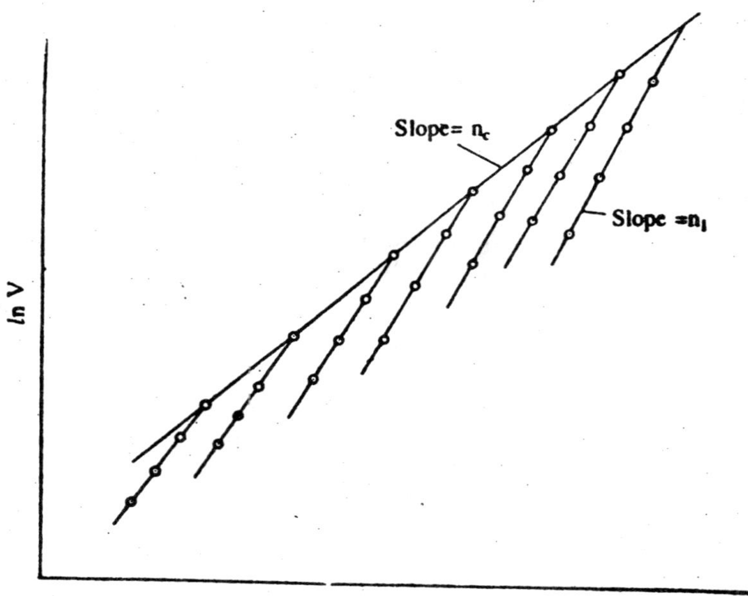

Fig, 2.3 : Plots of *l*n v *l*n c. The closed circles correspond to rates obtained in individual runs; the open circles are the initial rates.

Integration Method or Hit and Trial Method

This is a very simple procedure to calculate the rate constant. In this method, known quantities of the reaction are mixed in a reaction vessel and the progress of the reaction can be noted by determining the amount of reactant consumed after different intervals of time.

These values are then substituted in the equations for the first order, second order and third order reactions. The order of the reaction corresponds to that equation which gives the constant values of k. In this method, one equation after the other undergoes trial till the correct equation is known. *As this method involves the trial of different equations, it is usually called the hit and trial method.*

Limitations of the Method : Two main limitations are.

(a) This method can be used for simpler reactions only and not for complex reactions.

(b) In this method, the reaction should be studied over a wide time interval.

3. Fractional Change Method

It has already been proved that the time required to complete definite fraction of the reaction is independent of initial concentration of a first order reaction, inversely proportional to the initial concentration of the second order and inversely proportional to the second power of the initial concentration of the third order reaction and so on. This is only true when all the reactants taking pert in the reaction are at the same initial concentration.

For a first order reaction : $t \propto \frac{1}{a^0}$

For a second order reaction : $t \propto \frac{1}{a}$

For a third order reaction : $t \propto \frac{1}{a^2}$.

In general, the time required (t) to complete a definite fraction of a nth order reaction can be put as

$t \propto \frac{1}{a^{n-1}}$ where n is the order of the reaction.

Suppose we consider two experiments with initial concentrations a_1 and a_2. Let t_1 and t_2 be the times to complete the same fraction of the change. Then,

$$t_1 \propto \frac{1}{a_1^{n-1}} \text{ and } t_2 \propto \frac{1}{a_2^{n-1}}$$

$$\frac{t_2}{t_1} = \frac{a_1^{n-1}}{a_2^{n-1}} \text{ or } \frac{t_2}{t_1} = \left(\frac{a_1}{a_2}\right)^{n-1} \quad ...(1)$$

Taking logarithms of this equation, we get

$$\log\left(\frac{t_2}{t_1}\right) = (n-1)\log\left(\frac{a_1}{a_2}\right)$$

or $$(n-1)\log\left(\frac{a_1}{a_2}\right) = \log\frac{t_2}{t_1}$$

or $$(n - 1)(\log a_1 - \log a_2) = (\log t_2 - \log t_1)$$

or $$n - 1 = \frac{\log t_2 - \log t_1}{\log a_1 - \log a_2}$$

or $$n = 1 + \frac{\log t_2 - \log t_1}{\log a_1 - \log a_2} \quad ...(2)$$

By using the equation, the order of the reaction can be determined.

4. Graphical Method

As we have already discussed that the reaction velocity in a first order reaction is determined by the variation of one concentration term, in the second order by the variation of two concentration terms, and so on. Mathematically, it maybe put as

$$\frac{dx}{dt} = k_1 (a - x)^1 \text{ for the first order reaction}$$

$$\frac{dx}{dt} = k_2 (a - x)^2 \text{ for the second order reaction}$$

$$\frac{dx}{dt} = k_3 (a - x)^3 \text{ for the third order reaction}$$

....

....

$$\frac{dx}{dt} = k_n (a - x)^n \text{ for the nth order reaction.}$$

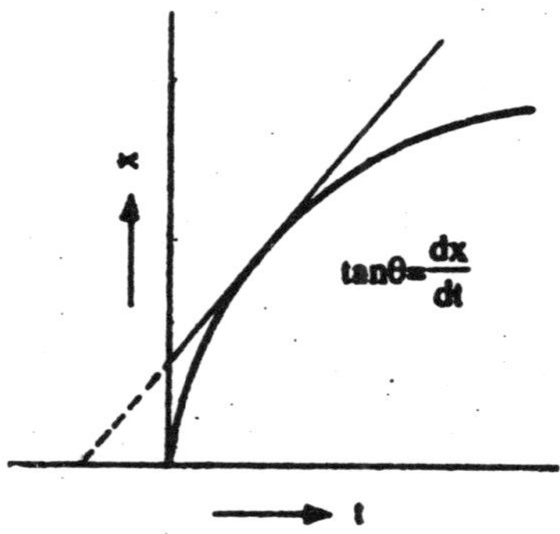

Fig. 2.4

If a curve is plotted between dx/dt and (a – x) at different time intervals, a straight line is obtained for the first order reaction. But if a straight lines obtained by plotting dx/dt against $(a - x)^2$, then the reaction is said to be of the second order. Similarly, if a straight line is obtained by plotting dx/dt against $(a - x)^n$, then the reaction is of the nth order.

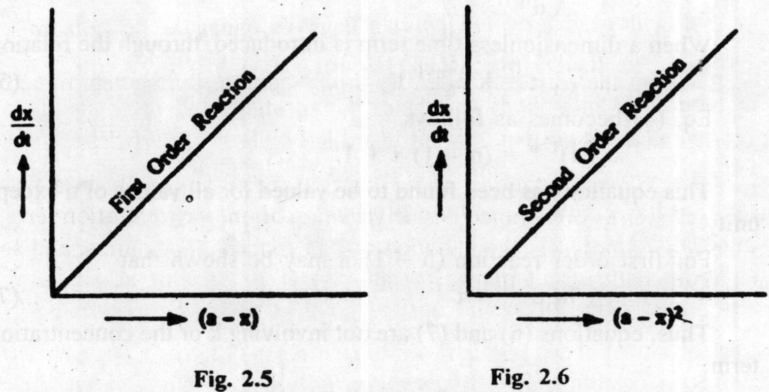

Fig. 2.5 **Fig. 2.6**

In this method, the value of dx/dt at different time intervals can be obtained by plotting x (the amount of substance decomposed) against time t. The value of tan θ at any time will give the value of $\frac{dx}{dt}$ at that time.

$$\frac{dx}{dt} = \tan\theta.$$

5. Powell Graphical Method

Powell (1961) introduced a unique graphical method for determining the order of a reaction. This method makes use of dimensionless parameters. For an nth order reaction (for equal concentration) we have,

$$kt = \frac{1}{(n-1)}\left\{\frac{1}{a^{n-1}} - \frac{1}{a_0^{\,n-1}}\right\} \qquad ...(1)$$

At any time during the progress of the reaction, the relative reactant concentration (α) may be put as follows.

$$a = \frac{a}{a_0} \qquad ...(2)$$

On substituting the value of a from Eq. (2), we get

$$\frac{1}{(n-1)\,a_n^{\,n-1}}\left(\frac{1}{\alpha^{n-1}}-1\right) = k\,t \qquad ...(3)$$

$$\left(\frac{1}{\alpha^{n-1}}-1\right) = k\,(n-1)\,a_0^{\,n-1}\,t \qquad ...(4)$$

When a dimensionless time term is introduced, through the relation

$$\tau = k\,a_0^{\,n-1}\,t \qquad ...(5)$$

Eq. (4) becomes as follows.

$$(\alpha)^{1-n} = (n-1)\,\tau + 1.$$

This equation has been found to be valued for all values of n except unity.

For first order reaction (n = 1) it may be shown that

$$\ln \alpha = -\tau \qquad ...(7)$$

Thus, equations (6) and (7) are not involving k or the concentration term.

For each order, thus a characteristic relation exists between α and τ. Plots of α against lot τ have been shown have been used for the determination of order and rate constant.

It reality, experimentally obtained values of a have been plotted against lot τ (since lot τ cannot be determined a priopri). These plots posses characteristic shapes as shown in Fig 2.7, but will be shifted along lot t axis. Amount of shift equals $-\log ka_0^{n-1}$, gives the value of k because n is determined by matching the a vs log t plots to those given above.

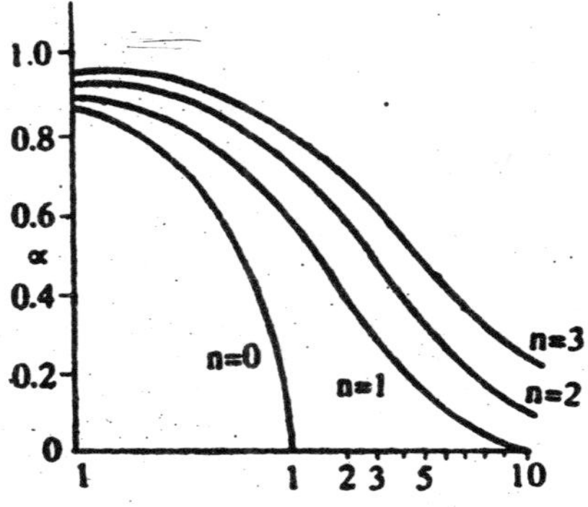

Fig. 2.7 : Plots of $\alpha = a/a_0$ vs $\log_{10} \tau$ for zero, first, second and third order reactions.

6. Vant Hoff's Differential Method

In 1884 vant Hoff suggested that the rate of the reaction of nth order is proportional to the nth power of concentration *i.e.,*

$$-\frac{dc}{dt} \propto c^n$$

or
$$-\frac{dc}{dt} = kc^n \qquad ...(1)$$

where c is the concentration of the reacting substance. Taking logarithms of equation (1), we get

$$\log\left(-\frac{dc}{dt}\right) = \log k + n \log c.$$

Suppose we start with experiments having initial concentration c_1 and c_2. Then

$$\log\left(-\frac{dc_1}{dt}\right) = \log k + n \log c_1 \qquad ...(2)$$

and
$$\log\left(-\frac{dc_2}{dt}\right) = \log k + n \log c_2 \qquad ...(3)$$

Substituting equation (3) from (2), we get

$$\log\left(-\frac{dc_1}{dt}\right) - \log\left(-\frac{dc_2}{dt}\right) = n (\log c_1 - \log c_2)$$

$$n = \frac{\log\left(\frac{dc_1}{dt}\right) - \log\left(-\frac{dc_2}{dt}\right)}{\log c_1 - \log c_2}.$$

This equation can be used for calculating the value of order of a reaction provided the values of dx_1/dt and dc_2/dt are known. These values can be measured by plotting the c against time t. Then the value of dc/dt can be measured from graphs by measuring angle θ at different places on the curves. The value of tan θ will give the value of dx/dt.

7. Isolation Method

This method id due to *Ostwald* (1902). In this method, all the reactants except one are taken in large quantities so that concentrations remain constant throughout this change. Thus, the order of the reaction is determined with respect to the isolated reactant which is not taken in large quantity. The experiment is repeated by isolating each reactant in turn. The total order of the reaction will be given by the sum of the order of isolated reactions.

This can be seen by considering the following general reaction.

$$n_1A + n_2B + n_2C \rightarrow \text{Products}$$

Reaction velocity is given by

$$\frac{dx}{dt} = C_{An_1} C_{Bn_1} C_{Cn_3}.$$

In the first experiment, the reactants B and C taken in large excess and the order of the reaction is to be determined with respect to A. Let this order be n_1 with respect to A.

In the second experiment, the reactants A and C are taken in large excess and the order of the reaction is to be determined with respect to B. Let this order be n_2. In the third experiment, the reactants A and B are taken in large excess and the order of this reaction is determined with respect to C. Let this order be n_3.

Total order of the reaction will be $n_1 + n_2 + n_3$ *i.e.,*

Total order = $n_1 + n_2 + n_3$.

8. Method of Ratio Variation

A complex reaction involving different substances may be studied by increasing greatly the concentration of the reactants, one at a time, and observing the reaction rates over a period in which the change in composition is not over 10 per cent. Consider, for instance, the reaction

$$A + B \longrightarrow C$$

the rate of which is given by

$$\frac{dx}{dt} = k\,[A]^m\,[B]^n.$$

The index m can be evaluated by doubing [A], while keeping [B] constant, and determining experimentally the change in the rate $\Delta x/\Delta t$, since

$$\frac{\left(\frac{\Delta x}{\Delta t}\right)_2}{\left(\frac{\Delta x}{\Delta t}\right)_1} = \frac{k[A]^m.2^m.[B]^n}{k.[A]^m.[B]^n} = 2n$$

where the subscripts 2 and 1 refer to the concentrations [2A] and [A] respectively. A simple procedure is adopted for the evaluation of n. The order of the reaction is then m + n.

By this method the kinetics of the following reaction was studied.

$$K_2C_2O_4 + 2HgCl_2 \longrightarrow HgCl_2 \longrightarrow Hg_2Cl_2 + 2KCl + 2CO_2$$

and the rate is seen to depend on the square of the concentration of potassium oxalate while actually there is only one molecule of it in the stoichiometric equation. However, the rate equation is

$$\frac{dx}{dt} = k\,[HgCl_2]\,[K_2C_2O_4]^2$$ and the overall order being 1 + 2 *i.e.*, 3.

9. Guggenheim's Method of Analysis

There are certain experiments in which it is necessary to take readings at t = ∞ before it can be analysed while this method does not involve this reading. The principle underlying this method is to take readings at equal intervals of time. The mathematics involving this method is as follows.

The equation for a first order reaction is as follows.

$$C_1 = C_0\,e^{-kt} \qquad ...(1)$$

$$C_2 = C_0\,e^{-k(t\,+\,\Delta t)} \qquad ...(2)$$

or $$C_1 - C_2 = C_0\,e^{-kt}\,(1 - e^{-k\Delta t})$$

or $$\ln\,(C_1 - C_2) = -\,kt + \ln\,C_0\,(1 - e^{-k\Delta t}) \qquad ...(3)$$

When ln $(C_1 - C_2)$ is plotted against t, the slope of the curve gives the value of k.

10. Comparison of Methods

The differential method is the most reliable one for investigating a reaction about which there is not much previous information. The main reason is that it does not require a tentative decision as to what the order should be. Instead, it determines directly how the rate depends on concentration. If the double logarithmic plots are linear, the reaction is having an order, which is given by the slope. If they are not, other procedures are used to determine the rate-concentration relationships.

A second reason for the superiority of the differential method is that it is able to distinguish clearly between the two orders of reaction, n_c

and n_t. This distinction is helpful in revealing information about the influence of products on the rates.

If the order of a reaction is reaction is well established, the method of integration may be useful for providing accurate rate constants. Otherwise, the method is having its limitations, for various reasons. In the first place, it creates a prejudice in favour of integral of half integral orders, and deviations from such order might escape notice. For example, a reaction having an order of 1-8 might fit the second-order integrated equation within the experimental error, and therefore the deviation from second-order kinetics would escape detection. The differential method would be more likely to indicate such a deviation.

Another difficulty with the method of integration is that the equations used, giving the variation of concentrations with time, are often quite similar for different kinetic laws. For example, the time course of a simple second order reaction order reaction is similar to that of a first-order reaction inhibited by products, and unless the experiments are done very carefully there is danger of confusion. The differential method, on the other hand, clearly reveals inhibition by products and avoids such confusion.

The method of integration gives rise in general to two difficulties when the two orders n_c and n_t are different. The analysis of a given run by the method of integration involves nt, the order with respect to time. If the work gets repeated at different reactant concentrations, the rate constants obtained will be the same only if the order with respect to concentrations, n_c is the same as n_t. If they are different, the rate constants vary with the initial concentration, and a careful analysis of the variation allows n_c to be obtained.

Kinetic data have sometimes been analyzed by seeing how half-lives depend on concentration. This method, however, is valid only when the two orders n_c and nt are the same. The half-life itself involves, n_t, while its variation with concentration involves nc in addition. Therefore, the order determined by this procedure will be a confusion mixture of n_c and n_t. The method cannot be recommended for reactions of previously unknown kinetics.

DISTURBING FACTORS IN THE DETERMINATION OF ORDER OF A REACTION

Only a few reactions are straight-forward, *i.e.,* first or second or third order reactions. Other reactions are complicated because the interpretation of their velocity data becomes difficult due to the occurrence of several reactions which take place simultaneously along the main reaction are known as *simultaneous reactions*. The simultaneous reactions may be divided into four types.

(a) **Side Reactions or Parallel Reactions :** When a reactant [A] undergoes two or more independent reactions at the same time, it is a case of side reactions. Each independent reaction gives rise to its own set of products.

$$A \begin{cases} \xrightarrow{\text{I}} B \; 90\% \\ \xrightarrow[\text{II}]{} C \; 10\% \end{cases}$$

Consider a reaction, where A is the reactant which undergoes two reactions I and II to from the products B and C. The reaction I is known as the *main reaction* and the reaction II known as *side reaction.* An example of side reactions is that when ethyl alcohol is made to react under different conditions, the following side reactions along the main reaction may occur.

$$C_2H_5OH \begin{cases} \xrightarrow{k_1} C_2H_4 + H_2O \\ \xrightarrow{k_2} CH_3CHO + H_2 \\ \xrightarrow{k_3} C_2H_5OC_2H_5 + H_2O \end{cases}$$

It is important to remark here that it is possible to convert any of the side reactions into the main reactions by adjusting the favourable experimental conditions,

Mathematical Treatment of Side Reactions : The side reaction is

$$A \begin{cases} \xrightarrow{k_1} B \text{ (I)} \\ \xrightarrow{k_2} C \text{ (II)} \end{cases}$$

Suppose a is the initial concentration of A. Let x of A decompose in time t to form partly as B and partly as C. If y and z are the amounts of B and C formed in time t, then we can write

Rate of disappearance of A $= -\,dx/dt$

Rate of formation of B and C together $= dx/dt$ ...(1)

Rate of formation of B $= dy/dt$ ...(2)

Rate of formation of C $= dz/dt$ ...(3)

As both B and C are formed from A, one can write

$$\frac{dx}{dt} = \frac{dy}{dt} + \frac{dz}{dt} \quad ...(3a)$$

If both the reactions I and II are of first order, the velocity for the formation of B from A is given by

$$dy/dt = k_1 (a - x) \quad ...(3b)$$

and the velocity for the formation of C from A is given by

$$dx/dt = k_2 (a - x) \quad ...(4)$$

where k_1 is the first order velocity constant for the formation of C from A. Substituting equations (3) and (4) in equation (2), we obtain

$$\frac{dx}{dt} = k_1 (a - x) + k_2 (a - x) = (a - x)(k_1 + k_2)$$

or
$$\frac{dx}{(a-x)} = (k_1 + k_2) \text{ or } \frac{dx}{a-x} = K\,dt \quad ...(5)$$

Integrating equation (5), we get

$$K \text{ (say)} = \frac{1}{t} \log \frac{a}{a-x} \quad ...(6)$$

where $K = k_1 + k_2$.

The value of K can be obtained by noting the concentration changes in A with time. But the separate evaluations of k_1 and k_2 require some relationship between them. Remembering that the amounts of B and C formed at any time will depend upon the rate of two reactions. Therefore, one can write

$$\frac{\text{Amount of B at any stage}}{\text{Amount of C at the same stage}} = \frac{\text{Rate of formation of B}}{\text{Rate of formation of C}} = \frac{dy/dt}{dz/dt}$$

or
$$\frac{\text{Amount of B at any stage}}{\text{Amount of C at the same stage}} = \frac{k_1 (a-x)}{k_2 (a-x)} = \frac{k_1}{k_2} = K' \text{ (say)} \quad ...(7)$$

[From equation (3a) and (4)]

Thus $$K' = \frac{k_1}{k_2}$$ ...(8)

Thus, the estimations of B and C at the end of the reaction would give the value of K'. Hence, the values of k_1 and k_2 can be evaluated separately by using equations (6) and (78) in terms of K and K' which can be measured experimentally.

From Equation (6) and (8), we have $k_1 + k_2 = k$

and $$\frac{k_1}{k_2} = K' \quad \text{or} \quad k_1 = K' k_2$$ ...(9)

Substituting equation (9) in (6), we get

$$K'k_2 + k_2 = K \quad \text{or} \quad k_2 (K' + 1) = K$$

or $$k_2 = \frac{K}{K'+1}$$ ...(10)

Substituting the value of k2 from equation (10) in (6), we get

$$k_1 = \frac{KK'}{1+K'}$$ ...(11)

Wegscheider's Test for Side Reactions : This test is used to distinguish side reactions from opposing or consecutive reactions. It states that

"*The ratio of the amount os substances formed in two side reactions is independent of time provided the order of two side reactions be the same.*"

This statement implies that this test is not applicable if the orders of two side reactions are different. This can be seen by considering the following cases.

Case I : *Side Reactions of the First order.* Consider the reaction,

$$A \begin{cases} \xrightarrow{k_1} B \\ \xrightarrow[k_2]{} C \end{cases}$$

where both the side reactions are of first order. Then

$$\frac{\text{Rate of formation of B}}{\text{Rate of formation of C}} = \frac{\text{Amount of B at any stage}}{\text{Amount of C at any stage}}$$

$$= \frac{k_1 (a - x)}{k_2 (a - x)} = \frac{k_1}{k_2} = K' \quad ...(13)$$

Thus, the constant K′ is independent of time.

Case II : *Side Reactions of nth Order.* Consider the case when both the side reaction are the nth order.

$$A \begin{cases} \xrightarrow{\text{nth order}} B \\ \xrightarrow[\text{nth order}]{} C \end{cases}$$

Rate of formation B = dz/dt = $k_1 (a - x)^n$ ...(14)

Rate of formation of C = dz/dt = $k_2 (a - x)^n$. ...(15)

$$\text{Thus, } \frac{\text{Rate of formation of B}}{\text{Rate of formation of C}} = \frac{\text{Amount of B at any stage}}{\text{Amount of C at any stage}}$$

$$= \frac{k_1 (a - x)^n}{k_2 (a - x)^n} = \frac{k_1}{k_2} = K' \quad ...(16)$$

Again, K' is independent of time.

Case III : *Side Reaction of Different Order.* Consider the case in which the first side reactions is of nth order and the second of mth order.

$$A \begin{cases} \xrightarrow{\text{nth order}} B \\ \xrightarrow[\text{nth order}]{} C \end{cases} \quad ...(17)$$

Rate of formation of B = dy/dt = $k_1 (a - x)^n$ ...(18)

Rate of formation of C = dz/dt = $k_2 (a - x)^m$ ...(19)

Therefore,

$$\frac{\text{Rate of formatioan of B}}{\text{Rate of formatioan of C}} = \frac{\text{Amount of B at any stage}}{\text{Amount of C at any stage}}$$

$$= \frac{k_1 (a - x)^n}{k_1 (a - x)^m} = K' (a - x)^{n-m} \quad ...(20)$$

K′ is not independent of the concentrations of A at any time and hence *Wegscheider's test fails in such cases.*

Applications of Wegscheider's Test : Skraup showed that the reaction between cinchonine and hydrochloric acid first forms cinchonine hydrochloride which then isonerises to form an isomeric form of cinchonine hydrochloride. Thus, he considered this reaction to be consecutive but not side reaction

Cinchonine + HCl + HCl → Cinchonine hydrochloride → isomer of cinchonine hydrochloride. Wegscheider studied this reaction experimentally and found that the amounts of ordinary cinchonine hydrochloride and its isomeric form were in constant ratio to one another and independent of the time chosen for the measurement. These observations clearly demonstrated that this reaction was case of side reaction but not the consecutive one.

$$\text{Cinchonine hydrochloride} \begin{cases} \nearrow \text{Cinchonine hydrochloride} \\ \searrow \text{Isomer of cinchonine hydrochloride} \end{cases}$$

(ii) Dehydration of alcohols:

$$C_2H_5OH \begin{cases} \nearrow C_2H_4 + H_2O \\ \searrow CH_3CHO + H_2 \end{cases}$$

(b) Opposing Reactions : *There are certain reactions in which the products of a chemical change react to form the original reactants. Such reactions are known as counter, reversible or opposing reactions.*

In opposing reactions, the net rate of the reaction will be influenced by both the forward and backward rates and thus causes a serious disturbance in the measurement of reaction velocity. Let us consider the various cases of opposing reactions.

Case I : Consider the simplest case when both the forward and backward reactions are of the first order

$$A \underset{k_2}{\overset{k_1}{\rightleftharpoons}} B \qquad \text{...(21)}$$

where k_1 and k_2 represent the *first order rate constants* for the *forward and backward* reactions respectively. Suppose a is the initial concentration of A. After a time t, x gm molecules of A have decomposed to form B. Then, the concentration of A at time t is a – x and the concentration of B at the same time is x. Thus,

Rate of the forward reactions = k_1 (a – x) ...(22)

Rate of the backward reaction = k_2x ...(23)

The net rate of production of B is

$$\frac{dx}{dt} = k_1 (a - x) - k_2x \qquad ...(24)$$

At equilibrium, the net rate is zero, *i.e.*, dx/dt = 0, thus

$$dx/dt = k_1 (a - x_e) - k_2x_e = 0$$

or $$k_1 (a - x_e) = k_2x_e \text{ or } k_2 = \frac{k_1 (a - x_e)}{x_e} \qquad ...(25)$$

where xe is the concentration at equilibrium.

Substituting the value of k_2 from equation (25) in eq. (24), we get,

$$\frac{dx}{dt} = k_1 (a - x) - \frac{k_1}{x_e} (a - x_e)\, x$$

$$= k_1a - k_1x - \frac{k_1a}{x_e} x + \frac{k_1x}{x_e} . x_e$$

$$\frac{dx}{dt} = k_1a - \frac{k_1a}{x_e} x = k_1a\left[1 - \frac{x}{x_e}\right] = \frac{k_1a}{x_e} [x_e - x]$$

or $$\frac{dx}{x_e - x} = \frac{k_1a}{x_e} dt \qquad ...(26)$$

Integrating this equation, we obtain

$$- \ln (x_e - x) \frac{k_1a}{x_e} t + \text{Integration constant.} \qquad ...(27)$$

When t = 0, x = 0, equation (27) becomes as

$$- \ln x_e = \text{Integration constant.} \qquad ...(28)$$

Substituting equation (28) in equation (27), we get

$$- \ln (x_e - x) = \frac{k_1a}{x_e} t - \ln x_e$$

or $$\ln x_e - \ln (x_e - x) = \frac{k_1a}{x_e} t$$

$$\ln \frac{x_e}{x_e - x} = \frac{k_1a}{x_e} t \text{ or } k_1 = \frac{x_e}{at} \ln \frac{x_e}{x_e - x} \qquad ...(29)$$

Thus, the value of k_1 can be calculated by using equation (29) if xe, x and t are known. The value of x can be calculated by noting the progress of the reaction at different time intervals 't' before the equilibrium is attained. When equilibrium is attained, the determination of concentration gives the value of xe. By substituting the value of k_1 (obtained as above) in eq. (25), the value of k_2 can be determined.

Examples 1:

The mutarotation of α-D-glucose into β-Dgluocose is an interesting example of a reversible first order reaction.

$$\alpha - \text{Glucose} \underset{k_2}{\overset{k_1}{\rightleftharpoons}} \beta - \text{Glucose}$$

(ii) Another interesting example is the conversion of γ-hydroxy butyric acid into lactone.

$$CH_2OHCH_2CH_2COOH \rightleftharpoons \begin{array}{c} O \text{———————} CO \\ | \qquad\qquad\quad | \\ CH_2\text{—}CH_2\text{—}CH_2 \end{array} + H_2O$$

(iii) Conversion of ammonium cyanate into urea

$$NH_4CNO \rightleftharpoons NH_2CONH_2$$

Case II : Let us consider the next case when the first order reaction is opposed by the reaction of the second order

$$A \underset{k_2}{\overset{k_1}{\rightleftharpoons}} B + C$$

where A is the reactant, B and C are products, k_1 is the rate constant for the froward reaction, and k_2 is the second order rate constant for the backward reaction.

Suppose a is the initial concentration of pure A, *i.e.*, the concentration of B and C being initially zero. Let x of A decompose in time t to form x of B and C. Then

Rate of the forward reaction $= k_1 (a - x)$

where $(a - x)$ is the concentration of A at time t

Rate of the forward reaction $= k_2x^2$.

At equilibrium,

Rate of the forward reaction = Rate of the backward reaction

$$\therefore \quad k_1 (a - x_e) = k_2x_e^2 \qquad ...(30)$$

where xe is the concentration of B and C at equilibrium, *i.e.*, $x = x_e$

From equation (30), we have

$$k_2 = k_1 \frac{(a - x_e)}{x_e^2} \qquad ...(31)$$

But the net rate of forward reaction is given by

$$\frac{dx}{dt} = k_1 (a - x) - k_2 x_2 \qquad ...(32)$$

Combining equation (31) and (32), we get

$$\frac{dx}{dt} = k_1 (a - x) - k_1 \frac{(a - x_e)}{x_e^2} x_2 \qquad ...(33)$$

On separating the variables and integrating the equation, we get

$$k_1 = \frac{x_e}{t(2a - x_e)} \ln \frac{ax_e + x(a - x_e)}{a(x_e + x)}$$

Examples 2:

(i) Hydrolysis of ethyl acetate.

(ii) Decomposition of alkyl ammonium halide in a solution.

Case III : Let us consider the case when the second order reaction is opposed by the reaction of the same order.

	A	+	B	$\rightleftharpoons$	C	+	D
Initially	a		b		c		d
	(a – x)		(b – x)		(c + x)		(d + x)

Rate of the forward reaction $= k_1 (a - x)(b - x)$ ...(34)

Rate of the backward reaction $= k_2 (c + x)(d + x)$ ...(35)

The net rate of the forward reaction is expressed as

$$\frac{dx}{dt} = k_1 (a - x)(b - x) - k_2 (c - x)(d + x).$$

On multiplying equation (36) by $1/k_2$, we get ...(36)

$$\frac{1}{k_2}\frac{dx}{dt} = \frac{k_1}{k_2} (a - x)(b - x) - (c + x)(d + x) \qquad ...(37)$$

$$= K (a - x)(b - x) - (c + x)(d + x) \left[\because K = \frac{k_1}{k_2}\right]$$

$$= K\,[ab + x^2 - (a + b)\,x^2] - [cd + x^2 + (c+d)\,x^2]$$
$$= [K - 1)\,x^2 - x\,[K\,(a + b) - (c+d)] + (K\,ab - cd)$$
$$= [x - A]\,[x - B]$$

$$A = \frac{Q+P}{2(K-1)'} \quad B = \frac{Q-P}{2(K-1)'}$$

$$Q = K\,(a + b) - (c + d)$$

$$P = \sqrt{[Q^2 - 4(K-1)(Kab - cd)]} \quad \text{and } K = \frac{k_1}{k_2}$$

Equation (38) is integrated to give

$$k_2 t = \frac{1}{A-B}\ln\frac{B(x-A)}{C(x-B)}.$$

Examples 3:

(i) Esterification of acids,

(ii) Dissociation of hydrogen iodide.

(c) Consecutive Reactions : *In such reactions, the products obtained in the first stage react with each other or with the reactants to form new products. This means that consecutive reactions proceed in stages.* It is important to remark that the velocity coefficient is being determined by such a stage which is the slowest. For example.

$$A \xrightarrow{k_1} B \xrightarrow{k_2} C \qquad ...(40)$$

This is the simplest case of a consecutive reaction in which both stages are of first order. Suppose a is the initial concentration of A and let x, y and z denote concentrations of A, B and C respectively at time t. Therefore, we can write

$$x + y + z = a \qquad ...(41)$$

The rate of disappearance of A is

$$-\,dx/dt = k_1 x \qquad ...(42)$$

where k_1 is the rate constant for the first stage (A → B).

Equation (42) can be written as

$$-\frac{dx}{x} = k_1\,dt \qquad ...(43)$$

Integrating equation (43), we get

$$-\ln x = k_1 t + I_c \qquad ...(44)$$

where I_c is the integration constant. When t = 0, x = a, equation (44) becomes as

$$-\ln a = I_c \qquad ...(45)$$

Substituting equation (45) in equation (44) we get

$$-\ln x = k_1 t - \ln a$$

$$\ln x - \ln a = -k1t$$

$$\ln \frac{x}{a} = -k_1 t \text{ or } x = are^{-k_1 t} \qquad ...(46)$$

Rate of formation of C is given by $\frac{dz}{dt} = k_2 y$ ...(47)

where k_2 is the rate constant for the formation of C from B.

Since the rate of formation of B is given by k_1x, and the rate of decomposition of B into C is given by k_2y, the rate at which B accumulates (dy/dt) in the system is written as

$$\frac{dy}{dt} = -\frac{dx}{dt} - \left(+\frac{dz}{dt}\right) = -\frac{dx}{dt} - \frac{dz}{dt} \qquad ...(48)$$

Substituting equations (42) and (47) in equation (48),

$$dy/dt = k_1 x - k_2 y$$

Substituting the value of x from equation (46) in equation,

$$dy/dt = k_1 a\, e^{-k_1 t} - k_2 y.$$

As this is a linear differential equation of the first order, the solution of this is

$$y = are^{-k_2 t}\left[\frac{k_1 e^{(k_2 - k_1)t}}{(k_2 - k_1)} - \frac{k_1}{k_2 - k_1}\right]$$

$$y = \frac{k_1 a}{k_2 - k_1}[e^{-k_2 t} - e^{-k_2 t}] \qquad ...(51)$$

Butz = a − x − y

Substituting equations (40) and (51) in equation (42),

or $$z = a - ae^{-k_1 t} - \frac{k_1 a}{k_2 - k_1}[e^{-k_1 t} - e^{-k_2 t}]$$

or $$z = \frac{a}{k_2 - k_1}[(k_2 - k_2 e^{-k_1 t}) - (k_1 - k_1 e^{-k_2 t})].$$

In Fig. (2.8) x, y, and z are plotted against functions of time for $a = 1$, $k_1 = 0.01$ and $k_1 = 0.02$. The curves corresponding to x, y and z are shown by A, B and C in the Fig. (2.8). From this Fig. (2.8), it is observed that

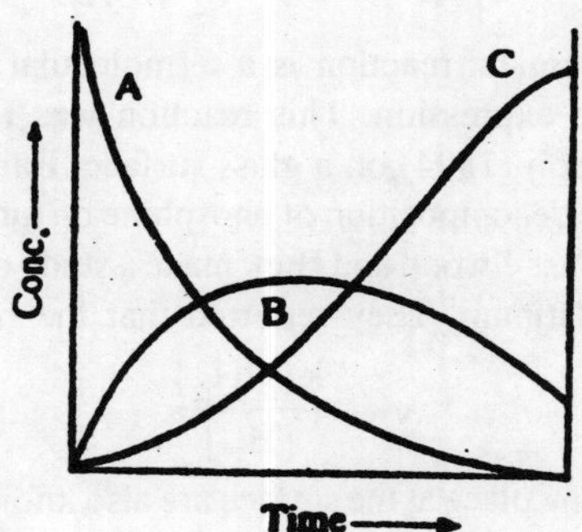

Fig. 2.8

(i) the concentration of A falls continuously.

(ii) the concentration of C rises continuously.

(iii) and the concentration of B rises to a maximum and then decreases with time.

Examples : (a)

Thermal decomposition of acetone occurs in two stages.

(i) $(CH_3)_2C{=}O \longrightarrow \underset{\text{ketene}}{CH_2 = C = O} + CH_4$.

(ii) $CH_2 = C = O \rightarrow \frac{1}{2}C_2H_4 + CO$

(b) The hydrolysis (acid) of ester of dibasic acid, (*e.g.*, diethyl succinate) occurs in two stages.

(i) $(CH_2COOC_2H_5)_2 + H_2O \longrightarrow$ CH_2COOH | $CH_2COOC_2H_5$ $+ C_2H_2OH$

H_2COOH

$$\text{(ii)}\ \begin{array}{l} CH_2COOH \\ | \\ CH_2COOC_2H_5 \end{array} + H_2O \quad \begin{array}{l} CH_2COOH \\ | \\ CH_2COOH \end{array} + C_2H_5OH$$

Surface Reaction : *Reactions which take place exclusively on the walls of the containing vessel are known as surface reactions.* Such reactions are mostly gaseous. Examples are the decomposition of AsH_3 and PH_3

$$4PH_3 \longrightarrow P_4 + 6H_2.$$

The above chemical reaction is a termolecular reaction but obeys the first order rate expression. This reaction was first investigated by Van't Hoff and Kooij (1894) on a glass surface. Barrer (1936) made an investigation of the decomposition of phosphine on tungsten order kinetics at high pressure. Hinshelwood and Burk made a study of the decomposition of ammonia on platinum. They reported that the rate law is

$$v = \frac{k[NH_3]}{[H_2]}$$

Reactions taking place at the surface are also known as *heterogeneous reactions.*

EFFECT OF TEMPERATURE ON REACTION RATES

It is a well established fact that the velocity of a chemical reaction increases with rise in temperature. In homogeneous reactions, the rate becomes doubled or trebled for each 10° rise of temperature. This increase in the reaction velocity with temperature is sometimes expressed in the form of temperature coefficient, which is defined as "*Temperature co-efficient of a chemical reaction is defined as the ratio of rate constants of a reaction at two different temperatures separated by 10°C. The two temperatures are generally taken as 35° and 25°C.*

Thus, the temperature co-efficient is expressed as.

$$\text{Temperature coefficient} = \frac{k_{35°}}{k_{25°}} \qquad ...(1)$$

In general the temperature co-efficient is expressed as

$$\textit{Temperature co-efficient} = \frac{k_{t+10°}}{kt} \qquad ...(2)$$

where k_t is the specific rate of the reaction at t°C and k_{t+10} is the specific rate of same reaction at t + 10°. In most of the homogeneous gaseous reactions, the value of temperature co-efficient lies between 2 and 3. *It means that the specific reaction rate becomes two fold or three fold for every 10°C rise of temperature. Some examples are.*

(i) The temperature co-efficient for the dissociation of hydriodic acid is 1.7.

(ii) The temperature co-efficient for the reaction of methyl iodide with sodium ethoxide is 2.9.

$$CH_3I + C_2H_5ONa \rightarrow CH_3OC_2H_5 + NaI$$

(iii) The temperature co-efficient for the decomposition of nitrogen pentoxide is 3.8 at room temperature.

Limitations

(i) From the above examples, it is evident that temperature co-efficient is an approximate method for expressing the effect of temperature on the reaction rates.

(ii) It fails to explain that the number of collisions in a reaction is increasing by 2% whereas the reaction velocity is increasing by 100% or 200% or more.

These two points led Arrhenius to suggest his own equation to express the influence of temperature on reaction velocity.

EXPLANATION OF TEMPERATURE COEFFICIENT ON THE BASIS OF SIMPLE COLLISION THEORY

Observations on rate experiments have indicated that rise in temperature increases the rate of reaction and fall in temperature decreases the rate, no matter whether the reaction is exothermic or endothermic. Following examples illustrate the effect of temperature on reaction rates.

(1) A piece of coal kept in air at room temperature is not affected at all. However, the same piece of coal if heated in air gets burnt up.

(2) Hydrogen and oxygen gases if mixed together do not react at ordinary temperature, inspite of the fact that this reaction is exothermic. However, if hydrogen and oxygen gaseous mixture

is ignited, there is an explosion resulting in the formation of water.

Now the effect of temperature on reaction rates can be explained on the basis of collision theory as follows.

(i) All collisions between the reacting molecules are not effective in producing a chemical change. There are only some collisions which lead to the reaction and to the formation of products.

(ii) The reactant molecules should first acquire a minimum amount of energy called *threshold energy* and then collisions will occur to result in a chemical reaction. The collisions between molecules which are associated with energy less than threshold energy do not result in a chemical reaction. It is observed that only a fraction of total molecules can have this energy.

(iii) In any collection of molecules there is a distribution of energies between the molecules of the system at a particular temperature as shown in Fig. (2.9).

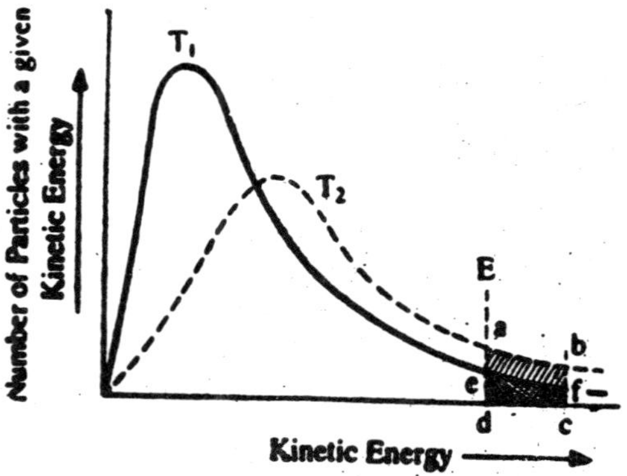

Fig. 2.9 : Energy distribution in molecules at different temperatures.

It is evident from the diagram that at given temperature T_1 some of the molecules possess very low kinetic energy while some others have high kinetic energy. However, most of the molecules possess intermediate kinetic energies. The molecules having energy greater than or equal to threshold energy are represented by the shaded portion c d e f.

As the temperature is raised, say from T°C to (T + 10)°C, *i.e.*, T_2, the whole distribution curve changes and is shifted to higher energies. Thus, the number of molecules having energy greater than or equal to threshold energy increases and this increases in Fig 2.9, is represented by the area a b e f. Thus, more molecules are effective in producing chemical changes. Hence the rate of reaction rises with rise in temperature.

(iv) Increase of temperature makes molecules to move faster. There is more frequent and more violent collisions with other molecules and hence more likely to cause reaction.

The above discussion answers the following questions.

(i) Why does a rise of temperature speed up a reaction ?

(ii) Why does a rise of temperature speed up a reaction to such a large extent ?

Arrhenius Equation : *Van't Hoof* proved that the value of equilibrium constant varied with the temperature. On the basis of this, *Van't Hoff* suggested that the logarithm of the specific rate of a chemical reaction must be linear function of the reciprocal of the absolute temperature.

Arrhenius : Extended the above suggestion of *Van't Hoff* and gave his own hypothesis. According to his hypothesis,

(i) *All the molecules of a system cannot take pari in the chemical reaction.*

(ii) *It is only a certain number of molecules which react. These reacting molecules are known as active molecules.*

(iii) *The molecules which do not take part in a chemical reaction are known as passive molecules.*

(iv) *There exists an equilibrium between passive and active molecules.*

$$M = (passive) \rightleftharpoons M\ (active)$$

(v) *When the temperature is raised, the equilibrium between active and passive molecules shifts towards the right. This increases the number of active molecules which are ready to take part in a reaction. Thus, the increase in reaction rate with rise in temperature is due to the increase in the number of active molecules than due to the number of collisions.*

Thus the basic concept of *Arrhenius* theory is that

'*The passive or non-active molecules can become active by absorption of heat energy.*'

DERIVATION OF ARRHENIUS EQUATION

Consider the following reversible reaction.

$$A + B \rightleftharpoons C + D$$

Rate of the forward reaction $= k_1 [A] [B]$ ...(3)

Rate of the backward reaction$= k_2 [C] [D]$

At equilibrium;

$$k1 [A] [B] = k_2 [C] [D]$$

or $$\frac{k_1}{k_2} = \frac{[C][D]}{[A][B]} = K_c = \frac{[C][D]}{[A][B]}$$

where $$K_c = \frac{k_1}{k_2} \quad ...(4)$$

From thermodynamics, Van't Hoff equation is

$$\frac{d \ln K_c}{dT} = \frac{\Delta E}{RT^2} \quad ...(5)$$

Substituting the value of K_c from equation (4) in (5), we get

$$\frac{d \ln \frac{k_1}{k_2}}{dT} = \frac{\Delta E}{RT^2} \quad ...(6)$$

or $$\frac{d \ln k_1}{dT} - \frac{d \ln k_2}{dT} = \frac{\Delta E}{RT^2}$$

Van't Hoff Proposed that equation (6) can be split up into two equations as follows

$$\frac{d \ln k_1}{dT} = \frac{E_1}{RT^2} + I \quad ...(7)$$

$$\frac{d \ln k_2}{dT} = \frac{E_2}{RT_2} + I \qquad ...(8)$$

where $\Delta E - E_1 - E_2$ and I is constant. It was observed that I was independent of temperature. Therefore, I must be equal to zero in equation (7) and (8).

$$\frac{d \ln k_1}{dT} = \frac{E_1}{RT^2} \qquad ...(9)$$

and
$$\frac{d \ln k_2}{dT} = \frac{E_2}{RT^2} \qquad ...(10)$$

Equations (9) and (10) are known as Arrhenius equations,

Integrating equations (9) and (10), we get

$$\ln k_1 = \frac{E_1}{RT} + \text{constant (A)} \qquad ...(1)$$

and
$$\ln k_2 = -\frac{E_2}{RT} + \text{constant (A)} \qquad ...(12)$$

Equations (11) and (12) can be put as

$$k_1 = A\ e^{-E_1/RT} \text{ and } k_2 = A\ e^{-E/RT} \qquad ...(12\ A)$$

In more generalk $= A\ e^{-E/RT}$ ...(13)

In equation (8). A is a constant. It is generally known as *frequency factor* of the reaction E is a term which has the dimensions of energy and is known as *energy of activation.* Equation (13) is known as *integrated form of Arrhenius equation.* Thus, equation (13) expresses the temperature dependence of k in terms of A and E. Both A and E are characteristics of the reaction.

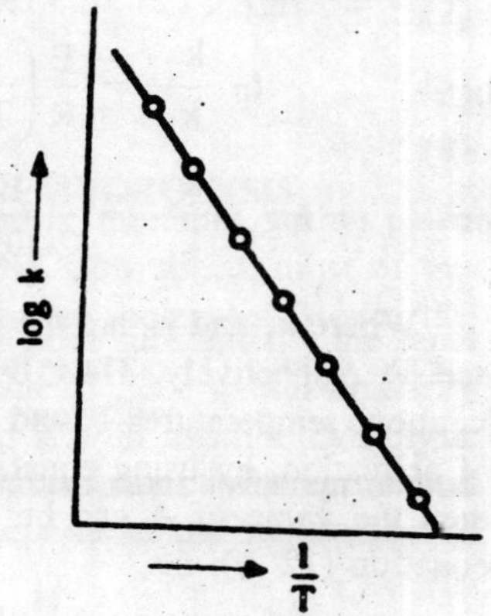

Fig. 2.10

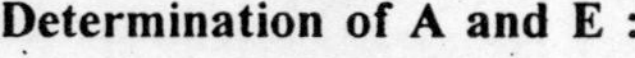

Determination of A and E :

First method

The logarithmic form of equation (13) is

$$\ln k = \ln A - \frac{E}{RT}$$

or since it is more convenient to use $\log_{10}$.

$$\log_{10} k = \log_{10} A - \frac{E}{2.303\,RT} \qquad ...(14)$$

If $\log_{10} k$ is plotted against 1/T, a straight line will be obtained. The slope of this line is given be

$$\text{Slope} = \frac{-E}{2.303\,RT} \qquad ...(15)$$

Thus, knowing the slope, the value of E can be easily calculated. The intercept of the line will give the value of log A. Such a plot is illustrated in the Fig. (2.10).

Second method. The logarithmic form of the equation (13) is

$$\ln k = \frac{E}{RT} + \text{Ln } A$$

Differentiating with respect to temperature, we obtain

$$\frac{d \ln k}{dT} = \frac{E}{RT^2}. \qquad ...(16)$$

Integrating equation (16) between the temperature limits T_1 and T_2, we get

$$\int_{k_1}^{k_2} d \ln = \int_{T_1}^{T_2} \frac{E}{RT^2}\, dT$$

or

$$\ln \frac{k_2}{k_1} = \frac{E}{R}\left(\frac{1}{T_1} - \frac{1}{T_2}\right)$$

or

$$\log_{10} \frac{k_2}{k_1} = \frac{E}{2.303\,R}\left(\frac{1}{T_1} - \frac{1}{T_2}\right) \qquad ...(17)$$

Where k_1 and k_2 are the velocity constants at two temperatures T_1 and T_2 respectively. Thus, by measuring the velocity constant k_1 and k_2 at two temperatures T_1 and T_2, it is possible to calculate E, the energy of activation, by using equation (17). When the value of E is known, then the value of A can be calculated by substituting value of E in equation (12 A), *i.e.,*

$$k_1 = Ae^{-E_1/RT} \text{ and } k_2 = A\,e^{-E_2/RT}.$$

The value of E is expressed in calories or better, kilocalories per gram molecule.

Activation Energy and Chemical Reaction : According to the concept of activation energy, all the molecules cannot take part in the chemical reaction. It is only a certain number of molecules that may be called active molecules which could take part in a chemical reaction. Thus, the reactants do not pass directly to the products but must first acquire necessary energy to pass over an energy barrier known as the activated state or transitions state. *The amount of energy which the reactants must absorb to pass over this activated energy barrier is known as the activation energy.*

The necessary energy of activation is acquired by the molecules as a result of interchanges occurring in collisions among the molecules.

SOLVED EXAMPLES

Example 1:

In the conversion of ammonium cyanate into urea, the following results were obtained by Humbly in 1885.

time (min)	0	45	72	157	312	600
(a – x) gm moles.	0.0916	0.0740	0.0656	0.0512	0.0348	0.0221

Find out the order of the reaction.

Solution:

In this problem,

$$a = 0.0916$$

t	a – x	x	$k_1 = \frac{2.303}{t} \log \frac{a}{a-x}$	$k_2 = \frac{1}{t}\frac{x}{a(a-x)}$
45	0.0740	0.0176	0.00574	0.0677
72	0.0656	0.0260	0.00462	0.0599
157	0.0512	0.0404	0.00370	0.0558
312	0.0348	0.0568	0.00308	0.0551
600	0.0228	0,0688	0.00231	0.0549.

Since the first order equation does not give a constant value of k while the second order equation yields a fairly constant value, the reaction is of the second order.

Example 2:

At a certain temperature, the half life periods for the catalytic decomposition of ammonia were found to be below.

Pressure (min)	50	100	200
Half-life period	3.52	1.92	1.0

Find out the order of reaction.

Solution:

We know

$$\frac{t_1}{t_2} = \left(\frac{a_2}{a_1}\right)^{n-1} = \left(\frac{p_2}{p_1}\right)^{n-1}$$

(a) In the present case.

$$p_1 = 50,\ p_2 = 100,\ t_1 = 3.52,\ t_2 = 1.92$$

$$\frac{3.52}{1.92} = \left(\frac{100}{50}\right)^{n-1}$$

or $(n - 1) \log 2 = \log \dfrac{3.52}{1.92}$

or $\qquad n = 1.88 \approx 2$

(b) With the second pair,

$$p_1 = 100 \qquad t_1 = 1.92$$

$$p_2 = 200 \qquad t_2 = 1.0$$

$$\frac{1.92}{1.0} = \left(\frac{200}{100}\right)^{n-1}$$

$$(n - 1) \log 2 = \log \frac{1.92}{1.0}$$

$$n = 1.95 \approx 2.$$

Hence the reaction is of second order.

Example 3:

From the work of L.T. Richer (1895) on the action of bromine on fumaric acid, the following data were obtained.

Ist Expt.		IInd. Expt.	
t (min.)	*Concentration*	*t (min.)*	*Concentration*
0	8.88	0	3.81
95	7.87	132	3.51

Mean concentration = 8.37   *Mean concentration* = 3.66

and $-\frac{dc_1}{dt} = 0.0106$ and $-\frac{dc_1}{dt} = 0.00227$

Find out the order of reaction.

Solution:

$$\frac{-dc_1}{dt} = 0.0106, \quad \frac{-dc_2}{dt} = 0.00227$$

$$c_1 = 8.37 \quad c_2 = 3.66$$

$$n = \frac{\log\left(\frac{-dc_1}{dt}\right) - \left(\frac{-dc_2}{dt}\right)}{\log c_1 - \log c_{22}}$$

$$= \frac{\log(0.0106) - \log(0.00227)}{\log 8.37 - \log 3.66}$$

$$= 1.87 \quad \text{or} \quad n \approx 2.$$

Example 4:

From the work of Dr. S.K. Anand on the action of acetone of iodine, the following data were obtained.

Exp. I	Exp. II
Concentration of I_2 *= 0.001 N*	*Concentration of* I_2 *= 0.0! N*
Concentration of HCl = 0.01	*Concentration of HCl = 0.01 N*

Concentration of Acetone = 3%Concentration of Acetone = 2%.

The progress of the reaction was examined by titrating 50 c.c. of reaction mixture at regular intervals against standard N/200 hypo solution. The following gives the value of hypo by the unreacted iodine.

Exp. I		Exp. II	
t (minutes)	*Volume of hypo used*	*time*	*Volume of hypro used*
0	10.0	0	10.0

10	8.5	10	9.0
20	7.0	20	8.0
30	5.5	30	7.0
50	2.5	50	5.0

Solution:

Exp. I			Exp. II		
Time (min.)	$C_0 - C$	$k_1 = \frac{C_0 - C}{t}$	Time	$C_0 - C'$	$k_2 = \frac{C_0 - C'}{t}$
10	1.5	$k_1 = 0.15$	10	1	$k_2 = 0.1$
20	3.0	$k_1 = 0.15$	20	2	$k_2 = 0.1$
30	4.5	$k_1 = 0.15$	30	3	$k_2 = 0.1$
40	6.0	$k_1 = 0.15$	40	4	$k_2 = 0.1$
50	7.5	$k_1 = 0.15$	50	5	$k_2 = 0.1.$

The overall order is given by

$$n = \frac{\log k_1 - \log k_2}{\log c_1 - \log c_2} = \frac{\log 0.15 - \log 0.1}{\log 3 - \log 2} = 1.$$

Hence the overall order of the reaction is one.

Example 5:

The chemical reaction between potassium oxalate and mercuric chloride is

$$2HgCl_2 + K_2C_2O_4 \rightarrow 2KCl + 2CO_2 + Hg_2Cl_2.$$

The weight of $HgCl_2$ *precipitated from different solutions in a given time at 100° C was as follows.*

$HgCl_2$ *Moles/litre*	$K_2C_2O_4$ *moles/litre*	*Time (min.)*	Hg_2Cl_2 *ppt. (moles)*
0.0836	0.404	65	0.0068
0.0836	0.202	120	0.0031
0.0418	0.404	60	0.0032

Calculate the order of the reaction.

Solution:

The differential rate equation can be written as

$$\frac{dx}{dt} = k\ [HgCl_2]m\ [K_2C_2O_4]n.$$

Thus, the order of the reaction is m + n.

From experiments (1) and (2), we have

$$\frac{\Delta x_1}{\Delta t_1} \Big/ \frac{\Delta x_2}{\Delta t_2} = \frac{k[HgCl_2]^m[2K_2C_2O_4]^n}{k[HgCl_2]^m[K_2C_2O_4]^n}$$

or $$\frac{0.0068}{65} \Big/ \frac{0.0031}{120} = 2^n$$

or $$4 = 2^n \text{ or } 2^2 = 2^n \text{ or } n = 2.$$

Similarly, from experiments (1) and (3), we have

$$\frac{\Delta x_1}{\Delta t_1} \Big/ \frac{\Delta x_2}{\Delta t_2} = \frac{k[2HgCl_2]^m[K_2C_2O_4]^n}{k[HgCl_2]^m[K_2C_2O_4]^n}$$

or $$\frac{0.0068}{65} \Big/ \frac{0.0032}{60} = 2^m \text{ or } 2 = 2^m \text{ or } 21 = 2^m \text{ or } m = 1.$$

Hence the order of the reaction = 2 + 1 = 3.

Example 6:

Determine the order of the reaction from the following data.

S. No.	*Conc. of A (moles/litre)*	*Conc. of B (moles/litre)*	*Rate of reaction (moles/litre) min^{-1}*
I	1.00	0.02	4.6×10^{-4}
II	1.00	0.04	9.2×10^{-4}
III	0.02	1.00	9.2×10^{-6}
IV	0.04	1.00	3.68×10^{-5}

Determine the true rate constant also.

Solution:

When the concentration of A is very large, rate depends upon the concentration of B. Therefore, nB, (the order of reaction with respect to B) is calculated using following equation from sets I and II.

$$n_B = \frac{\log\left(\frac{-dc_1}{dt}\right) - \log\left(\frac{-dc_2}{dt}\right)}{\log c_1 - \log c_2}$$

$$= \frac{\log(4.6\times10^{-4}) - (9.2\times10^{-4})}{\log 0.02 - \log 0.04} = 1.$$

When B is in large quantity as in sets III and IV, n_A is obtained using equation,

$$n_A = \frac{\log 9.2\times10^{-6} - \log 3.68\times10^{-5}}{\log 0.02 - \log 0.04} = 2.$$

Therefore, the order of reaction in $n_A + n_B$, *i.e.*, 2 + 1 = 3

or The true rate constant $= \dfrac{\text{Rate}}{C_A C_B}$.

From the four sets of data

(I) $k = \dfrac{4.6\times10^{-4}}{(1.00)^2\,(0.02)} = 2.3\times10^{-2}$ (litre/mole) min^{-1}.

(II) $k = \dfrac{9.2\times10^{-4}}{(1.00)^2\times0.04} = 2.3\times10^{-2}$

(III) $k = \dfrac{9.2\times10^{-6}}{(0.02)^2\times1.00} = 2.3\times10^{-2}$

(IV) $k = \dfrac{3.68\times10^{-5}}{0.04\times1.00} = 2.3\times10^{-2}$.

Example 7:

Now we consider the same reaction with different set of data

$A \rightarrow B + C$

Time	*0*	*t*
Total pressure of A + B + C	P_1	P_2

Find k.

Solution:

In this we are given total pressure of the system at these time intervals. The total pressure obviously includes the pressure of A, B and

C. AT $t = 0$, the system would any have A. Therefore the total pressure at $t = 0$ would be the initial pressure of A \ P_1 is the initial pressure of A. AT time t let us assume that same moles of A decomposed to give B and C because of which is pressure is reduced by an amount x while that of B and C is increased by x each. That is

$$A \rightarrow B + C$$

Initial P_1 0 0

At time t P_{1-x} x x

$\therefore$ total pressure at time $t = P_1 + x = P_2$

Example 8:

We are given a first order reaction

$A \rightarrow B + C$ where we assume that A, B and C are gases. The data given to us is

$A \rightarrow B + C$

Time	*0*	*t*
Partial pressure of A	P_1	P_2

Find k.

And we have to find the rate constant of the reaction.

Solution:

Since A is a gas and assuming it to be ideal, we can state that $P_A = [A]RT$ [From $PV = nRT$]. \ at $t = 0$, $P_1 = [A]_o RT$ and at $t = t$, $P_2 = [A]_t RT$. Thus the ratio of the concentration of A at two different time intervals is equal to the ratio of its partial pressure at those same time intervals.

$$\therefore \frac{[A]_o}{[A]_1} = \frac{P_1}{P_2}$$

$$\therefore k = \frac{1}{t} \ln \frac{P_1}{P_2}$$

$$\Rightarrow x = P_2 - P_1$$

Now the pressure of A at time t would be

$$P_{1-x} = P_1 - (P_2 - P_1) = 2P_1 - P_2$$

$$\therefore \ l = \ln \frac{[A]_0}{[A]_t} = \ln \frac{P_1}{(2P_1 - P_2)}.$$

Example 9:

→ B + C

Time	*t*	*∞*
Total pressure of (B + C)	P_2	P_3

Find k.

Solution:

	A →	B + C	
At t = 0	P_1	0	0
At t = t	$P_1 - x$	x	x
At t = ∞	0	P_1	P_1

$\therefore \ 2P_1 = P_3$

Þ $P_1 = \frac{P_3}{2}$

$2x = P_2$.

Example 10:

A → B + C

Time	*0*	*t*
Volume of reagent	V_1	V_2

Find k.

The reagent reacts with A, B and C. Find k.

Solution:

As we had seen in the previous case that the result of illustration 5 matches that of illustration of 1, we can use a similar logic for solving this one. That is

	A →	B + C	
At t = 0	V_1	0	0
At t = t	$V_1 - x$	x	x

At t = ∞ 0 V_1 V_1

where x is the volume of the reagent for those moles of A which have been converted into B and C

$$\therefore V_1 + x = V_2$$

$$\Rightarrow x = V_2 - V_1$$

$$\Rightarrow V_1 - x = 2V_1 - V_2$$

$$\therefore \quad k = \frac{1}{t}\ln\frac{[A]_0}{[A]_t} = \frac{1}{t}\ln\frac{V_1}{(2V_1 - V_2)}$$

This result looks perfectly OK. The problem is that this is true only if we make an assumption in the beginning of this problem (which you may have thought about). The assumption is that the 'n' factor of A, B and C with the reagent is same.

This can be made clear as follows. Let y moles of A be converted to B and C. If y moles of A require x ml of reagent and if the 'n' factor of A is n_1 with the reagent n_1y equivalents of the reagent are present. Now since the same y moles of B and C are formed and we find the same volume of reagent used for B and C, it can be calculated that y moles of B and C also contain n_1y equivalents, each, which implies that 'n' factor of B and C are also 'n_1'.

Example 11:

Now let us assume that A, B and C are substances present in a solution. From a solution of A, a certain amount of the solution (small amount) is taken and titrated with a suitable reagent that reacts with A. The volume of the reagent used is V_1 at t = 0 and V_2 at t = t

Time	*0*	*T*
Volume of reagent	V_1	V_2

The reagent reacts only with A. Find k.

Solution:

It can be understood that the volume of the reagent consumed is directly proportional to the concentration of A. Therefore the ratio of volume of the reagent consumed against A at t = 0 and t = t is equal to $[A]_0/[A]_t$

$$\therefore\ k = \frac{1}{t}\ln\frac{[A]_0}{[A]_t} = \frac{1}{t}\ln\frac{V_1}{V_2}.$$

Example 12:

$$A \xrightarrow{D} B + C$$

Time	*0*	*t*	*∞*
Volume of reagent	V_1	V_2	V_3

The reagent reacts with only C and D. Find K.

Solution:

$V_1 = V_D$

$V_2 = x + V_D$

$V_3 = V_A + V_D$

$V_3 - V = V_A$

$V_3 - V_2 = V_A - x$

$$\therefore\ k = \frac{1}{t}\ln\frac{[A]_0}{[A]_t} = \frac{1}{t}\ln\frac{(V_3 - V_1)}{(V_3 - V_2)}.$$

3

ACTIVATION ENERGY AND CHEMICAL REACTIONS

INTRODUCTION

We know that for a chemical reaction to occur, collisions between reactant molecules must occur. It has been postulated that only those collisions between reactants molecules which are associated with a certain minimum amount of energy can causes a chemical reaction. The minimum amount of energy which must be associated with molecules so that natural collisions may result in chemical reaction is termed as the should energy. Molecules possessing energy less then threshold energy will no react on collision Molecules possessing energy equal to or in excess of threshold energy constitute only a small fraction of the total number of molecules. This explains why a small fraction of the total number collisions result in chemical reaction. The number of molecules possessing energy equal to or in excess of threshold energy increases appreciably even with a small rise in temperature. So, the rate of reaction increase considerably even with a small increase of temperature.

As already explained, there is a certain minimum energy (or threshold energy) which the colliding molecules must a quire before they are in a position of react. Most of the molecules have much less kinetic energy then the threshold energy. The excess energy that the reactant molecules having energy less than the threshold energy must acquire in order to react to give the final products in known as activation energy. Therefore,

Activation energy = Threshold energy – Energy possessed by the molecules initially.

In others words passive or non-active molecules, *i.e.*, molecules possessing energy less than threshold energy, can be activated by absorption of excess energy known as activation energy.

As already stated, the reactant molecules have to acquire a minimum amount of energy before they can yield products on mutual collisions, *i.e.*, there is an energy barrier placed between reactants and products. This barrier has to be crossed before the reactants can yield products. If determines the magnitude of threshold energy which the reactant molecules must acquire before they can react to give products.

Further, different reactions require different amounts of activation energy. The concept of activation energy gives us an idea whether a given reactions is slow or fast at a give temperature. A reaction which has lower activation energy will proceed at a faster rate at a given temperature or vice versa. The differences in activation energy are mainly responsible for observed difference in rates of reactions.

The concept of activation energy as applied to chemical reaction can be explained by plotting energy against the progress of a reaction $(A \rightleftharpoons B)$ as shown in figure. The point A represent the initial energy (E_1) of the reactants., AC represents the energy barrier which the reactants have to cross and B represent the products of energy. E_2. At the beginning of rising position of the curve AC, the molecules have come very close to each other, but they cannot react as they possess energy less than the threshold energy. If however, sufficient energy is given to them, they can go up and reach the summit, C of the energy barrier. The molecules are then said to be in an activated state. In this state, molecules are under conditions of a cute strain. The bonds between atoms of the reacting molecules become very fable. The conditions now is such that the probability of formation of new bonds between atoms of the molecules of the reactant is fairly strong. This probability is shown by the curve B moving down the barrier.

In the above case, the reactants are shown to possess a higher total energy than the products. So, according to the law of conservation of energy, there will be a release of energy, *i.e.*, heat will be evolved. The reaction is exothermic and the amount of heat evolved gives the heat of reaction. This release of energy is shown by ΔE. Evidently $\Delta E = E_1 - E_2$, where E_1 is the normal energy of the reactants and E_2 of the products. In this case. ΔE is negative as energy is released and not absorbed. An exothermic reaction has this a lower activation energy or a faster rate of reaction. Now, consider the converse reaction $(B \rightleftharpoons A)$. Since the energy (E_2) of molecules of B is less than the energy (E_1) of A, as shown in figure, energy will be absorbed and not released, *i.e.*, the reverse reaction will be endothermic. An endothermic reaction has always a greater activation energy and hence has a slower rate of reaction than the opposing reaction.

Effect of Catalyst : A catalyst is a substance which can increase the reaction velocity but itself remains unchanged in amount and chemical composition at the end of the reaction. The reactants in a reaction have to cross an energy barrier before they can react to give products. The higher the energy barrier the slower is the rate of reaction. When a catalyst is added, a new reaction path with a lower energy barrier is obtained. As the energy barrier is reduced in magnitude, a large number of reactant molecules can get over it. This increases the reaction rate.

"*When two molecules, having the necessary energy of activation, come together, they must first form an activated complex or transitions state*". For example,

$$A + BC = AB + C.$$

When A is relatively far from BC, the atoms of BC are vibrating about their mean positions and thus the potential energy of the system is unaffected. As the distance of A is decreasing towards BC, the nuclei of atoms B and C of molecule BC are forced apart and this results an increase in the potential energy. This increase in potential energy is contained till a configuration A – B – C is attained. This A – B – C is known as the *activated complex* or *transitions state* and has the maximum potential energy. The activated complex A – B – C can decompose to yield the products of the reaction (AB and C) or *vice-versa*.

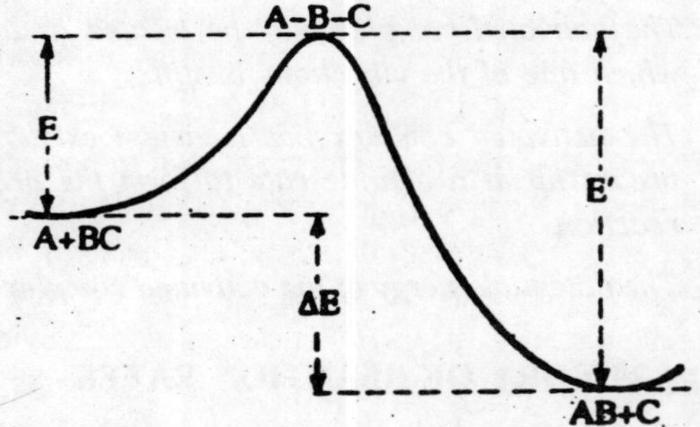

Fig. 3.1 : Change of P.E. in a chemical reaction.

$$\underset{\text{Initial State}}{A + BC} \rightleftharpoons \underset{\text{Activated Complex}}{[A - B - C]} \rightleftharpoons \underset{\text{Final State}}{AB + C}.$$

The change of potential energy is shown in Figure (3.1).

The energy of activation for the forward reaction is given by

$$E = E_{A-B-C} - E_{A+BC}.$$

Where E_{A-B-C} is the potential energy of the activated complex and E_{A+BC} is that of the reactants (A + BC).

Similarly, the energy of activation for the backward or reverse direction is given by

$$E' = E_{A-B-C} - E_{AB+C}$$

where E_{A+BC} is the potential energy of the products (AB + C), this $\Delta E = E - E'$ where ΔE gives the heat of the reaction at constant volume.

The activated complex of the bimolecular decomposition of hydrogen iodide might be represented as follows.

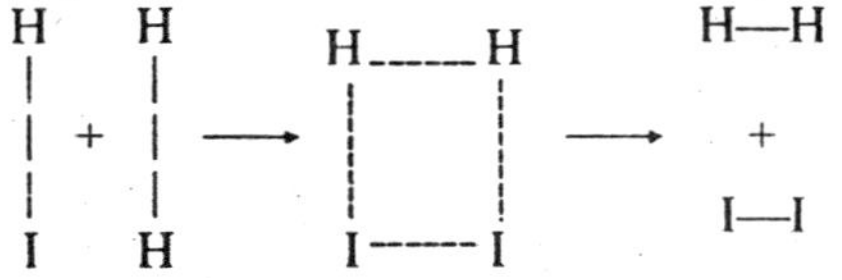

$$\begin{matrix} H & & H \\ | & + & | \\ I & & H \end{matrix} \longrightarrow \begin{matrix} H & \cdots & H \\ \vdots & & \vdots \\ I & \cdots & I \end{matrix} \longrightarrow \begin{matrix} H-H \\ + \\ I-I \end{matrix}$$

Activated Complex

CHARACTERISTICS OF AN ACTIVATED COMPLEX

(i) The formation of the activated complex is regarded as the characteristic of all chemical changes.

(ii) The activated complex may be treated as a molecule in which one of the vibrations is stiff.

(iii) The activated complex has transient existence only and breaks up at a definite rate to form the products of the reaction.

(iv) The potential energy of the activated complex is maximum.

COLLISION THEORY OF REACTION RATES

1. Chemical Collision Theory : According to the classical collision theory, molecules must collide to react and the rate of a chemical reaction is proportional to the number of reacting molecules, *i.e.*,.

Rate $\propto$ number of colliding molecules per litre/per second. The collision theory accounts for the dependence of the rate on the product of concentration terms. Let us consider that molecule A combines directly with a molecule of B to form AB, *i.e.,* A + B → AB. Visualise four molecules each of A and B in a box. In how many ways can be collision between A and B occur ? One A molecule can collide with any of four B molecules, hence it has four opportunities for collision.

The same is true for each of four A molecules, making 16 possible collisions in all. Therefore, assuming the reaction rate is proportional to the number of collisions, *i.e.,* collisions per second. Thus, the rate of the reaction is proportional to the product of the concentrations.

$$\text{rate} \propto [A]\,[B]$$

Mathematical Treatment of Classical Collision Theory : The rate of the reaction between two identical molecules is given by

$$dx/dt = ze^{-E/RT} \qquad ...(1)$$

where z is the number of binary collisions per second between two identical molecules in 1 ml. of a gas and E is the energy of activation of the process. According to the kinetic theory of gases, the reaction velocity is related to the concentrations by the relation

$$\frac{dx}{dt} \propto n^2 \quad \text{or} \frac{dx}{dt} = kn^2 \qquad ...(2)$$

where k is the specific rate constant and n is the concentration of the reactant. From equations (1) and (2), we have

$$k = \frac{z}{n^2} e^{-E/RT} \qquad ...(3)$$

From the simple kinetic theory of gases, the value of z is given by

$$z = 4n^2\sigma^2\left(\frac{\pi RT}{M}\right)^{1/2} \qquad ...(4)$$

where s is the collision diameter. Substituting eq. (4) in (3), we get

$$k = 4\sigma^2\left(\frac{\pi RT}{M}\right)^{1/2} e^{-E/RT} = Ze^{-E/RT} \qquad ...(5)$$

where $Z\left\{= 4\sigma^2\left(\frac{\pi RT}{M}\right)^{1/2}\right\}$ is known as *collision number* and is defined as

" The number of collisions per second when there is only one reactant molecule per ml of the gas."

On taking logarithm of equation (5), we get

$$\ln k = \ln \left\{4\sigma^2\left(\frac{\pi RT}{M}\right)^{1/2}\right\} - \frac{E}{RT}.$$

Differentiating the expression with respect to temperature,

$$\frac{d \ln k}{dT} = \frac{1}{2T} + \frac{E}{RT^2} \quad ...(5A)$$

The expression 1/2T can be neglected in comparison with E/RT^2 Therefore, equation (5A) becomes as

$$\frac{d \ln k}{dT} = \frac{E}{RT^2} \quad ...(6)$$

Equation (6) is identical with the Arrhenius equation. If the reaction involves two different molecules, then the value of collision number (Z) is given by

$$Z = \sigma^2 \left[8\pi RT \frac{(M_1 + M_2)}{M_1 M_2}\right]^{1/2} \quad ...(7)$$

where M_1 and M_2 are molecular weights and σ is the mean of the molecular collision diameters of the two reactants A and B.

Test of Simple Collision Theory : A test of simple collision theory is that the values of Z calculated from equation (4) or (7), are compared with the experimental values for a number of as reactions. Alternatively, it may be done by comparing the specific rate of a reaction which is calculated from equation (5) with the observed rate. As an example of the calculation, the following reaction is considered

$$2HI \rightleftharpoons H_2 + I_2.$$

The experimental specific rate at 556°K is 3.5×10^{-7}. We will now calculate the specific rate at 556°C by utilizing equation (5).

E = 4400 cal/mole, R = 2 cal/mole

T = 550°K, a 3.5×10^{-8} cm, M = 128.

Substituting these values in equation (5), we get

$k = 5.4 \times 10^{-7}$ per mole per sec.

Thus, there is a good agreement between experimental value (3.3×10^{-7}) and the value (5.4×10^{-7}) calculated from equation (5). The success of the collision theory can also be seen from the following examples.

(i) $2N_2O \rightarrow 2N_2 + O_2$

(ii) $2HI + H_2 + I_2$

(iii) $H_2 + I_2 \rightarrow 2HI$

Failures of the Theory : The theory fails in a number of cases such as.

(i) The bimolecular polymerisation of ethylene occurs at a rate which is slower by a factor of 5×10^{-4} than the theoretical value calculated from simple collision theory [equation (5)]

(ii) The polymerisation of 1 : 3 butadiene proceeds at rate which is slower by a factor of 10^{-4} than the value calculated from equation (5).

(iii) The reaction between ethyl alcohol and acetic anhydride vapours occurs at a rate which is slower by a factor of 10^{-5} than the value calculated from simple collision theory equation (5).

(iv) The chain reactions proceed at a very fast rate which is much higher than the value obtained from equation (5).

Modified Collision Theory : The number of colliding molecules calculated from the kinetic theory of gases is of the order of magnitude of 1032 molecules per litre per second at standard conditions. Hence, if collisions were the only requirement, for reaction all gaseous reactions should proceed at practically the same explosive rate. However, for the same reactant concentration at the same temperature, 300°C, the decomposition rate of gaseous hydrogen iodide is 4.4×10^{-3} mol/litre-hour, while the decomposition rate of gaseous dinitrogen peroxide is 9.4×10^{-6} moles. litre/hour. Clearly, *collisions between molecules cannot be the only factor involved in determining the rate of a reactions.* There, the earlier theory suffered a serious set back. But the concept of Arrhenius modified the simple collision theory. The new modified theory is

(i) The collisions between all molecules do not lead to chemical reaction.

(ii) The collisions between such molecules which possess energy of activation lead to the chemical reaction. It means that if the energy of the colliding molecules is less than activation energy, no reaction occurs, if the energy of the colliding molecules is equal to or greater than activation energy, reaction occurs. Thus, in order to account for the derivations from the collisions theory, equation (5) is modified to take the form

$$k = PZ\ e^{-E/RT}$$

where P is the probability factor or steric factor to make allowance for such effect which decreases or increases the speed of reactions. The value of P varies from 1 to 10^{-8}. In certain reactions, the value of P is as large 10^{-8} have been reported.

WEAKNESSES OF THE COLLISION THEORY

(i) Various attempts of correlate the value of P with the characteristics of reacting molecules have not been very successful.

(ii) The theory does not advance any explanation for the abnormally high rates that are sometimes observed.

(iii) When simple collision theory is applied to reversible reactions, a logical weakness is observed. This can be understood by considering the following reversible reaction of general type.

$$x + y \quad x + y \underset{k_2}{\overset{k_1}{\rightleftarrows}} xy$$

The rate constant for the forward reaction is given by

$$k_1 = P_1Z_1\ e^{-E/RT} \qquad ...(9)$$

and the rate constant for the backward reaction is

$$k_2 = P_2Z_1\ e^{-E/RT} \qquad ...(10)$$

$$k_2 = P_2Z_2\ e^{-E/RT}$$

But $$K = \frac{k_1}{k_2} \qquad ...(11)$$

Substituting equations (9) and (10) in (11), we get

$$K = \frac{P_1Z_1\, e^{-E_1/RT}}{P_2Z_2\, e^{-E_2/RT}} = \frac{P_1Z_1}{P_2Z_2}\, e^{-(E_1-E_2)/RT} \qquad ...(12)$$

From thermodynamics, the value of equilibrium constant is given by

$$\Delta G = \Delta H \ T\Delta S$$

or $\Delta G = -\Delta H + T\Delta S$

$$\therefore \quad K = e^{-(T\Delta S - \Delta H)/RT}\, e^{-\Delta S/RT}\, e^{-\Delta H/RT} \quad ...(13)$$

Comparing equations (13) and (12), we get

$$\frac{P_1 Z_1}{P_2 Z_2} e^{-(E_1 - E_2)/RT} = e^{\Delta S/R}\, e^{-\Delta H/RT} \quad ...(14)$$

But $E_1 - E_2 = \Delta H$ (change in energy). Therefore equation (14) becomes

$$\frac{P_1 Z_1}{P_2 Z_2} e^{-\Delta H/RT} = e^{-\Delta S/R}\, e^{-\Delta H/RT}$$

or
$$\frac{P_1 Z_1}{P_2 Z_2} e^{-\Delta S/R} \quad ...(15)$$

If the dimensions of the reactants and products are roughly of the same sizes, then $Z_1 = Z_2$. Hence equation (15) is simplified

$$\frac{P_1}{P_2} = e^{\Delta S/R} \quad ...(16)$$

But simple collision theory does not give any physical explanation of relationship (16) of the probability factor with entropy change.

(iv) Simple collision theory cannot explain unimolecular reactions unless Lindemann's hypothesis is taken into account.

Causes of Weakness of the Collisions Theory : Following are the main causes for the weakness of simple collision theory.

(i) This theory is entirely based on classical mechanics which is very crude and needs modification.

(ii) Now account is taken of the internal motion of the reactants.

(iii) Numerical calculations are usually made in terms of elastic hard spheres, although such models are cases of ideal cases, *i.e.,* for from the reality.

(iv) The fact that every combination of initial and final states must be characterised by a different reaction, and cross-section, which are not considered.

From the above facts it follows that simple collision theory must be a case of over simplification which needs modification. Therefore, a more modern theory, known as transition state theory, has replaced the older collisions theory.

ABSOLUTE REACTION RATE OF TRANSITIONS STATE THEORY

Introduction : A more modern theory, first proposed by *Pelzer and Minger* in 1932 and later extensively developed by *Eyring* and collaborators (1935) has replaced the older collision theory. In this theory details of collisions processes are not considered. Instead, and attempt is made to consider a reaction in terms of the changes in the potential energy of the reactants as they approach each other from infinite separation and then separate as different and distinguishable product molecules when they return to a state of zero potential energy. This theory is known as the *transitions state theory* or occasionally, as the *theory of absolute reaction rates* because it uses fundamental physical properties such as dimensions, vibrational frequencies, masses, etc., of the reacting molecules to calculate the rates of chemical reactions.

This theory can be conveniently treated in two stages.

(a) The formation of the rate equation using statistical mechanics to describe the equilibrium between reactants and the activated complex.

(b) Formulation of the rate equation in terms of thermodynamical state functions characterising the transition state complex and the reactants.

Let us discuss these two facts one by one.

(a) Statistical-mechanical Derivation of the Rate Equation : The various postulates of this theory are.

(i) In order for any chemical reaction to take place, it is necessary that the reactants, possessing sufficient energy, must approach one another to form the activated complex. These complexes will be in equilibrium with the reactants

$$\text{Reactants} \rightleftharpoons [\text{Activated complex}]$$

(ii) Activated complex is regarded as a normal molecule in every respect except that in addition to having three translational degrees of freedom, it has a fourth degree of freedom along the reaction co-ordinate.

(iii) Activated complex decomposes along the fourth degree of freedom to form the products. Then the rate of reaction is given by the rate of decomposition of a activated complex which is to be calculated.

[Activated complex] → Product.

It is assumed that the activated complex has one vibrational cegree of freedom less than a normal molecule.

Schematically, the above postulates can be put as

Reactants ⇌ [Activated Complex] → Products.

The activation energy of a reaction in the light of transitions-state theory, is defined as *the additional energy which the reacting molecule should acquire before forming the intermediate activated complex for the reaction.* This is how in the Fig. (3.2). This diagram is true for endothermic reactions.

In Fig. (3.2), there is a small initial fall in the potential energy of reactants which is due to the approaching of the reactant molecules close to one another before reacting to yield the activated complex. Similarly, there is small final rise in the energy of the products which is due to the falling apart of the molecules after the reaction.

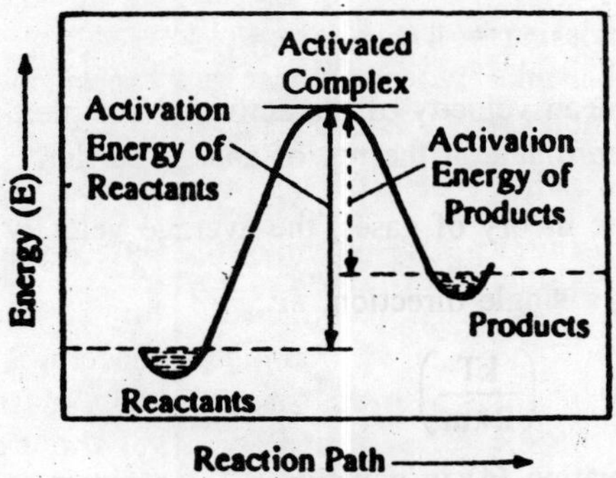

Fig. 3.2 : Activation energy required for the formation of activated complex.

From Fig. (3.2), it also follows that the activation energy of the back reaction may be defined as the additional energy which the product molecules should acquire before forming the activated complex. In Fig. 1, it is clear that the activation energy of the backward reaction is less than that of the forward reaction. The reason for this is that the products possess higher energy than the reactants, *i.e.*, the reaction has been assumed to be endothermic. If a reaction is exothermic the reverse thing is there, *i.e.*, the products possess less energy then the reactants and the activation energy of the back-reaction will be higher than that of the forward reaction.

Let us consider bimolecular reaction between a molecule A and a molecule B. The mechanism for this is

$$A + B \rightleftharpoons [AB]^* \rightarrow \text{Products} \qquad ...(1)$$

Here $[AB]^*$ is the activated complex and its concentration will now be calculated by using simple means of statistical mechanics. As earlier stated that the rate of the reaction is equal to the rate of decomposition of the activated complex. Therefore,

Rate of the reaction = frequency of crossing the energy barrier × concentration of activated complexes at the top of energy barrier × probability of crossing the barrier. ...(2)

We will consider all the above factors one by one.

(i) The frequency of crossing the barrier means the number of activated complexes crossing the energy barrier per unit time. This is given by

$$\text{Frequency of crossing} = \bar{u}/\delta . \qquad ...(3)$$

Here $\bar{u}$ is the mean velocity of the activated complex in a length δ in the reaction coordinate at the top of energy barrier.

From the kinetic theory of gases, the average velocity $\bar{u}$ is given by $\sqrt{\left(\frac{kT}{2\pi m}\right)}$ along a single direction, *i.e.*,

$$\bar{u} = \left(\frac{kT}{2\pi m}\right) \qquad ...(4)$$

Substituting equation (4) in equation (3), we obtain

$$\text{Frequency of crossing the barrier} = \frac{1}{\delta}\sqrt{\left(\frac{kT}{2\pi m}\right)} \qquad ...(5)$$

(ii) The concentration of activated complex means the number of activated complexes per unit volume lying in the length d.

Let the concentration of activated complexes = C^*

(iii) As pointed out earlier, the activated complex is treated as a normal molecule in which one of the vibrational degree of freedom is replaced by transitional motion in the reaction co-ordinate. The value of this translational partition function along the length δ in reaction co-ordinate is given by.

$$= \frac{(2\pi mkT)^{1/2}\,\delta}{h}$$

where h is the Planck's constant, k the *Boltzmann constant* and m is the effective mass of the activated complex in the reaction co-ordinate.

The factor $(2\pi m\,kT)^{1/2}\delta/h$ may be taken as a measure of the probability of activated complex at the top of energy barrier. Therefore, the probability of crossing the barrier is

$$= \frac{(2\pi mkT)^{1/2}\,\delta}{h} \quad ...(8)$$

Substituting equations (5), (6) and (8) in (2), we get.

Rate of the reaction = Rate of breaking up of the activated complexes.

$$= \frac{1}{\delta} \times \sqrt{\left(\frac{kT}{2\pi m}\right)} \times C^* \times \frac{(2\pi m\,kT)^{1/2}\,\delta}{h}$$

$$= C^* \frac{kT}{h}. \quad ...(9)$$

Here kT/ h is known as the *universal frequency* having the dimensions of time^{-1}. The value kT/h *depends* only on the *temperature* and is *independent* of the *nature* of the reactants and the *type of the reaction.* If k_r is the specific reaction rate, the rate of formation of activated complex is given by

Rate of formation of activated complex = $k_r C_A C_B$. ...(10)

where C_A and C_B are the concentrations of A and B respectively.

For steady reaction,

Rate of formation of activated complex = Rate of decomposition of activated complex. Therefore, equation (9) and (10) give

$$k_r C_A C_B = \frac{C^* kT}{h} \text{ or } k_r = \frac{C^*}{C_A C_B} \times \frac{kT}{h} \quad ..(11)$$

Applying the law of the mass action of the first equilibrium $A + B \rightleftharpoons [AB]^*$ of expression (1), we obtain

$$K^* = \frac{C^*}{C_A C_B} \quad ...(12)$$

where K* is the equilibrium constant and C* is the concentration of activated complex. From equations (11) and (12), we have

$$k_r = \frac{kT}{h} K^* \quad ...(13)$$

From statistical mechanics, the value of equilibrium constant K* in terms of partition functions is given by

$$K^* = \frac{\theta^*}{\theta_A \theta_B} e^{-\Delta E_0 / RT} \quad ...(14)$$

where θ_A and θ_B are the partition functions of the reactants A and B θ^* is the partition function of the activated complex [AB]*, and ΔE_0 is the change in the zero point energy in passing from the reactants to the activated complex state, *i.e.*, heat of activation. Substituting the value of K* from equation (14) in (15), we get,

$$k_r = \frac{kT}{h} \cdot \frac{\theta^*}{\theta_A \theta_B} e^{-\Delta E_0 / RT} \quad ...(15)$$

Equation (15) is regarded as a basis for the calculation of the specific rate of a reaction from the physical properties of the reacting substances.

The Thermodynamical Formulation Reaction Rates : The equations of the theory of absolute reaction rates may be expressed in terms of the thermodynamical functions instead of partition functions. This formulation of reaction rates is known as the *thermodynamical formulation of reaction rates.*

The basis for the thermodynamical formulation of rate constant is that the equilibrium between reactants and activated complexes may be expressed in terms of partition functions as well as thermodynamical functions.

Again, consider the general scheme,

$$A + B \rightleftharpoons [AB]^* \rightarrow \text{products.}$$

The equilibrium constant for the $A + B \rightleftharpoons [AB]^*$ may be stated as

$$K^* = \frac{\theta^*}{\theta_A \theta_B} e^{-\Delta E_0 / RT} \quad ...(16)$$

From equation (15), we have

$$k_r = \frac{kT}{h}.\frac{\theta^*}{\theta_A\theta_B}e^{-\Delta E_0/RT} \quad ....(17)$$

comparison of equation (16) with (17) leads to

$$k_r = \frac{kT}{h} K^*. \quad ...(18)$$

K*, the equilibrium constant, is related to the free energy of formation DG* of the activated complex by the following thermodynamic expression.

$$-\Delta G^* = RT \ln K^* \quad ...(19)$$

Therefore, equation (191) may be written as

$$-(\Delta H^* - T\Delta S^*) = RT \ln K^* \quad ...(20)$$

Solving equation (20), we get

$$\ln K^* = \frac{-(\Delta H^* - T\Delta S^*)}{RT}$$

or $$K^* = e^{-(\Delta H^* - T\Delta S)/RT} \quad ...(21)$$

Substituting equation (12) in equation (18), we get

$$k_r = \frac{kT}{h}e^{-\Delta(H^* - T\Delta S)/RT}$$

$$k_r = \frac{kT}{h}e^{-\Delta H^*/RT}.e^{\Delta S^*/R} \quad ...(22)$$

Equation (22) was derived by *Wyne-Jones B and Erying (1935).*

This equation (22) involves ΔH* (the heat of activation) which is not an experimental measured quantity. Therefore, equation (22) is modified to involve $E_{exp,}$ in place of ΔH*. Here E_{exp}, can be measured experimentally and is known as *experimental energy of activation,* We know

$$\frac{d \ln k^*}{dt} = \frac{\Delta E^*}{RT^2} \quad ...(23)$$

where DE* is the increase in energy in passing from the initial state to the activated state. Taking logarithm of eq. (18) we get

$$\ln k_r = \ln k - \ln h + \ln T + \ln K^*$$

Differentiating it, we obtain

$$\frac{d \ln k_r}{dT} = \frac{1}{T} + \frac{d \ln k^*}{dT} \qquad ...(24)$$

Substituting equation (23) in equation (24), we obtain

$$\frac{d \ln k_r}{dT} = \frac{1}{T} + \frac{\Delta E^*}{dT^2} = \frac{RT + \Delta E^*}{RT^2}$$

As we know, $E_{exp} = RT + \Delta R^*$ ...(25)

But $\Delta H^* = \Delta E^* + P\Delta V^*$ ...(26)

Substituting equation (26) in equation (25), we obtain,

$$E_{exp} = RT + (\Delta H^* - P\Delta V^*) \qquad ...(27)$$

When equation (27) is applied to unimolecular reactions, to reactions in solution and to the reactions in gas phase, different equations are obtained.

(i) *Unimolecular Reactions* : For unimolecular reactions, the change in the number of molecules as the activated molecule formed is zero. Therefore, ΔV^* is zero. Thus, equation (27) becomes as

$$E_{exp} = RT + \Delta H^*$$

$$\Delta H^* = E_{exp} - \Delta RT \qquad ...(28)$$

Substituting equation (28) in equation (22), we get

$$k_r = \frac{kT}{h} e^{-\Delta S^*/R} e^{-(E_{exp} - RT)/RT}$$

$$= \frac{kT}{h} e^{-\Delta S^*/R} e^{-E_{exp} - RT} e^{RT/RT}$$

or $$k_r = e \frac{kT}{h} e^{\Delta S^*/R} e^{-E_{exp} - RT} \qquad ...(29)$$

(ii) *Reaction in solution* : For this ΔV^* is zero.

Therefore, equation (27) becomes as

$$E_{exp} = \Delta H^* + RT$$

$$\Delta H^* = E_{exp} = E_{exp} - RT.$$

Again, substituting this equation in (22), we get

$$k_r = \frac{kT}{h} e^{\Delta S^*/R} e^{-(E_{exp} - /RT)/RT}$$

or $$k_r = e. \frac{kT}{h} e^{\Delta S^*/R} e^{-E_{exp}/RT} \qquad ...(30)$$

(iii) *Reactions in gas phase* : The following relation holds good

$$P\Delta V^* = \Delta n^* RT \qquad ...(31)$$

where Δn^* denotes the increase in the number of molecules when the activated complex is formed from the reacting species.

Substituting equation (31) in (27), we get

$$E_{exp} = RT + \Delta H^* - \Delta n^* \times RT$$

or

$$\Delta H^* = E_{exp} + RT \times \Delta n^* - RT$$

$$= E_{exp} + RT(\Delta n^* - 1) \qquad ...(32)$$

Substituting equation (32) in equation (22), we get

$$k_r = \frac{kT}{h} e^{-[E_{exp} + RT(\Delta n^* - 1)]/RT} e^{\Delta S^*/R}$$

$$= \frac{kT}{h} e^{-E_{exp}/RT} . e^{\Delta S^*/R} . e^{-(\Delta n^* - 1)} \qquad ...(33)$$

For a bimolecular reaction in gas phase, Dn* is – 1. Therefore, equation (33) becomes

$$k_r = e^2 \frac{kT}{h} e^{-E_{exp}/RE} . e^{\Delta S^*/R} \qquad ...(34)$$

Thus, we see that.

Equation (12) is the general thermodynamical formation of reaction rate.

Equation (29) is applicable to unimolecular reactions.

Equation (30) is applicable to reactions in solution,

Equation (33) is applicable to bimolecular reactions in gas phase.

Comparison of Collision Theory and Absolute Reaction Rate Theory

From equation (15), we have

$$k_r = \left(\frac{kT}{h}\right) \frac{\theta^*}{\theta_A \theta_B} e^{-E_0/RT}$$

Replacing ΔE_0 by E, energy of activation, we get

$$k_r = \left(\frac{kT}{h}\right) \frac{\theta^*}{\theta_A \theta_B} e^{-E/RT} \qquad ...(35)$$

$$k_r = Ae^{-E/RT} \qquad ...(36)$$

Comparing the two equations, we have

$$A = \frac{kT}{h}\frac{.\theta^*}{\theta_A\theta_B} \qquad ...(37)$$

Reaction between atoms : The value of A can be calculated by using equation (37) for a reaction between atoms A and B. From thermodynamics, we have

$$\theta_A = \frac{(2\pi m_A kT)^{3/2}}{h^3} \qquad ...(38)$$

$$\sigma_B = \frac{(2\pi m_B kT)^{3/2}}{h^3} \qquad ...(39)$$

$$\theta^* = \left[\frac{2\pi(m_A + m_B)}{h^3}\right]^{3/2}\left[\frac{8\pi^2 IkT}{h^2}\right] \qquad ...(40)$$

where I is the moment of inertia and its value is given by

$$I = \sigma^2\left[\frac{m_A m_B}{m_A + m_B}\right] \qquad ...(41)$$

where s is the mean diameter of the activated complex. Substituting equations (38), (39), (40) and (41) in equation (37), we get

$$A = \frac{kT}{h} \times \frac{\dfrac{[2\pi(m_A + m_B)kT]^{3/2}}{h^3} \cdot \dfrac{8\pi^2 m_A m_B \sigma^2 kT}{(m_A + m_B)h^2}}{\dfrac{(2\pi m_B kT)^{3/2}}{h^3} \cdot \dfrac{(2\pi m_A kT)^{3/2}}{h^3}}$$

$$= \left[8\pi\frac{(m_A + m_B)}{m_A m_B}RT\right]^{1/2} \sigma^2$$

$$= \left[8\pi\frac{(Nm_A + Nm_B)}{(Nm_A\ Nm_B)}NkT\right]^{1/2} \sigma^2$$

$$= \left(8\pi\frac{M_A + M_B}{M_A M_B}RT\right)^{1/2} \sigma^2 \qquad ...(42)$$

where R = Nk and M = mN.

Both the statistical and collision theories lead to the same result, provided the reactants are both atoms.

Now we will extend the above treatment to simple molecules.

Reactions between molecules. The partition function θ of a molecule may be written as

$$q = q_T^{\ t}\ qrt\ q_V^{\ v}$$

where t, r and v are the numbers of degrees of fieedom.

Consider the reaction between two non-linear molecules A and B containing a and b atoms respectively. Then,

$$q_A = q_T^{\ 3}\ q_R^{\ -3}\ q_V^{\ 3\ a\ -\ 6} \qquad \text{...(43)}$$

$$q_B = q_T^{\ 3}\ q_R^{\ 3}\ q_V^{\ 3\ b\ -\ 6} \qquad \text{...(4)}$$

and $$q^* = q_T^{\ 3}\ q_R^{\ 3}\ c_V\ (3a + 3b) - 7 \qquad \text{...(45)}$$

(Noting that the activated complex has one vibrational degree of freedom less than a normal molecule). Substituting equations (43), (44) and (45) in equation (37), we get

$$A = \frac{kT}{h}\frac{q_T^{\ 3}\ q_R^{\ 3}\ q_V^{\ (3a+3b)-7}}{q_T^{\ 3}\ q_R^{\ 3}\ q_V^{\ 3a-6}\ q_T^{\ 3}\ q_R^{\ 3}\ q^{3b-5}}$$

$$= \frac{kT}{h}\cdot\frac{q_V^{\ 5}}{q_T^{\ 3}\ q_R^{\ 3}}.$$

Compare this equation with the following equation

$$R = PZ.$$

$$\because \qquad PZ = \frac{kT}{h}\cdot\frac{q_V^{\ 5}}{q_T^{\ 3}\ q_R^{\ 3}} = \frac{kT}{h}\frac{q_R^{\ 2}}{q_T^{\ 3}}\cdot\frac{q_V^{\ 5}}{q_R^{\ 5}} \qquad \text{...(46)}$$

But $$Z = \frac{kT}{h}\frac{q_R^{\ 2}}{q_T^{\ 3}}$$

$$\therefore \qquad PZ = \frac{q_V^{\ 5}}{q_R^{\ 5}}\ Z \text{ or } P = \frac{q_V^{\ 5}}{q_R^{\ 5}}.$$

At ordinary temperature, q_V is unity and q_R may be about 10 – 100 in the case of complex molecules. Thus, the value of θ_V/θ_R is 10^{-1} to 10^{-2} so that$(\theta_V/\theta_A)5$ is 10^{-5} to 10^{-10}. Thus, the value of P is 10^{-5} to 10^{-10}. It is expected that such reactions should proceed more slowly than calculated on the basis of collision theory. This is actually found both in gas phase and in solution. Thus, *the absolute reaction rate theory gives the physical significance of the probability factor.*

Transmission Coefficient : It is not necessary that every activated complex is converted into one of the reaction products. To allow for this possibility, it is convenient to introduce a transmission co-efficient K in equation (15).

$$k_r = K \frac{kT}{h} \frac{\theta^*}{\theta_A \theta_B} e^{-\Delta E_0/RT} \quad ...(47)$$

For many reactions the transmission co-efficient is unity. It means that every activated complex becomes a product.

The value of K has been found to less than unity in the following types of reactions.

(i) In bimolecular atom recombination in the gas phase and the reverse decomposition of atomic molecules.

(ii) In such reactions in which there is change from one type of electronic state to another.

In certain cases, the value of K may be greater than unity. This may be the case when there is quantum mechanical tunnelling.

Entropy of a Activation : From equation (22), we have

$$k_r = \left(\frac{kT}{h}\right) e^{\Delta S^*/R} e^{-\Delta H^*/RT} \quad ...(48)$$

which may be compared with collision theory expression.

$$k_r = PZe^{-E/RT} \quad ...(49)$$

and we get

$$PZ\ e^{-E/RT} = \left(\frac{kT}{h}\right) e^{\Delta S^*/R} . e^{-\Delta H^*/RT}$$

But $E = DH^*$

$$\therefore \ PZ\ e^{-DH^*/RT} = \left(\frac{kT}{h}\right) e^{\Delta S^*/R} . e^{-\Delta H^*/RT}$$

or $$PZ = \frac{kT}{h} e^{\Delta S^*/R}$$

$$\therefore \ P = \frac{kT}{Zh} e^{\Delta S^*/R} \quad ...(50)$$

Thus, the probability factor P is related to the entropy of activation DS*.

Advantages of Thermodynamic Formulation over Absolute Theory of Reaction Rates

(i) Thermodynamic formulation provides a quantitative treatment of the steric factor P in bimolecular reactions.

(ii) Thermodynamic formulation (22) is not restricted to bimolecular processes but is equally applicable to the rate constants of unimolecular and termolecular processes.

(iii) Thermodynamic formulation (22) can be applied to reactions in solution.

Discussion of Transitions-State Theory : The concept of entropy of activation in transition state theory is very useful for quantitative purposes and one notable advantage over the collision theory is in the field of bimolecular reactions. In the collision theory the reactant molecules, whatever their atomicities and shapes, are merely regarded as spherical particles, and the entropy change in passing from the reactants to the activated collision pair is merely that to be associated with the replacement of three translational degrees of freedom by two of rotation.

In the transition state theory, however, an account is taken of the internal degrees of freedom of the reactants and the changes these undergo on reaction, and we find for instance that in bimolecular association reaction between non-linear polyatomic molecules the formation of the activated complex generally involves the net replacement of three translational degrees of freedom by five vibrational degrees of freedom.

At temperatures in the range 0° to 500°C such a change ins accompanied by a decrease of entropy and we, therefore, expect that P factor for such reactions will be very small, as is found to be the case.

Validity of the Transition State Theory : In view of the nature of the assumptions in the various parts of the transition state theory, it is important to remember that the transition state theory should be utilised carefully. Because the theory is useful in describing and correlating rate studies particularly in solution, it does not follow that its microscopic (molecular) details correspond to the actual situation in nature.

Until theories are verified by experiment, one should be prepared to reject today a theory which appeared to be satisfactory yesterday. Unfortunately, in the case of the transition state theory, an adequate experimental test of the theory is not possible at present.

UNIMOLECULAR REACTIONS

Introduction : *These are the reactions in which the activated complex is formed from a single reactant molecule.* These are two types of unimolecular reactions.

(i) **Elementary Unimolecular Reactions :** *These are the reactions in which the products are obtained from reactants in one stage i.e.,* via a single activated complex. For example : isomerisation of cyclopropane into propylene.

$$\underset{CH_2 - CH_2}{\overset{CH_2}{\wedge}} \longrightarrow CH_3-CH=CH_2$$

(ii) **Unimolecular Reactions in Free Radical Mechanisms.** *In most of the decompositions it has been observed that they are not occurring via single activated complexes but follow the free radical mechanisms.* In such mechanisms, the first step is a unimolecular decomposition. For example, in the decomposition of ethane, the first step is the dissociation of ethane into two methyl radicals which is a unimolecular decomposition.

$$C_2H_6 \longrightarrow 2CH_3.$$

Unimolecular reactions are of the first order under certain conditions, but they become of the second order at low pressures.

THEORIES OF UNIMOLECULAR REACTIONS

It is assumed that activation of the molecules in unimolecular reactions is due to the collisions, it means that the reaction should follows the second order kinetics because the number of collisions is proportional to the square of the concentration.

(a) **Perrin Theory :** In 1919, *Perrin* suggested that activation of molecules in unimolecular reactions had been due to the absorption of infra-red radiations emitted by the sides of a containing vessel.

Failures of the theory.

(i) Experimental evidences could not be traced out in support of this theory.

(ii) *Perrin's* main suggestion was that the unimolecular reaction rate is independent of pressure. But in actual practice it has been observed that the rate constant does not remain constant at very low pressures.

(b) Lindemann's Theory : Lindemann suggested that :

(i) Reactant molecules are activated by collisions with one another. Some of the molecules may possess sufficient energy to pass into the final products without receiving additional energy. Such molecules are usually termed as *activated molecules*.

(ii) If the rate of activated molecules to convert into the products in slow as compared with the rate with which activated molecules are deactivated by collisions, a *stationary concentration* of activated molecules may be formed.

As the activated molecules are in equilibrium with the normal molecules, the concentration of activated molecules will be proportional to that of normal molecules.

It, therefore, follows that the rate of the reaction is proportional to the concentration of activated molecules which in turn is proportional to the first power of the concentration of normal molecules as the two types are in equilibrium with each other. It means that the reaction is of the first order.

(iii) At low pressures, the collisions cannot maintain a supply of activated molecules. Therefore, the rate of the reaction will depend upon the rate of activation and hence is proportional to the square of the concentration of reacting molecules. Thus, the reaction becomes of the second order at low pressures.

MATHEMATICAL FORMULATION OF LINDEMANN'S THEORY

(i) Activation by collision. The process of energisation by collision is represented by

$$A + A \longrightarrow A^* + A$$

where A is a normal molecule and A^* is an activated molecule which possesses sufficient energy to pass into the products.

$\therefore$ Rate of activation $= k_1 \, [A]^2$

where k_1 is the specific rate constant for the second order reaction.

(ii) *Deactivation by collision. The* process of deactivation may be represented as

$$A^* + A \xrightarrow{k_2} A + B$$

$\therefore$ Rate of deactivation $= k_2$ [A*] [A]

where the rate of deactivation is slow as compared with the rate of activation.

(iii) *Decomposition.* The decomposition of activated molecules may be represented as

$$A^* \xrightarrow{k_3} \text{Products}$$

where k_3 is the first order constant.

As the concentration of A* is so small that it is not changing with time, it means that according to steady state treatment $\frac{d[A^*]}{dt}$ is zero *i.e.*,

$$\frac{d[A^*]}{dt} = 0 \quad ...(1)$$

But the rate of formation of A* is given by

$$\frac{d[A^*]}{dt} = k_1 [A]^2 - k_2 [A] [A^*] - k_3 [A^*]. \quad ...(2)$$

Applying condition of equation (1) to equation (2), we get

$$k_1 [A]^2 - k_2 [A] [A^*] - k_3 [A^*] = 0$$

or $$k_1 [A]^2 = k_2 [A] [A^*] + k_2 [A^*]$$

$$= [A^*] (k_2 [A] + k_2)$$

or $$[A^*] = \frac{k_1 [A]^2}{k_2 [A] + k_3} \quad ...(3)$$

But the rate of reaction will be proportional to the concentration of the activated molecules, one may write

$$v = k_3 [A^*] \quad ...(4)$$

Substituting equation (3) in equation (4), we get

$$v = \frac{k_3 k_1 [A]^2}{k_2 [A] + k_3} \quad ...(5)$$

(i) **At High Pressure :** At high pressure, k_2 A is far greater than k_3. It means that k_3 (A) can be neglected in comparison with k_2 (A).

$$v = \frac{k_3 k_1 [A]^2}{k_2 [A]} = k' [A]$$

Thus, the reaction is of the *first order*.

(ii) **At Low Pressure :** At low pressure, k_3 is far greater than k_2. It follows that k_2 can be neglected in comparison with k_3.

Therefore,

$$v = \frac{k_1 k_3}{k_3} [A]^2 = k_1 [A]^2.$$

Hence the reaction is of the *second order*.

Criticism of the Lindemann's Theory : According to this theory there is a change from first-order to second order kinetics as the pressure is lowered. This has been observed experimentally in a number of instances. Thus, the evidence is against the radiation theory and in favour of Lindemann's theory.

Although this theory gives a satisfactory qualitative interpretation of unimolecular reactions, yet quantitatively it is not completely satisfactory and, therefore, certain modifications are required. However, there are two main difficulties of Lindemann's theory.

(i) Suppose that a first-order coefficient k is defined by the equation

$$v = k [A] \quad \text{...(6)}$$

Equations (5) and (6) give rise to,

$$k [A] = \frac{k_3 k_1 [A]^2}{k_2 [A] + k_3} = \frac{k_\infty}{1 + \dfrac{k_3}{k_2 [A]}} \quad \text{...(7)}$$

The concentration $[A]_{1/2}$ at which k should become equal to $k_\infty/2$ has been defined by the equation

$$k_1 [A]_{1/2} = k_3 \quad \text{...(8)}$$

If follows from this equation that when $k = k_\infty/2$.

$$k_1\ [A]_{1/2} = \frac{k_1 k_3}{k_2} = k_\infty$$

or $$[A]_{1/2} = \frac{k_\infty}{k_1} \qquad \text{....(9)}$$

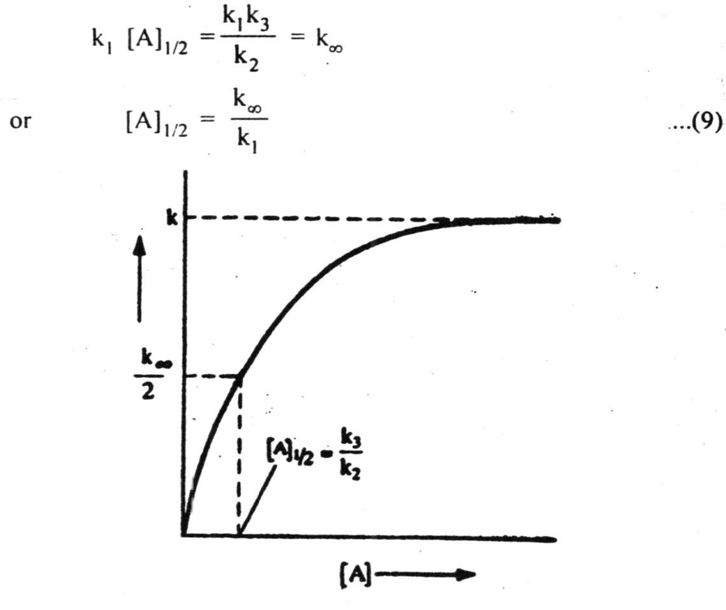

Fig. 3.3

The value of k_∞, the first order rate constant at high pressures, can be obtained from experiment and the value of k_1 can be obtained from Ze^{-E^*} RT, where E* is the energy of activation. However, for several reasons, this procedure gave rise to the result that the first-order rate constant should fall at a much higher pressure than was actually observed. Since there can be no doubt about k_∞, the error must be in the estimation of k_1. It becomes, therefore, necessary for the collision theory to be modified in such a manner as to give large values for k_1. This difficulty was overcome by Hinshelwood.

(ii) The second difficulty with the simple Lindemann's theory of unimolecular reactions becomes apparent when one plots the experimental results. Equation (7) may be written as

$$\frac{1}{k} = \frac{k_2}{k_2 k_1} + \frac{1}{k_1\,[A]} \qquad \text{...(10)}$$

and a plot of 1/k against the reciprocal of the concentration should give a straight line. However, deviations from linearity have been found, of the kind shown schematically in. The deviations are explained, as will be seen according to the theories of Kassel, Rice and Ramsperger, and Slater.

(b) Hinshelwood's Theory : Thus first difficulty with the Lindemann's theory, that the first order behaviour is maintained down to lower concentrations than the theory appeared to permit, requires that k_1 must have a larger value than $Ze^{-E^*/}RT$. It was pointed out that this expression is actually only applicable to a molecule having one degree of vibrational freedom, a molecule having more degrees of vibrational freedom has a greater probability of acquiring the energy E*, since this energy may now be distributed amongst all the degrees of freedom in the molecule. The expression for k_1 that applies to the case of S-degrees of vibrational freedom is

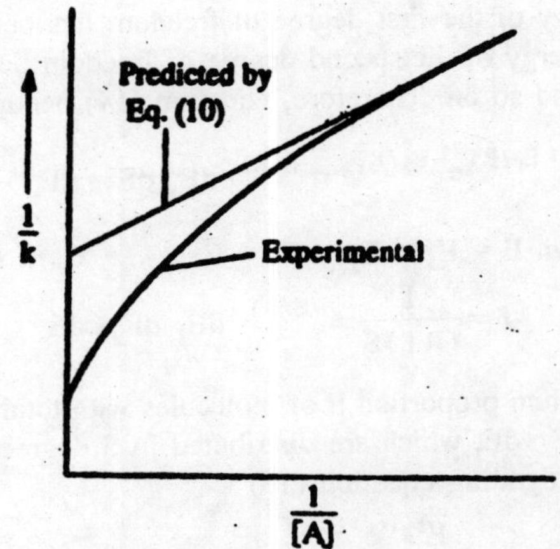

Fig. 3.4

$$k_1 = \frac{Z}{S-1!}\left(\frac{E^*}{RT}\right)^{s-1} e^{-E^*/RT} \quad ...(11)$$

where k is the Boltzmann constant.

Expression (11) can be proved by the application of statistical mechanics.

According to the statistical mechanics, the fraction of molecules, f, having an amount of energy between E and E + dE in one degree of freedom is proportional to $e^{-E/RT}$ *i.e.,*.

$$f = A\, e^{-E/RT}\, dE \quad ...(12)$$

where A is the proportionality constant. As the total fraction of molecules having an amount of energy between zero and infinity should be unit, it means that equation (12) becomes as

$$A \int_0^\infty e^{-E/RT} \, dE = 1.$$

This equation on simplification becomes as

$$A = \frac{1}{RT} \qquad ...(13)$$

Substituting equation (13) in (12), we get

$$f = \frac{1}{RT} e^{-E/RT} \, dE \qquad ...(14)$$

If energy is distributed in s degrees of vibrational freedom in such a way, the energy of the first degree of freedom lies between E_1 and $E_1 + dE_1$, theenergy of the second degree of freedom lies between E_2 and $E_2 + dE_2$ and so on. Therefore, equation (14) becomes as

$$f = \frac{1}{(RT)S} . e^{-E_1/RT} e^{-E_2/RT} ... e^{-S/RT} \, dE_1 \, dE_2 ... dE_s \qquad ...(15)$$

As we known $E = E_1 + E_2 + E_2 + E_3 + ... + E_s$, it means

$$f = \frac{1}{(RT)S} . e^{-E/RT} \, dE_1 \, dE_2 ... dE_s \qquad ...(16)$$

The equilibrium proportion F of molecules with total energy lying between E and E +dE, which are distributed in S degrees of freedom isobtained by integrating equation (16)

$$F = \frac{E^{s-1} e^{-E/RT} \, dE}{(S-1)!(RT)^S}$$

$$= \frac{1}{(S-1)!} \left(\frac{E}{RT}\right)^{s-1} \frac{1}{RT} e^{-E/RT} \, dE$$

or $$\frac{dk_1}{dk_2} = \frac{1}{(S-1)!} \left(\frac{E}{RT}\right)^{s-1} \frac{1}{RT} e^{-E/RT} \, dE$$

where k_1 and k_2 have same significance as discussed in Lindemann's Theory.

If E* represents the minimum energy which a molecule must possess so as to decompose into products, it means

$$\frac{k_1}{k_2} = \frac{[A^*]}{[A]} = \int_{E^*}^{\infty} \frac{1}{(S-1)!}\left(\frac{E}{RT}\right)^{s-1} \frac{1}{RT} e^{-E/RT} dE$$

Changing the differential to $de^{-E/RT}$, and integrating by parts, we obtain

$$\frac{k_1}{k_2} = \frac{[A^*]}{[A]}$$

$$= \left[\frac{1}{(S-1)!}\left(\frac{E^*}{RT}\right)^{s-1} + \frac{1}{(S-1)!}\left(\frac{E^*}{RT}\right)^{s-2} \ldots\right] e^{-E/RT} \quad \ldots(17)$$

In order for the first term is to be greater than the second, one must have

$$E^* > (S-1)\,RT \quad \ldots(18)$$

Under the above restriction, equation, (17) becomes as

$$\frac{k_1}{k_2} = \frac{[A^*]}{[A]} = \frac{1}{(S-1)!}\left(\frac{E^*}{RT}\right)^{s-1} e^{-E/RT} \quad \ldots(19)$$

As k_2 is a collision frequency, it means that it may be written as Z_2 and, thus, equation (19) for k_1 becomes as

$$k_1 = Z_2 \frac{1}{(S-1)!}\left(\frac{E^*}{RT}\right)^{s-1} e^{-E/RT} \quad \ldots(20)$$

This equation is same as equation (11). By inserting numerical values it may be shown that this expression can give rise to much higher rates or activation than corresponds to the expression $Ze^{-E^*/RT}$. If the experimental activation energy is taken as 40 k cal per mole and S as 12, equation (20) gives rise to a value of 9.9×10^{-12} for k_1. The simple expression $Z\,e^{-E^*/RT}$ gives rise to 3.1×10^{-18}. The difference in this case is of the order of 106, and equation (20) will therefore lead to the prediction that the first-order rate coefficient will begin to fall off at very much lower pressures than predicted by simple theory. In practice S is usually found by a method of t rail and error.

Limitation of Hinshelwood's Theory : Although the Hinshelwood's theory of unimolecular reactions represents and analytical formulations the time lag theory of Lindemann, it does not in general account satisfactorily for the observed variation of the rate constant with pressure. This failure may in large measure be attributed to the rather unrealistic assumptions that the activation energy per molecule may be distributed among all or many of the vibrational modes of the molecule and that excess of energy in these bonds above the activation energy does not increase the probability of reaction.

(c) Kassel and Rise and Ramsperger Theory : Kassel and Rice and Ramsperger attempted to explain the observations shown in Figs. 1 and 2 by developing a theory which is based on the assumption that k1 is a function of the energy possessed by the energized molecule A. Though the theories given by Kassel and Rise and Ramsperger have been essentially similar yet the quantum version of Kassel has been more realistic and can be shown to give rise to the classical version under suitable conditions. Discussed below is the Kessel theory.

Kassel postulated a model of a molecule as a complex of s vibrations which are having the same frequency of vibration v. The vibrations have been quantized and can exchange energy among them in the multiples of hv. Only such a molecule is regarded to be capable of monomolecular decomposition in which m quanta *i.e.,* an energy equal to mhv, get concentrated on a definite vibrational degree of freedom of it. The probability of decomposition of an active molecule can be regarded to be taken to be proportional to the probability of a definite concentration of the quanta on one of it.

In this model it is assumed that the reacting molecule is a system of s identical harmonic oscillators, each of frequency v. These oscillators are said to get loosely coupled, they interact enough so that energy can flow from one too another, but not strongly enough to perturb each other's energy levels. Thus, the reactant gas is taken to be a double statistical assembly, at the *intermolecular* level molecules interchange energy at every collision, while at the *intermolecular* level, energy gets interchanged randomly among the oscillators, between the times of molecular collisions. The total energy of the energized molecule may be given as follows.

$$e_j = jhv \qquad ...(1)$$

where k denotes the total number of quanta.

The fraction of the molecules in the jth level may be put as follows

$$P_j = \frac{N_j}{N} = \frac{g_j \exp(-jhv/kT)}{\sum_{i=0}^{\infty} g_i \exp(-ihv/kT)} \qquad ...(2)$$

The statistical weight g_j, can be determined by the degeneracy of the jth energy level, it refers to the number of different ways the j quanta can be distributed among the s oscillators and is given by

$$g_j = \frac{(j+s-1)!}{j!(s-1)!} \qquad ...(3)$$

The statistical weight for states in which the s oscillators are having j quanta among them and a particular one has m quanta, may be given as follows.

$$\frac{(j-m+s-1)!}{(j-m)!(s-1)!} \qquad ...(4)$$

The probability that a particular oscillator is having m quanta and all s oscillators are having j quanta is the ratio of these

$$\frac{(j-m+s-1)!\,j!}{(j-m)!(j+s-1)!} \qquad ...(5)$$

When j is very large, Eq (5) can be approximated to

$$\left(\frac{j-m}{j}\right)^{s-1} \qquad ...(6)$$

As the total energy of the molecule, e, is proportional to j and the energy of the oscillators corresponding to the bond is breaking , ε_C, is proportional to m, the above expression may be put as follows.

$$\left(\frac{\varepsilon-\varepsilon_C}{\varepsilon}\right)^{s-1} \qquad ...(7)$$

As the rate with which the required energy passes into the particular degree of freedom is proportional to this quantity, then we can write

$$k_1 = k^{\ddagger}\left(\frac{\varepsilon-\varepsilon_C}{\varepsilon}\right)^{s-1} \qquad ...(8)$$

$k^{\ddagger}$ is the rate constant corresponding to the free passage of the system over the potential-energy barrier, when e is sufficiently large, the energized molecule is essentially an activated molecule and therefore can pass immediately into the final state. The variation of k_1 with energy (ε) is shown in Fig. 3.5.

The experimental rate constant at high pressure (k_∞) may be obtained

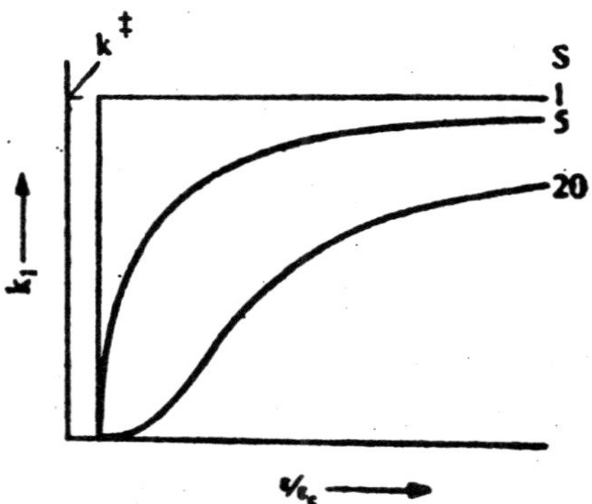

Fig. 3.5 : The variation of k_1 with ϵ/ϵ_c, (Eq. 8) for different values of s.

$$k_\infty = k_1\left(\frac{k_2}{k_{-2}}\right) = \int_{\varepsilon_C}^{\infty} k_1\left(\frac{dk_2}{k_{-2}}\right) \qquad \text{...(9)}$$

If we use equation (8), then Eq. (9) may become as follows.

$$k_\infty = \int_{\varepsilon_C}^{\infty} k\ddagger\left(\frac{\varepsilon-\varepsilon_C}{\varepsilon}\right)^{s-1} \cdot \frac{1}{(s-1)!}\left(\frac{\varepsilon}{kT}\right)^{s-1}$$

$$\frac{1}{kT}\exp(-\varepsilon/dT)\,d\varepsilon \qquad \text{...(10)}$$

This equation on solving gives,

$$k_\infty = k\ddagger \exp(-\varepsilon_C/kT) \qquad \text{...(11)}$$

The $k\ddagger$ refers to the order of vibrational frequency in many cases. We also know

$$dk_{uni} = \frac{k_1(dk_2/k_{-2})}{1+k_1/k_{-2}[A]} \qquad \text{...(12)}$$

On inserting the value of k_1, and dk_2/k_{-2} from Eq. (8) and integrating from ε_C to ∞ we get

$$k_{uni} = \int_{\varepsilon_C}^{\infty} \frac{k\ddagger\left(\frac{\varepsilon-\varepsilon_C}{\varepsilon}\right)^{s-1}\frac{1}{(s-1)!}\left(\frac{\varepsilon}{kT}\right)^{s-1}\exp(-\varepsilon_C/kT)\frac{d\varepsilon}{kT}}{1+\frac{k\ddagger}{k_{-2}[A]}\left(\frac{\varepsilon-\varepsilon_C}{\varepsilon}\right)^{s-1}}$$

...(13)

Eq. (13) gives the variation of rate constant with pressure concentration of reactant A. The above relationship can also be expressed as

$$\text{kuni} = \frac{k\ddagger \exp(-\varepsilon_C / kT)}{(s-1)!} \int_0^{\infty} \frac{x^{s-1} \exp(-x)\, dx}{1 + \frac{k\ddagger}{k_{-2}[A]}\left(\frac{x}{b+x}\right)^{s-1}} \quad ...(14)$$

where $x = \frac{\varepsilon - \varepsilon_C}{kT}$, $b = \frac{\varepsilon_C}{kT}$.

On using $k_\infty = k\ddagger \exp(-\varepsilon_c / kT)$, Eq. (14) becomes as follows.

$$\frac{k_{uni}}{k_\infty} = \frac{1}{(s-1)!} \int_0^{\infty} \frac{x^{s-1} \exp(-x)\, dx}{1 + \frac{k\ddagger}{k_{-2}[A]}\left(\frac{x}{b+x}\right)^{s-1}} \quad ..(15)$$

Thus, Eq. (15) for a fixed value of s describes the concentration dependence of the reaction rate constant. In practice, the value of s can be determined by trial and error, what value of is will predict the observed variation of kuni with the pressure. The value of s that is needed generally corresponds to about the half the total number of normal vibrational modes in the molecule.

(d) The PRKM Method : In 1952 Marcus refined the RRK theory by taking into consideration all vibrations and rotations of the energized molecule (A*) in terms of the actual vibrational frequencies and rotation constants under the basic frame work of the activated complex theory.

This improved theory known as PRKM has been regarded as the most successful and refined statistical theory of reaction kinetics of the day.

The PRKM theory has been based on the scheme

$$A + A \underset{k_{-2}}{\overset{k_2}{\rightleftharpoons}} A + A^*$$

$$A^* \xrightarrow{k_1} A^\ddagger$$

$$A^\ddagger \xrightarrow{A^\ddagger} \text{Products}.$$

Here A* refers to an energized species as before and must attain critical configuration of the activated complex, $A^\ddagger$, *enrout* to reaction.

The individual vibrational frequencies of A* and $A^{\ddagger}$ are regarded explicitly, the way normal mode vibrations contribute to reaction is taken into consideration and an allowance is made for zero point energies of all species, PRKM treatment yields.

$$\frac{dk_2}{k_{-2}} = \frac{N_c(\varepsilon_C)\exp(-\varepsilon_C/kT)\,d\varepsilon_C}{fc} \qquad ...(16)$$

where N_C (ε_C) refers to the energy density of the degrees of freedom which are contributing to the breaking of bonds, ε_C the critical energy in the activated complex, somewhat above the complex's zero point energy, and f_C the partition function which is corresponding to critical energy contribution.

It is possible to calculate k_1 by considering the relationship between the concentration of activated complexes and the energized molecules.

$$k_1 = f^{\ddagger} \frac{\int_0^{\varepsilon} N_1\left(\varepsilon_n^{\ddagger}\right)}{f_R N_C(\varepsilon_C)\,h} d\varepsilon\ddagger\ n \qquad ...(17)$$

The critical energy ε_C is equal to the sum of$\left(\varepsilon_n^{\ddagger}\right)$ denotes the energy density of quantum sates at energy $\varepsilon_n^{\ddagger}$ for vibrational modes. $f^{\ddagger}$ R/fR refers to the ratio of portion functions which allow for changes in moments of inertia in the non-contributing rotations on forming the activated state from the energized molecule. Substitution of these terms into Eq. (12) and integration over the appropriate energy range yields the vibration of k_{uni} with [A] once again as a complicated function, which under high pressure gets reduced to

$$k_{\infty} = \frac{kTF^{\ddagger}}{hF_A}\exp[-\varepsilon_0/kT] \qquad ...(18)$$

The above equation is of the from which is similar to that obtained from the activated complex theory.

(e) The Slater's Treatment : As opposed to statical theories with energy flow between the vibrations, Slater proposed a *dynamic* theory without energy flow. Vibrations are required to be simple harmonic, which does not allow energy flow between modes, but when different modes come into phase, the vibrational amplitude is changed. For suitable phase relationships the vibrational amplitude may be extending beyond the critical length corresponding to the reaction. This is illustrated in Fig. 3.6.

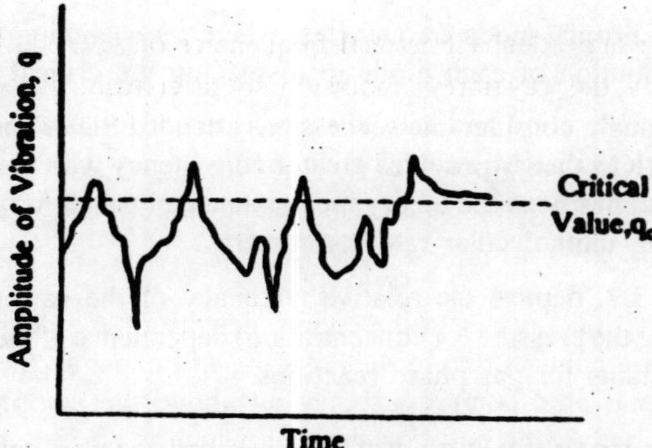

Fig. 3.6 : The vibration of amplitude of an internal coordinate, q with time q_c is the critical value at which unimolecular reaction occurs.

Slater provided a classical treatment in 1939 and later a quantum mechanical derivation. The mathematical treatment is again complex but the variation of k_{uni} with [A] can be calculated. The number of vibrational modes that may contribute to the critical reaction coordinate is selected to give the best fit with the experimental data, and the result is approximately half the maximum possible modes of vibration.

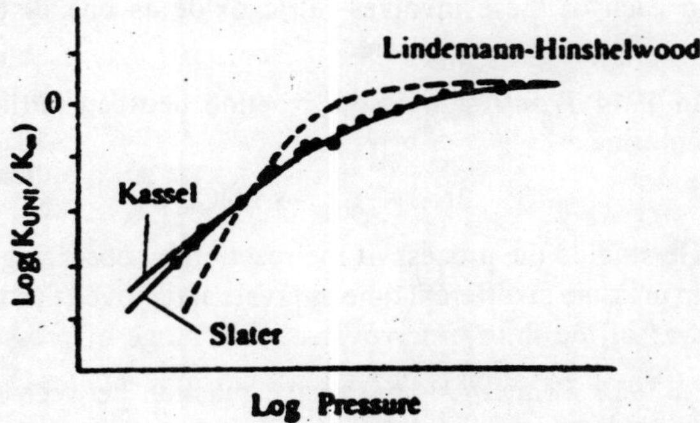

Fig. 3.7 : Comparison of various theories for describing the pressure dependence of the gas phase reaction.

Under high pressure conditions, the rate constant k∞ may be expressed as

$$k_\infty = v \exp(-\varepsilon^*/kT) \quad \text{...(19)}$$

where e* refers to the minimum energy for the process as defined by the theory, and v an average vibration frequency – a weighed root mean square of normal mode frequencies, where the weighing factors denote the contribution of each mode to displacing the critical coordinate.

Although, considerable success has attended Slater's method, on the whole PRKM theory provides greater consistency with the experimental results and has been the theory most extensively used for predicting and correlating unimolecular rate parameters.

Fig. 3.7. depicts the relative adequacy of the various theories in describing the pressure (or concentration) dependence of the unimolecular rate constants for gas phase reactions.

TRIMOLECULAR REACTIONS OR TERMOLECULAR REACTIONS

Introduction : *A termolecular reaction is one in which the activated complex is formed from three reactant molecules.* The concentration of this activated complex can be calculated by using the methods of *statistical mechanics.* All the known termolecular reactions are found to be free from side reactions and, therefore, may be considered as elementary.

The number of elementary termolecular reactions is very small. Only five homogeneous gas reactions are definitely known to be of third order and each of these involves nitric oxide as one of the reacting molecules. Various examples are.

(i) In 1914 *Trautz* studied the reaction between nitric oxide and chlorine.

$$2NO + Cl_2 \rightarrow 2NCl$$

He studied the process of the reaction by observing the change in pressure at different time intervals and proved that this reaction was of the third order over a wide range of pressures.

(ii) In 1918 *Bodenstein* studied the reaction between nitric oxide and oxygen to form nitrogen dioxide.

$$2NO + O_2 \rightarrow 2NO_2.$$

He studied this reaction by immersing the reaction vessel in a constant temperature bath and observing the change in pressure as a function f the time by using a manometer containing bromonaphthalene. Again this reaction was found to be of the third order.

(iii) *M. Trautz* and *V. P. Dalal* (1918) reported a similar type of reaction between nitric oxide and bromine.

(iv) In 1926, *C. N Hinshelwood* and *Green* reported another termolecular reaction

$$2NO + H_2 \rightarrow N_2O + H_2O$$

This reaction takes place at high temperature.

(v) In 1926. *C. N. Hinshelwood* and others reported another reaction between heavy hydrogen and nitric oxide,

$$2NO + D_2 \rightarrow N_2O + D_2O$$

SIMPLE COLLISION THEORY

According to this theory, an activated complex is formed by a collision which involves simultaneous contact of three molecules. But the chance of such a collusion between three molecules is almost zero. Thus, the formation of activated complex could not be explained on the basis of simple collision theory and needs a good deals of discussion.

Trautz's Theory : In 1916 *Trautz* suggested that termolecular reactions are taking place in two stages. In 1918, he proposed the following mechanism for the reaction between NO and Br_2 to form NOBr.

(i) A molecule of nitric oxide and one of the bromine molecules undergo a sticky collision to form a complex $NOBr_2$ which has a short life time.

$$NO + Br_2 \underset{K_2}{\overset{K_1}{\rightleftharpoons}} NOBr_2 \qquad ...(1)$$

(ii) This complex $[NOBr_2]$ then reacts with NO to form NOBr.

$$NOBr_2 + NO \xrightarrow{k'} 2NOBr \qquad ...(2)$$

As the complex $[NOBr_2]$ has a short life time, a small concentration of $[NOBr_2]$ is maintained in equilibrium with NO and Br_2 and therefore, the law of mass action can be applied to step (10) *i.e.,*

$$K = \frac{k_1}{k_2} = \frac{[NOBr_2]}{[NO][Br_2]}$$

or $\quad [NOBr_2] = K\,[NO]\,[Br_2] \qquad ...(3)$

where K is the equilibrium constant for the step (1).

The rate of the reaction is determined by the step (2) *i.e.*,

$$\frac{d}{dt}[NOBr] = k'[NO][NOBr_2] \qquad ...(4)$$

Substituting equation (3) in (4), we get

$$\frac{d}{dt}[NOBr] = k'K\{NO\}^2[Br_2] = k_3[NO]^2[Br_2]$$

where k_3 is another constant.

Thus, the reaction is of the third order. Similarly, the mechanism of other termolecular reactions can be written as.

(a) (i) $NO + O_2 \rightleftharpoons NO_3$

(ii) $NO_3 + NO_2 \longrightarrow 2NO_2$

(b) (i) $NO + Cl_2 \rightleftharpoons NOCl_2$

$NOCl + NO_2 \longrightarrow 2NOCl.$

Objection : If the collision frequency of a termolecular reaction is calculated by using Trautz's theory, this value comes out to be much larger than the actual frequency factor of termolecular reactions. Thus, Trautz's theory cannot be accepted.

BODENSTEIN'S THEORY

In 1922 *M. Bodenstein* gave a different but more general treatment for explaining termoleculear reactions. Her did not specify the particular pair of molecules between which a sticky collision is taking place.

Objection : He estimated the ratio of ternary to binary collisions and this value was found to be about 1/1000. Thus, value was much larger than the frequency factor of a termolecular reaction by a factor of 10^5. Thus, this theory cannot be accepted.

ABSOLUTE THEORY

The reason for the above discrepancy is that the simple kinetic theory, by treating the molecules as hard spheres, takes no account of the considerable loss of entropy when three molecules unit to form the activated complex.

The absolute theory of reaction rates takes into account this loss of entropy. This theory also makes assumptions to the structure of the activated complex.

The application of absolute theory to termolecular reactions was worked out by Gershinowitz and Eyring.

For the reactions

$$2NO + X_2 \rightarrow 2NOX$$

the rate constant can be put as follow.

$$k = \frac{kT}{h} \frac{q\ddagger}{q^2 NoqX_2} e^{-E_0/RT} \quad \text{...(5)}$$

the statistical factor being unity. If there occurs no free rotation in the activated complex, this equation becomes as follows.

$$k_1 = \frac{kT}{h} \frac{\frac{(2\aleph m\ddagger kT)^{3/2}}{h^3} \frac{8\aleph^2 (I_A I_B I_C)^{1/2} (kT)^{3/2}}{h^3} \prod^{11} q\ddagger(v)}{\prod_t^3 \frac{(2\aleph m_i kT)^{3/2}}{h^3} \prod_t^3 \frac{8\aleph^2 I_i kT}{h^2} \prod_t^3 q_i(v)} e^{-E_0/RT} \quad \text{...(6)}$$

where the subscript i refers to the initial state. The activated complex is having six atoms, so that there are 3 × 6 – 7 = 11 vibrational factors. As these vary only slightly with temperature, the temperature dependence of the rate constant, to a good approximation, may be given as follows

$$k \propto T^{-3.5} e^{E_0/RT} \quad \text{...(7)}$$

The pre-exponential factor for this reaction, of the type L + L + L varies at T–3.5.

Gershinowitz gave the following configuration of the activated complex

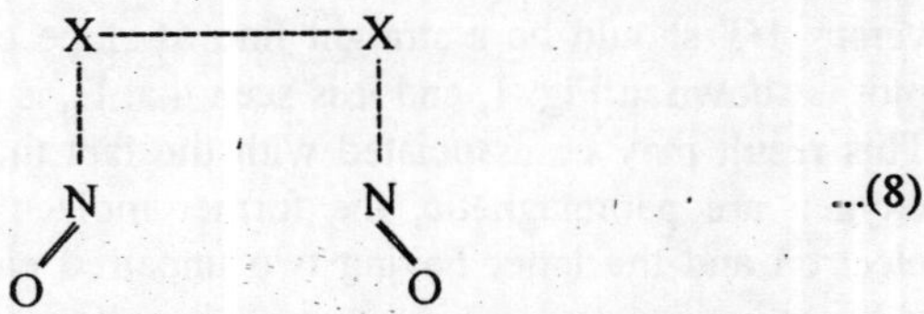

Owing to the face rotation about Br – Br bond, they replaced one of the vibrational factors in the activated complex by the following expression.

$$\frac{(8\pi^3 I_D kT)^{1/2}}{h} \quad ...(9)$$

where ID is the moment of inertia for the free rotation about the X...X bond. They took value of E_0 as 4.89 k cal *i.e.*,

$$E_0 = 4.8 \text{ k cal.}$$

By using equations (6), (7) and (9) they found out the value of kr which was in close agreement with experimental values over a considerable temperature range. Similar agreement has been reported for the reaction between nitric oxide and oxygen in which the value of E0 was taken as zero. Thus, the absolute theory has been adequate for interpreting termolecular reaction and avoids the difficulties that have been reported in simple kinetic theory of collisions.

In the case of the 2NO + O2 reaction, E0 is zero, and the rate constant itself, according to CTST, should decrease with increasing temperature being proportional to T–3.5.

Alternatively, if the activated complex is having the structure

```
            O . . . O
           /         \
      O=N             N=O
```

one of the vibrational partition functions in Eq. (6) must be replaced by the expression for free rotation. The effect of this has been to replace the term T–3.5 in Eq. by T–3.0. The resulting rate expression can be written as follows.

$$\ln k = \ln (\text{constant}) - 3 \ln T + f(T) \quad ...(10)$$

where the constant can be evaluated and f (T) is the contribution from the vibrational partition functions. It follows that a plot of

$$\ln k + 3 \ln T - f(T) \quad ...(11)$$

versus 1/T should be a straight line of slope equal to $- E_0/R$. Such a plot is shown in Fig. 1, and it is seen that E_0 is extremely close to zero. This result may be associated with the fact that both nitric oxide and oxygen are paramagnetic, the former molecule having one unpaired electron and the latter having two unpaired electrons.

The rate constant therefore is calculated with E_0 taken as zero. The moments of inertia are derived from spectroscopic measurements, but the evaluation of those of the activated complex requires a knowledge of its configurations and dimensions. The relevant values are not very critical and estimates were made on the basis of the normal interatomic distances in other molecules.

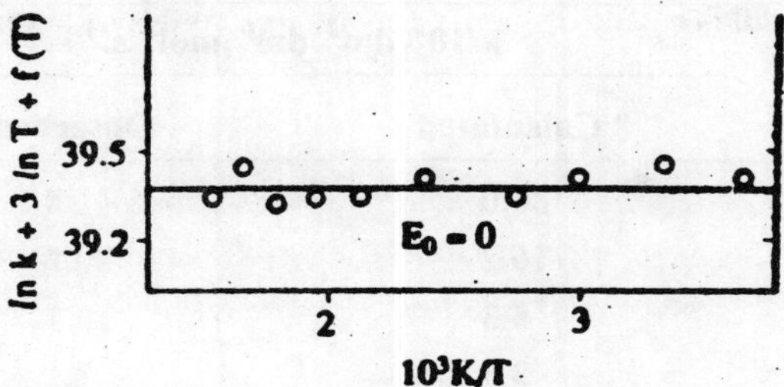

Fig. 3.8 : Plot of *ln* k + 3 T = f'(T) [see eq. (11)] versus 1/T for the reaction between nitric oxide and oxygen.

The observed results and those calculated by the methods outlined above are shown in Table 3.1. The agreement is very satisfactory, especially in that both theory and experiment show a minimum rate at about the same temperature. The contributions of the vibrational partition functions becomes greater, the higher the temperature and eventually overcome the decrease due to the factor $T^{-3.0}$.

The reaction between nitric oxide and chlorine has been treated in a similar manner, the activated complex being assumed to have the configuration shown for N_2O_4‡, with a Cl_3 molecule replacing the O_2 molecule. A plot of the same function against 1/T again gave a straight line, but the slope corresponded to an energy of activation of 20.1 kJ mol^{-1}. Using this value, one can calculate the rate constants given in Table 3.1. The agreement between calculated and observed values is again quite satisfactory, especially in view of the assumptions made in the theory and the fact that the data are not very accurate.

The treatment of trimolecular gas reactions is one of the outstanding successes of conventional transition-state theory. Using very straight forward procedures, we can account for the unusual temperatures dependencies of these reactions in a simple way and we can calculate reliable values for the pre-exponential factors.

As was noted, a simple version of collision theory gave results in error by several orders of magnitude, and it gave no interpretation of the temperature dependence. Other theories can give satisfactory results only at the cost of considerable labour.

Table 3.1 : Calculated and observed rate constant for the nitric oxide-oxygen reaction.

	$k/10^3$ dm^3 dm^6 mol^2 s^{-1}	
T/K	**Calculated**	**Observed**
80	86.0	41.8
143	16.2	20.2
228	5.3	10.1
300	3.3	7.1
413	2.2	4.0
564	2.0	2.8
613	2.1	2.8
662	2.0	2.9

Table 3.2 : Calculated and observed rates for the nitricoxide chlorine reaction.

	k/dm^2 mol^{-2} s^{-1}	
T/K	**Calculated**	**Observed**
273	1.4	5.5
333	2.2	9.5
355	8.6	27.2
401	18.3	72.2
451	25.4	182
506	64.5	453
566	120.2	11.30

Kinetic data for some trimolecular reactions studied in the mass spectrometer have been compiled and discussed and are listed in Table 3.3. All these reactions appear to occur with zero activation energy, so that the temperature dependence of the rates is attributed to that of the pre-exponential factors. The predicted temperature dependences for linear and nonlinear complexes are shown in table. In view of the simplicity of the model and the possible experimental error, the agreement between theory and experiment is very satisfactory. These trimolecular reactions occur such more slowly than bimolecular reactions having zero activation energy, because of the considerable entropy loss when the activated complex is formed.

Table 3.3 : Temperature dependences of rates of trimolecular ion-molecule reactions

Reaction	Experimental temperature dependence	Theoretical temperature dependence	
		Linear complex	Nonlinear complex
$He^+ + He + He \rightarrow He_2^+ + He$	T^{-1}	T^{-1}	$T^{-1.5}$
$N^+ + N_2 + He \rightarrow N_3^+ + He$	$T^{-1.7}$	T^{-2}	$T^{-1.5}$
$N_2^+ + N_2 + He \rightarrow N_4^+ + He$	$T^{-1.6}$	T^{-3}	$T^{-2.5}$
$N^+ + N_2 + N_2 \rightarrow N_3^+ + N_2$	$T^{-2.5}$	T^{-3}	$T^{-2.5}$
$O_2^+ + O_2 + N_2 \rightarrow O_4^+ + O_2$	T^{-3}	T^{-4}	$T^{-3.5}$
$N_2^+ + N_2 + N_2 \rightarrow N_4^+ + N_2$	$T^{-4.4}$	T^{-4}	$T^{-3.5}$
$H_3^+ + H_2 + H_2 \rightarrow H_5^+ + H_2$	$T^{-.46}$	$T^{-4.5}$	T^{-4}
$H_3O^+ + H_2O + CH_4 \rightarrow H_5O_2^+ + CH_4$	$T^{-4.2}$	$T^{-5.5}$	T^{-5}

CHAIN REACTIONS

Introduction : *A chain reaction is one in which the products of the reaction carry on the reaction on the part of reacting molecules and thus a long series of self-repeating steps is started.*

These reactions were first proposed by *Bodenstein* to account for the high yields of photochemical reaction between hydrogen and chlorine. *Christiansen* and *Kramers* independently proposed chain reactions to describe the kinetics of unimolecular reactions.

Some of the processes involved in chain reactions may, and occasionally do take place at interfaces between phases, *e.g.*, termination and initiation of chain may occur at the walls of the reaction vessel or on the surface of catalytic or inhibiting powers; however, the greater proportion of the product molecules are usually formed in the gas or liquid phase and in this sense chain reactions may be regarded as homogeneous. However, chain reactions differ in many respects from the homogeneous reactions.

DISTINGUISHING FEATURES OF CHAIN REACTIONS

(i) The probability factor P of chain reactions is generally greater than unity. It means that the reaction chain may be the cause if the value of P is found to be very large.

(ii) Sometimes, the chain reactions possess excessive speeds which may lead to explosion. It is noteworthy, that the transition from steady reaction rate to explosion at some parameter such as pressure or composition is very abrupt, *i.e.,* the explosion boundary is well defined. Such explosion boundaries possess varied shapes and may have further explosion regions which differ from the first in properties such as the colour and temperature of the flame and the products formed.

(iii) In all non-chain processes, the highest rate is observed at the beginning and the rate of reaction falls off with time. On the other hand, chain reactions begin at zero rate. It then rises to maximum and then falls off with time. This is illustrated in the Fig. 3.9.

In this figure, curve 2 indicates a very rapid increase of rate in which very high values are momentarily attained, and this corresponds to an ignition. Such reaction systems may momentarily reach very high temperatures, be luminous, be accompanied by an audible click and possible cause explosion. Curve represents much slower chain reactions in which the maximum rate is not reached until after a considerable time interval, but the maximum rate is maintained for an appreciable period before slowly falling off. Curve 1 represents a typical non-chain reaction.

(iv) As the chain reactions begin at zero rate, it therefore requires enough time so that the rate of the reaction could be detected experimentally. This time interval is called the induction period. *This induction period* is regarded as the striking feature of chain reactions. In the oxidation of hydrogen sensitized by NO_2 or NOCl, induction period of 2–3 minutes are observed during which no detectable pressure change occurs and after which either an explosion or a slow reaction may ensue.

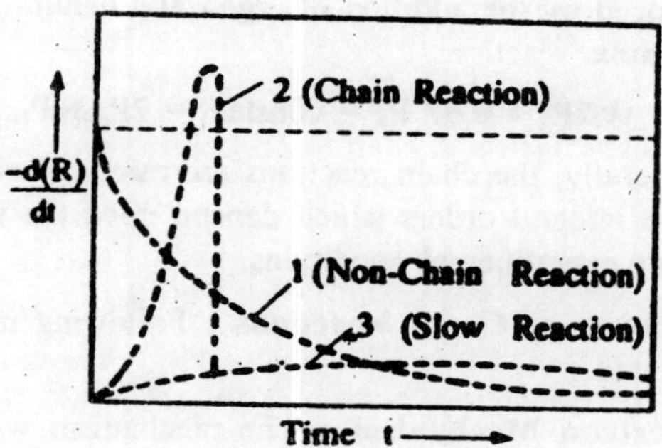

Fig. 3.9

(v) The speed of chain reactions is retarded or accelerated by traces of other substances. As an example of positive catalysis we have the lowering of the spontaneous ignition temperature of a mixture of 100 mm. of hydrogen and 50 mm. of oxygen from 580°C to 330°C by the addition of 0.1 mm. of nitrosyl chloride. Moreover, the same substance may be a positive catalyst for one group of reactions and a negative catalyst for others. Thus, 0.1 m m. nitrosyl chloride added to 400 mm. of an equimolecular mixture of carbon monoxide and chlorine will reduce the rate of photosynthesis of phosgene at room temperature by a factor of thirty five.

(vi) The rate of chain reactions is influenced by the changes in shape of the containing vessel. Generally, the rate of chain reactions is reduced if the radius of the reaction vessel is reduced. For equimolecular mixtures of hydrogen and chlorine of total pressure less than 30 mm. Hg pressure, it was found that the rate of photo-chemical reaction is proportional to the square of the diameter of the cylindrical reaction vessel. At higher pressures those surfaces effects are much less marked and there is evidence that in some cases a chain reaction cannot proceed in the absence of a surface.

(vii) Foreign gases which are chemically unchanged in a chain reacting gas mixture often modify the reaction kinetics. At low pressure, the action is frequently catalytic but at high pressures they generally exert a retarding action. In the hydrogen-oxygen reaction it was established that under certain conditions the upper limit (P_u) of $2H_2 + O_2$, mixture in a cylindrical reaction vessel was reduced by the addition of argon and helium according to the formula

$$2P_u + 0.67\ P_A = \text{constant} = 2P_u + P_{H0}.$$

(viii) Generally, the chain reactions are rarely of simple orders but have integral orders which depend upon the vessel shape and other experimental conditions.

Mechanism of Chain Reactions : Following mechanisms have been suggested :

(a) Electron Mechanism : The mechanism was suggested by *Bodenstein*. According to him, electrons act as chain carriers. These electrons are generated when chlorine exposed to light.

$$Cl_2 + h\nu \longrightarrow Cl_2^+ + e^-$$

$$e^- + Cl_2 \longrightarrow Cl_2^-$$

$$Cl_2^- + H_2 \longrightarrow 2HCl + e^-$$

This mechanism could not be accepted because no ionisation of chlorine was detected when exposed to light.

(b) Activation Mechanism : Again, this mechanism was proposed by *Bodenstein*. According to him, the chlorine molecules become activated when exposed to light.

$$Cl_2 + h\nu \longrightarrow Cl^*$$

$$Cl_2^* + H_2 \longrightarrow 2HCl^*$$

$$HCl^* + Cl_2 \longrightarrow Cl_2^* + HCl$$

where* represents the activated substance.

In this mechanism, an activated molecule of hydrogen chloride can transfer its energy to chlorine molecule but not to the hydrogen molecule. Thus, the main drawback of this theory is that it could not explain transfer of energy from hydrogen chloride to hydrogen.

(c) Free Atoms and Radicals Mechanism : According to this theory atoms and radicals act as chain carries. For example.

$$Cl_2 + h\nu \longrightarrow Cl + Cl$$

$$Cl + H_2 \longrightarrow HCl + H$$

$$H + Cl_2 \longrightarrow HCl + Cl.$$

The presence of free atoms and free radicals in a reaction system has been well established. The main methods which are used for detecting free radicals are discussed in the next section.

Detection and Estimation of Atoms and Radicals in Chain reactions : In order to study the kinetics of gas reactions, it is important to employ methods which will show the presence of atoms and radicals, and will provide an estimate of their concentrations. The important methods are.

1. **Spectroscopic Methods :** Emission and absorption spectroscopy have both been used for the detection and estimation of atoms and radicals in reaction systems. Emission spectra are employed when the reaction is accompanied by a flame, while absorption spectra is applied in studies using the technique of flash photolysis, which yields very high concentration of radicals.

   Indirect spectroscopic methods are increasingly used for the identification of radicals and atoms.

2. **Electron Spin Resonance Spectroscopy :** The underlying idea in this method is that substances such as atoms and free radicals which contain unpaired electrons produce a splitting of energy levels in a strong magnetic field. The atoms or radicals to be studied are introduced into a quartz tube between the poles of a magnet, and at right angles to the broad face of a wave-guide; the power is provided by a magnetron or klystron. A crystal detector is used to measure the absorption of radiation. As the intensity of absorption depends upon the number of unpaired electrons, the method may be employed for measuring radical concentrations. The region of absorption varies for different species and hence the atoms and radicals can be detected.

3. **Mass Spectrometry :** This technique is now being employed to a considerable extent for the identification and estimation of free radicals in reaction systems. *Eltenton* demonstrated the presence of methyl and ethyl radicals in the decomposition of ethane and of methylene in the decomposition of diazomethane. More recently a large number of free radicals have been detected by this method,

4. **Trapping of Free Radicals :** In recent years a large amount of work has been carried out on the "trapping" of free radicals. In this, the reaction system is passed over a surface that has been

cooled, say, by liquid air. The solid that condenses is frequently of a vivid colour, and then the technique like electron spin spectroscopy (E.S.R.), mass spectrometry are employed for study. But this technique has not been used for determining radical concentrations.

5. **Chemical Methods :** A number of molecules for the detection and estimation of atoms and free radicals are based upon their high chemical reactivity. For example, the introduction of metallic oxide into reaction system has been used for the detection of hydrogen atoms. Hydrogen atom concentration has also been determined by introducing Para hydrogen into the reaction system and following the rate of the reaction.

$$H + p\text{-}H_2 \longrightarrow 0\text{-}H_2 + H$$

A disadvantage of the method is that the conversion is also catalysed by paramagnetic substances such as other radicals that may be present in the reaction system.

Kinetics of Chain Reactions : There are two approaches to the problems of the mathematical representation of chain reactions.

(i) Non-Steady Treatment : This method is followed by *Russians.* It involves highly complicated mathematical equations.

(ii) Steady Treatment : This method is followed by *British* and *American workers*. According to this treatment, the concentration of a free radical or intermediate or chain carrier is constant at any instant.

$$\frac{d[R]}{dt} = 0.$$

where R is the free radical or intermediate or a chain carrier.

Let us now apply steady state treatment to the general chain reaction. This general scheme is.

(i) $M \xrightarrow{k_1} R$ chain initiation

(ii) $M + R \xrightarrow{k_2} \alpha R + M'$. Branching chain

(iii) $M + R \longrightarrow \text{Product}$ Final reaction

(iv) $R \xrightarrow{k_4}$ destruction at surface

(v) $R \xrightarrow{k_5}$ destruction in gas phase .

Applying the steady treatment to the radical R, we get

$$\frac{dR}{dt} = k_1 [M] + k_2 [\alpha - 1] [R] [M] - k_3 [R] [M]$$

$$- k_4R - R - k_5R = 0$$

$$\text{or} \quad R = \frac{k_1 [M]}{k_3 [M] - [k_4 + k_5] - k_2 [\alpha - 1] M} \quad ...(1)$$

The overall rate of the reaction is given by step (iii)

$$v = k_1 [R] [M] \quad ...(2)$$

Substituting the value of [R] from equation (1) in equation (2), we obtain

$$v = \frac{k_3 k_1 [M]^2}{k_3 [M] + [k_4 + k_5] - k_2 [\alpha - 1] M} \quad ...(3)$$

Chain Length : *It is defined as the rate of the over-all reaction divided by the rate of the initiation reaction.*

Rate of the initial reaction (i) = k_1 [M] ...(4)

The rate of the over-all reaction = v

$$\text{where} \quad v = \frac{k_3 k_1 [M]^2}{k_3 [M] + [k_4 + k_5] - k_2 [\alpha - 1] M} \quad ...(5)$$

∴ chain length

$$= \frac{v}{k_1 M}$$

$$= \frac{k_3 k_1 [M]^2}{(k_3 [M] + [k_4 + k_5] - k_2 [\alpha - 1] M) k_1 [M]}$$

$$= \frac{k_3 [M]}{k_3 [M] + [k_4 + k_5] - k_2 [\alpha - 1] M} \quad ...(6)$$

EXAMPLES OF THE STEADY TREATMENT

(a) **Decomposition of Ozone :** *Chapman* and his *co-workers* investigated the decomposition of ozone in the presence of excess of oxygen. They reported that the reaction was of second-order in ozone.

In 1906 John studied this reaction and reported that the rate of the reaction was inversely proportional to the oxygen concentration. The mechanism proposed by the him is

(1) $O_3 \underset{k_2}{\xrightarrow{k_1}} O_2 + O$

(2) $O + O_3 \xrightarrow{k_3} 2O_2$.

The decomposition of simple molecules like ozone must be of second order. Therefore, *Benson* and *Axworthy* (1975) gave the following mechanism.

(1) $O_3 + M \underset{k_2}{\overset{k_1}{\rightleftharpoons}} O_2 + O + M$

(2) $O + O_3 \xrightarrow{k_3} 2O_2$.

By applying steady-state treatment to the oxygen atoms, we get

$$\frac{d[Q]}{dt} = k_1 [O_3] [M] - k_2 [O_2] [O] [M] - k_3 [O] [O_3] = 0$$

or $$k_2 [O_2] [O] [M] + k_3 [O] [O_3] = k_1 [O_3] [M]$$

or $$[O] (k_2 [O_3] [M] + k_3 [O_3]) = k_1 [O_3] [M]$$

or $$[O] = \frac{k_1 [O_3][M]}{k_2 [O_3][M] + k_3 [O_3]}.$$

The overall rate is given by

$$\frac{-d[O_3]}{dt} = k_1 [O_3] [M] - k_2 [O_2] [O] [M] + k_3 [O] [O_3] \quad ...(7)$$

Combining equations (6) and (7), we get

$$\frac{-d[O_3]}{dt} = \frac{2k_1k_3 [O_3]^2 [M]}{k_2 [O_2][M] + k_3 [O_3]} \quad ...(8)$$

All the experimental results were consistent with eq. (8).

(b) **Reaction Between Hydrogen and Bromine :** This reaction was first studied by *Budenstein* and *Lind* (1907) over the temperature range from 205°C. Unlike the analogous hydrogen-iodine reaction, they found that this reaction was not a simple bimolecular process but the rate of such a reaction might be expressed by the following complicated empiric l equation.

$$\frac{-d[HBr]}{dt} = \frac{k[H_2][Br_2]^{1/2}}{1+[HBr]/m[Br_2]} \quad ...(1)$$

where K and m are constants.

Christiansen, Herzfeld and *Polayi* (1920) suggested the correct mechanism. They applied the steady state t eatment to this reaction mechanism and obtained the empiri al rate eq. (1). This reaction is initiated by the dissociation of brc nine molecule to atoms

Chain Initiation :

(i) $Br_2 \xrightarrow{k_1} 2Br$

This dissociation is taking place to a small extent.

Chain Propagation :

The initiation reaction (i) is followed by a slow reaction

(ii) $Br + H_2 \xrightarrow{k_2} HBr + H$

and the rapid reaction

(iii) $H + Br_2 \xrightarrow{k_3} HBr + Br$

Chain Inhibition :

(iv) $Br + HBr \xrightarrow{k_4} H_2 + Br$

(v) $Br + Br \xrightarrow{k_5} Br_2$.

Here k_1, k_2, k_3, k_4, and k_5 represent the specific rates of all the five reaction respectively. In this mechanism, the rate of the formation of HBr is given bysteps (i) and (iii). Hence

$$\frac{d[HBr]}{dt} = k_2\,[H_2]\,[Br] + k_3\,[H]\,[Br_2].$$

The rate of consumption of HBr is given by step (iv)

$$\frac{-d[HBr]}{dt} = k_4\,[H]\,[HBr].$$

Therefore, the net rate of formation of HBr will be given by

$$\frac{d[HBr]}{dt} = k_2 [H_2] [Br] + k_3 [H] [Br] - k_4 [H] [HBr] \quad ...(2)$$

As equation (2) involves concentrations of bromine and hydrogen atoms which are too small quantities to be measured directly, it is required that their concentrations must be expressed in terms of measurable quantities. This is done by writing down and solving steady state equations for [H] and [Br]. The steady-state equations for [H] and [Br] are,

$$\frac{d[H]}{dt} = k_2 [Br] [H_2] - k_3 [H] [Br_2] - k_4 [H] [HBr] \quad ...(3)$$

$$\frac{d[Br]}{dt} = 2k_1 [Br_2] - k_2 [Br] [H_2] + k_3 [H] [Br_2] + k_4 [H] [HBr] - 2k_5 [Br]^2 \quad ...(4)$$

As the concentrations of hydrogen and bromine atoms are relatively small, it is, therefore regarded that a stationary state is reached soon after the reaction has started. Therefore, the time derivative of atom concentration can be taken to be zero over a short interval of time. In simple words, it means

$$\frac{d[H]}{dt} = 0 \text{ and } \frac{d[Br]}{dt} = 0.$$

Applying this condition of stationary states of equations (3) and (4), we get

$$k_3 [H] [Br_2] + k_4 [H] [HBr] - k_3 [Br] [H_2] = 0 \quad ...(5)$$

and $2k_1 [Br_2] + k_3 [H] [Br_2] + k_4 [H] [HBr] - 2k_5 [Br]^2 - k_2 [Br] [H_2] = 0 \quad ...(6)$

Equations (5) and (6) can be written as

$$k_2 [Br] [H_2] = k_3 [H] [Br_2] + k_4 [H] [HBr] \quad ...(7)$$

and $k_2 [Br] [H_2] = 2k_1 [Br_2] + k_3 [H] [Br_2] + k_4 [H] [HBr] - 2k_5 [Br]^2 \quad ...(8)$

Subtracting equation (7) from (8), we get

$$2k_1 [Br_2] = 2k_5 [Br]^2$$

or $$[Br]^2 = \frac{k_1}{k_5} [Br_2]$$

or $$[Br] = \left(\frac{k_1}{k_5}\right)^{1/2} [Br_2]^{1/2} \qquad ...(9)$$

Substituting the value of [Br] from Eq. (9) in (7), we obtain

$$k_2\left(\frac{k_1}{k_5}\right)^{1/2} [Br_2]^{1/2} [H_2] = k_3 [H] [Br_2] + {}^*k_4 [H] [HBr]$$

or $$k_2\left(\frac{k_1}{k_5}\right)^{1/2} [Br_2]^{1/2} [H_2] = [H] (k_3[Br_2] + k_4 [HBr])$$

or $$[H] = \frac{k_3 (k_1/k_5)^{1/2} [Br_2]^{1/2} [H_2]}{k_3 [Br_2] + k_4 [HBr]} \qquad ..(10)$$

Inserting the values of [H] and [Br] from equations (10) and (9) in (2), we obtain the rate expression for the formation of hydrogen bromide which is given by

$$\frac{d[HBr]}{dt} = k_2\left(\frac{k_1}{k_5}\right)^{1/2} [Br_2]^{1/2} [H_2] + k_3 \frac{k_2 (k_1/k_5)^{1/2} [Br_2]^{1/2} [H_2]}{k_3 [Br_2] + k_4 [HBr]} [Br_2] - k_4 \frac{k_2 (k_1/k_5)^{1/2} [Br_2]^{1/2} [H_2]}{k_3 [Br_2] + k_4 [HBr]} [HBr]$$

$$\frac{d[HBr]}{dt} = k_2\left(\frac{k_1}{k_5}\right)^{1/2} [Br_2]^{1/2} [H_2] \left[1 + \frac{k_3 [Br_2]}{k_3 [Br_2] + k_4 [HBr]} - \frac{k_4 [HBr]}{k_3 [Br_2] + k_4 [HBr]}\right]$$

$$= k_2\left(\frac{k_1}{k_5}\right)^{1/2} [Br_2]^{1/2} [H_2] \left[\frac{k_3 [Br_2] + k_4 [HBr] + k_3 [Br_2] - k_4 [HBr]}{k_3 [Br_2] + k_4 [HBr]}\right]$$

$$= k_2\left(\frac{k_1}{k_5}\right)^{1/2} [Br_2]^{1/2} [H_2] \left[\frac{2k_3 [Br_2]}{k_3 [Br_2] + k_4 [HBr]}\right]$$

$$= \frac{2k_2 (k_1/k_5)^{1/2} [Br_2]^{1/2} [H_2]}{k_3 [Br_2]/k_3 [Br_2] + k_4 [HBr]/k_3 [Br_2]}$$

$$= 2k_2 = \frac{[k_1/k_5]^{1/2} [Br_2]^{1/2} [H_2]}{1 + k_4 [HBr]/k_3 [Br_2]}.$$

This equation is identical with the empirical equation (1).

(c) Photochemical Combination of Hydrogen and Bromine:

Please see the chapter on Photochemistry.

(d) Hydrogen-Chlorine Reaction:

Please see the chapter on Photochemistry.

Comparative Study of Hydrogen-Halogen Reactions : The thermal reactions of hydrogen with chlorine, bromine and iodine have been found to be kinetically very different.

For example, the hydrogen-iodine reaction is a molecular reaction while others being free-radical reactions. The kinetic differences between the hydrogen chlorine and hydrogen bromine reactions have been explained in terms of the above mechanism, the main differences are.

1. The inclusion of the reactions $H + HBr \rightarrow H_2 + Br$ and exclusion of $H + HCl \rightarrow H_2 + Cl$,
2. The different chain-ending steps, $Br + Br \rightarrow Br_2$ being assumed for bromine, and the three processes $H + O_2$, $Cl + O_2$ and $Cl + X$ for chlorine.

The differences between the reactions are to be explained in terms of differences between the activation energies for the reactions Table 3.4. The activation energy for the molecular reaction $H_2 + I_2 \rightarrow 2HI$ is 40 K. cal.

In order for reaction ($H_2 + I_2$) to occur by an atomic mechanism, the overall activation energy would be 51 J cal and since this is so large, hence, the reaction proceeds by the molecular mechanism. But in the case of chlorine and bromine the position is different.

For example with bromine, the initial step requires 23 Kcal, and second one 18 Kcal, making a total of 41 Kcal which is appreciably less than the estimated value for the molecular reaction.

Table 3.4 : Activation Energies for the Hydrogen-Halogen Reactions

Reaction	*Activation energy (k cal) with Y equal to*		
		Br	*I*
$H_2 + Y_2 \rightarrow 2HY$	50	45	40
$1/2\ Y \rightarrow Y$	28	23	17
$Y + H_2 \rightarrow YH + H$	6.0	17.6	34.2
$H + Y_2 \rightarrow HY + Y$	2–3.6	1.2	0
$H + HY \rightarrow H_2 + Y$	5.0	1.2	1.5

The reason that the reaction $H + HCl \rightarrow H_2 + Cl$ is unimportant, whereas the reaction $H + HBr \rightarrow H_2 + Br$ plays a role, is also seen from the above table. The reaction $H + HCl \rightarrow H_2 + Cl$ has an activation energy of 5 K, cal, and therefore occurs at a negligible speed as compared to $H + Cl_2 \rightarrow HCl + Cl$, the activation energy for which is about 2K. cal The values for the corresponding bromine reactions are very close to one another (1.2 K. cal).

The reaction of $H + O_2$ is important in the chlorine reaction because its speed in 1/20 that of $H + Cl_2$. In case of $H + Br_2 \rightarrow HBr + Br$, rate is much higher, however, in comparison with it that of $H + O_2 \rightarrow HO_2$ is quite negligible. In case of bromine reaction the main chainending step is $2Br \rightarrow Br_2$, while the reactions involving oxygen are usually predominant in the hydrogen-chlorine reaction.

(e) Photolysis of Acetaldehyde : Please see chapter on Photochemistry.

(f) Auto-oxidations:

The reactions which take place between molecular oxygen and other substances are usually referred to as auto-oxidations. An example for an auto-oxidation is the reaction of hydrogen with oxygen. Let us now discuss this reaction.

THE HYDROGEN–OXYGEN REACTION

The simplified mechanism of this reaction is due to *Hinshelwood*. The mechanism is.

(1) $H_2 \rightarrow 2H$

(2) $H + O_2 \rightarrow OH + O$

(3) $O + H_2 \rightarrow OH + H$

(4) $H + O_2 + M \rightarrow HO_2 + M$

(5) $HO_2 \rightarrow$ removal at surface

(6) $HO_2 + H_2 \rightarrow H_2O + OH$

(7) $OH + H_2 \rightarrow H_2O + H$

(8) $H \rightarrow$ removal at surface

(9) $OH \rightarrow$ removal at surface.

The reactions (2) and (3) are chain-braching processes. The HO_2 radical is fairly stable. This an diffuse to the surface and be removed. Another possibility is that HO_2 can react with hydrogen (Reaction 6). This reaction is important at higher pressure in the neighbourhood of the third limit.

At the lower limit, the important processes are (8) and (9) but the reactions (4), (5) and (6) are not important.

We will now consider the reaction at *low pressures.* The steady state equation for hydrogen atoms is

$$v_t - k_2 [H] [O_2] + k_3 [O] [H_2] + k_2 [OH] [H_2] - f_8 [H] = 0 \quad ...(11)$$

Similarly, the steady state equations for oxygen atoms and for hydroxyl radicals are given below.

$$k_2 [H] [O_2] - K_3 [O] [H_2] = 0 \quad ...(12)$$

$$k_2 [H] [O_2] + K_3 [O] [H_2] - k_7 [OH] [H_2] - f_8 [OH] = 0 \quad ..(13)$$

In equation (11), v_t is the rate of initiation reaction (1) and f_8 is the coefficient for reaction (8)

In equation (13), f_9 is the rate coefficient for reaction (9).

By adding equations (11) and (12), we get

$$v_i \; k_7 [OH] [H_2] - f_8 [H] = 0 \quad ...(14)$$

By adding equations (12) and (13), we get

$$2k_2[H] [O_2] - k_7 [OH] [H_2] - f_9 [OH] = 0 \quad ...(15)$$

From equation (14), we have

$$f_8\,[H] = v_t + k_7\,[OH]\,[H_2]$$

$$[H] = \frac{v_t + k_7[OH][H_2]}{f_8} \quad ...(16)$$

Substituting the value of [H] in equation (15), we obtain

$$2k_2\left[\frac{v_t + k_7[OH][H_2]}{f_8}\right][O_2] - k_7[OH]\,[H_2] - f_9[OH] = 0$$

$$\text{or} \quad 2k_2\,v_t\,[O_2] + 2k_2\,k_7\,[OH]\,[O_2] - k_7\,f_8\,[OH]\,[H_2] - f_8f_9\,[OH] = 0$$

$$\text{or} \quad 2k_2\,v_t\,[O_2] = f_8\,f_9\,[OH] + k_7\,f_8\,[OH]\,[H_2] - 2k_2k_7\,[OH]\,[H_2]\,[O_2])$$

$$= [OH]\,)f_8\,f_9 + k_7\,f_8\,[H_2] - k_2k_7\,[H_2]\,[O_2])$$

$$\text{or} \quad [OH] = \frac{2k_2\,v_t[O_2]}{f_8f_9 + k_7f_8[H_2] - 2k_2k_7[H_2][O_2]} \quad ...(17)$$

At low pressures the rate of the formation of water is given by reaction (7),

$$v = k_7\,[OH]\,[H_2] \quad ...(18)$$

Combining equations (17) and (18), we get,

$$v = \frac{2v_t\,k_2\,k_7[H_2][O_2]}{f_8f_9 + k_7f_8[H_2] - 2k_2k_7[H_2][O_2]} \quad ...(19)$$

In equation (19), the sum $f_7f_8\,[H_2] + f_8\,f_9$ is related to the surface removal procésses. When the pressure is increased, the value of $k_7f_8\,[H_2] + f_8\,f_9$ decreases. A limit may be reached when $k_7\,k_8\,[H_2] + f_8\,f_9$ becomes equal to $2k_2k_7\,[H_2]\,[O_2]$. Thus, the denominator becomes zero and so the rate becomes infinite.

UPPER AND LOWER EXPLOSION LIMITS

This was studied by *Hinshelwood and Semenoff.*

The reaction between hydrogen and oxygen proceeds at a measurable speed if the temperature is between 450° and 600° C. Below this range, the reaction becomes slow but above this range, explosion takes place.

Consider a mixture of O_2 and H_2 in the ratio of 1 : 2 which is maintained at 550°C and at a pressure of about 2 mm. When the pressure is slowly increased, the rate of the reaction also increases slowly, as shown in Fig. (3.10). At a certain critical pressure of 50 mm or so the mixture explodes. The exact value of this critical pressure will depend upon the size and shape of the vessel. The limit at which it occurs is known as *first explosion limit or lower explosion limit.*

Now, maintain the mixture of O_2 and H_2 at 200 mm, pressure. Again, the reaction proceeds at a steady rate. If the pressure is reduced, the rate of the reaction is also reduced. But at about 100 mm, the reaction mixture explodes. Thus, there is a pressure region between 50 and 100 mm within which explosion takes place. Above and below this pressure region, reaction proceeds at a normal rate. The pressure limit of 100 mm at which the explosion occurs is known as *upper explosion limit or second limit.*

In Fig. (3.10) there is a third limit of explosions which is at still higher pressure. Here explosions are due to the rise in the temperature of the reaction system. Such explosions are known as *thermal-explosions.*

In Fig. (3.11), the variation of explosion limits with the temperature is shown. From this figure, it is seen that explosion will take place at all pressures above 600°C but no explosion will occur under any pressure conditions below 460°

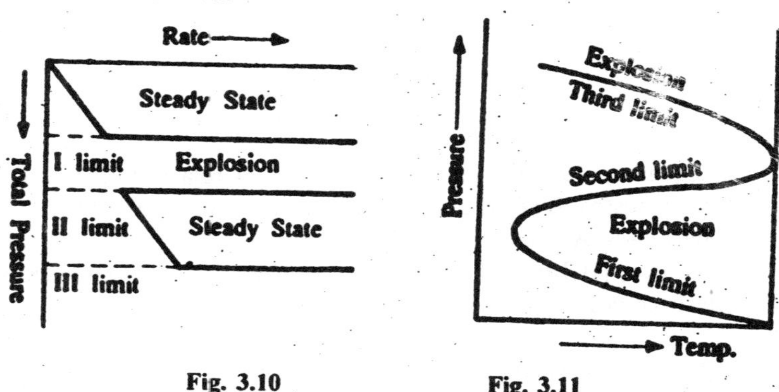

Fig. 3.10 Fig. 3.11

POLYMERISATION REACTIONS

These are the reactions in which is number of molecules react to form a large molecule. In certain cases, molecule of water is eliminated and in such cases the polymerisation is said to be of the *condensation type.* In other cases there is simple addition of molecules to one another, and the process is then termed as *addition polymerisation.*

Polymerisation reactions in the gas phase occur by complex mechanisms that may be either of the *molecular type or may involve free radical processes.* The kinetic features of these two types of polymerisation reactions will now be considered.

(a) Molecular Mechanism : Suppose a polymer Mm + n is formed.

$$M_m + M_n \longrightarrow M_{m+n}$$

where Mm and Mn represent a chain containing m and n monomers respectively, m and n can have any value from unity upwards.

The monomer M_1 may be removed by reaction with itself and with any other molecule present, its net rate of disappearance is therefore

$$-\frac{d[M_1]}{dt} = k[M_1]^2 + k[M_1][M_2] + \ldots \qquad \ldots(1)$$

This may be put as

$$\frac{d[M_1]}{dt} = -k[M_1]\sum_{n=1}^{\infty}[M_1] \qquad \ldots(2)$$

A molecule having two monomer molecules may be formed by reaction between two monomers, and disappears by reaction with a molecule of any length, the net rate of production of dimer is therefore given by

$$\frac{d[M_2]}{dt} = \frac{1}{2}k[M_1]^2 - k[M_2]\sum_{n=1}^{\infty}[M_n] \qquad \ldots(3)$$

The factor 1/2 is required because the formation of one dimer requires the reaction of two monomers. In general, the net rate of formation of n dimers is given by

$$\frac{d[M_n]}{dt} = \frac{1}{2}k\sum_{S=1}^{S=n-1}[M_S][M_{n-S}] - k[M_n]\sum_{S=1}^{\infty}[M_n] \qquad \ldots(4)$$

Addition of all of these equations gives rise to

$$\frac{d\sum_{n=1}^{\infty}M}{dt} = -\frac{1}{2}k\left\{\sum_{n=1}^{\infty}[M_n]\right\}^2 \qquad \ldots(5)$$

If $[M_1]_0$ represents the total concentration of monomers at the beginning of the reaction, and $\sum_{n=1}^{\infty} [M_n]$ that of all the molecules at time t, the fraction of reaction that has occurred at time t is given by

$$f = \frac{[M_1]_0 - \sum_{n=1}^{\infty} [M_n]}{[M_1]_0} \qquad ...(6)$$

Equation (5) and (6) gives rise

$$\frac{df}{dt} = \frac{1}{2} [M_1]_0 \, k \, (1 - f)^2$$

which integrates to

$$f = \frac{[M_1]_0 kt}{2 + [M_1]_0 kt}.$$

The concentration of n-mers is given by

$$[M_n] = [M_1]_0 \, f^{r-1} \, \{1 - f)^2 \qquad ...(7)$$

This equation is useful in that it gives the distribution of polymers corresponding to any extent to reaction.

(b) Free-Radical Mechanisms : Olefinic substances, such as ethylene and styrene usually polymerise by free-radical mechanisms. The reactions sometimes occur in the gas phase and sometimes in solution or in the liquid phase. The general reaction scheme for this type of polymerisation reaction may be put is

(i) $? \longrightarrow R_1$ Initiation

(ii) $R_1 + M \longrightarrow R_2$

(iii) $R_2 + M \longrightarrow R_3$ Chain propagation

....

(iv) $R_{n-1} + M \longrightarrow R_n$

(v) $R_n + R_m \longrightarrow R_{n+m}$ Termination.

Application of the steady-state treatment leads to set of equations. The first of these, applying to R1, is

$$v_t - k_p [R_1] [M] k_t [R_1])[R_1] + [R_2] + ...) = 0 \qquad ...(8)$$

where M = a monomer molecule,

R_1 = an initiating free radical,

R_n = a radical consisting of R_1 added to a chain of (n − 1) monomer molecules.

v_t = the rate of the initial molecules.

k_p = rate constant for the propagation and

k_t = rate constant for termination.

The final term is for the removal of radical R_1 by reaction with R_1, R_2, R_3, etc. The equation (8) may be written as

$$v_t - k_p [R] [M] - k_t [R_1] \sum_{n=1}^{\infty} [R_n] = 0 \qquad ...(9)$$

Similarly for R_2,

$$k_p [R_1] [M] - k_p [R_2] [M] - k_t [R_2] \sum_{n=1}^{\infty} [R_n] = 0 \qquad ...(10)$$

In general for R_n

$$k_p [R_{n-1}] [M] - k_p [R_n] [M] - k_t [R_n] \sum_{n=1}^{\infty} [R_n] = 0 \qquad ...(11)$$

There is an infinite number of such equations, and the sum of all the them is simplified as

$$v_i - k_t \left(\sum_{n=1}^{\infty} [R_n] \right)^{1/2} = 0 \qquad ...(12)$$

Thus $\sum_{n=1}^{\infty} [R_n] = \left(\frac{v_i}{k_t} \right)^{1/2}$ ...(13)

The rate of disappearance of monomer is

$$\frac{-d[M]}{dt} = k_p [M] \sum_{n=1}^{\infty} [R_n] = k_p \left(\frac{v_i}{k_t} \right)^{1/2} [M] \qquad ...(14)$$

Various special cases of this may be considered, corresponding to different initiation processes. Thus,

(i) The rate of thermal initiation may sometimes the proportional to the square of the monomer concentration.

$$v_t = k_i [M]^2$$

and in this case $-\frac{d[M]}{dt} = k_p \left(\frac{v_i}{k_t} \right)^{1/2} [M]^2.$

The overall process is thus of second-order. This type of behaviour has been observed in the polymerisation in the gas phase.

(ii) Initiation may sometimes involve a second-order reaction between a catalyst C and the monomer, the rate is given by

$$v_t = k_i\,[C]\,[M].$$

The rate of polymerisation is then

$$-\frac{d[M]}{dt} = k_p\left(\frac{v_i}{k_t}\right)^{1/2} [M]^{3/2}\,[C]^{1/2}.$$

The law is obeyed in the polymerisation of styrene catalysed by benzoyl peroxide,

(iii) In photochemical initiation, the rate of the first step is given by

$$v_t = I$$

where I is the intensity of light absorbed in mole quanta per second. The rate of polymerisation is then

$$-\frac{d[M]}{dt} = k_p\left(\frac{I}{k_t}\right)^{1/2} [M]^{3/2}.$$

KINETICS IN LIQUID SOLUTIONS

When a chemical reaction is occurring in a solution, the solvent is generally present in such a large quantity, that its concentration is not expected to change during the course of the reaction and the rate expression accordingly will remain the same. However, some cases in their stochiometric equation s for chemical reaction, do not involve the solvent whereas some other reactions are known in which solvent enters into chemical reactions and would not be generated at the end of the reaction, in the former cases, solvent has not any effect on the chemical reaction whereas in the latter cases, solvent would be said to exert a chemical effect on the reaction.

Comparison Between Gas-phase and Solution Reactions : In reactions that do occur in the gas-phase as well as in solution, the solvent has only a minor influence on the rate of reaction. An example of such a case is the thermal decomposition of nitrogen pentoxide. In the nitric acid solution on the other hand, the rate constant is significantly lower and the activation energy higher, indicating that this solvent plays a more active role in the reaction.

The former case has been found to be true in the case of ideal solutions where as the latter case has been found to be true in the case of reactions between ions. So we will apply the theory of absolute reaction rates to the case. (a) reactions in ideal solutions, and (b) reactions between ions. We will discuss these on one by one.

(I) Theory of Absolute Reaction Rates Applicable to Reactions in Ideal Solutions : The application of the theory of absolute reaction rates to reactions in solution is simple in principle but is difficult to carry out quantitatively owing to the imperfect knowledge of the liquid state. As with gas reaction, the rate can be expressed in terms of partition functions for the reacting species and the activated state, but they should contain terms for the influence of the solvent environment. However, since there is so much uncertainty about partition functions in solution, it is better to treat rates in solution in a less fundamental manner, by using activity coefficients. Let us consider the reaction,

$$A + B \rightarrow \text{products} \qquad ...(1)$$

The fundamental postulate of the theory of absolute reaction rate is that the reactants A and B will form the activated complex M* which corresponds energetically to the top of the energy barrier. M* is always in equilibrium with the reactants A and B.

$$A + B \rightleftharpoons M^* \qquad ...(2)$$

The rate of the reaction is the rate at which the activated complex decompose into products.

$$M^* \rightarrow \text{products.}$$

The activated complex is an aggregate of atoms which may be thoughtof as being similar to an ordinary molecule except that it has one specialvibration with respect to which it is unstable. This vibration leads todissociation of the complex in to products. If the frequency of this vibrationis v, then the rate at which products are formed is

$$\text{rate} = vc^*. \qquad ...(3)$$

where c* is the concentration of activated complex. But in solution,

$$K^* = \frac{a^*}{a_A a_B} = \frac{\gamma^* c^*}{\gamma_A \gamma_B c_A c_B}$$

or $$c^* = K^* \left(\frac{\gamma_A \gamma_B}{\gamma^*}\right) c_A c_B \qquad ...(4)$$

where the a*s are activities and γ's are activity co-efficients, K* is the equilibrium constant for the reaction. Then the rate becomes by using equations (3) and (4)

$$\text{rate} = vK^* \left(\frac{\gamma_A \gamma_B}{\gamma^*}\right) c_A c_B \qquad ...(5)$$

But the elementary reaction, A + B → products, has the rate,

$$\text{rate} = k_r c_A c_B \qquad ...(6)$$

Equating equations (5) and (6) we get

$$k_r\, c_A\, c_B = vK^* \left(\frac{\gamma_A \gamma_B}{\gamma^*}\right) c_A c_B \ \text{ or } k_r = vK^* \left(\frac{\gamma_A \gamma_B}{\gamma^*}\right) \qquad ...(7)$$

where k_r is the velocity constant for this reaction leading to products. For an ideal gaseous system, the activity coefficients are unity, so that

$k_r = k_g$ and equation (7) reduces

$$k_g = vK^* \qquad ...(8)$$

Dividing eq. (7) by (8), we get $\left(\frac{\gamma_A \gamma_B}{\gamma^*}\right)$ ...(9)

Equation (9) can be used for relating the rate constant in solution not only to that in the gas phase but also to that in very dilute solution, where the substance behaves ideally. If the rates in solution and in the gas phase are to be equal, then $\gamma_A\gamma_B/\gamma^*$ must be equal to unity. This may easily arise for a unimolecular reaction. It has been seen that this similarity in rates is approximately true for the decomposition of nitrogen pentoxide in a number of solvents.

If the solvent lowers the free energy of the reactant more than it does that of the activated complex, then γ_A and Y_B will be small, while γ^* will not be so small. The rate constant in this case will be smaller in solution than in the gas. Conversely, if the activated complex is strongly solvated while the reactants are not, the rate constant will be larger in solution than the gas.

(II) Theory of Absolute Rates Applicable to the Reaction Between ions in Solution : For reactions between ions of opposite sign the frequency factors are much higher than "normal" ($\sim 10^{10}$ litres mole^{-1} sec^{-1}) whereas if the ions are of the same sign, the frequency factors are abnormally low. The effects of electrostatic attraction or repulsion are obviously important. The situation can be considered in terms of lision theory; if the ions are of opposite signs, the frecuency of collision will be increased by attractive forces, but if they are of the same sign, the frequency of collisions will be decreased. One may again use absolute rate theory by first considering the influence of the solvent on the rate and frequency factors of reactions in solution; afterwards, with the aid of the Debye-Huckel theory the influence on rates of the ionic strength, *i.e.*, salt effects, will be treated-

(a) Application of absolute theory to the reaction between ions (influence of solvent) : The procedure that will be utilised involves making estimates of the free energy change in going from the initial state in a reaction to the activated state. According to the theory of absolute reaction rates the rate constant is related to the free energy of the activation by the equation.

$$k_r = \frac{RT}{h} e^{-\Delta G^* / RT} \qquad ...(1)$$

In this theory, the charged ions are considered to be conducting spheres and the solvent is regarded as a continuous dielectric, having a fixed dielectric constant ε. The simplest situation is represented in Fig. 3.12. The reacting molecules are regarded as conducting spheres; the radii are r_A and r_B and the charges Z_Ae and Z_Be. Initially the ions are at an infinite distance from one another, and in the activated state they are considered to be intact and they are at a distance d_{AB} apart. The force acting between the ions, which are at a distance x apart, is equal to

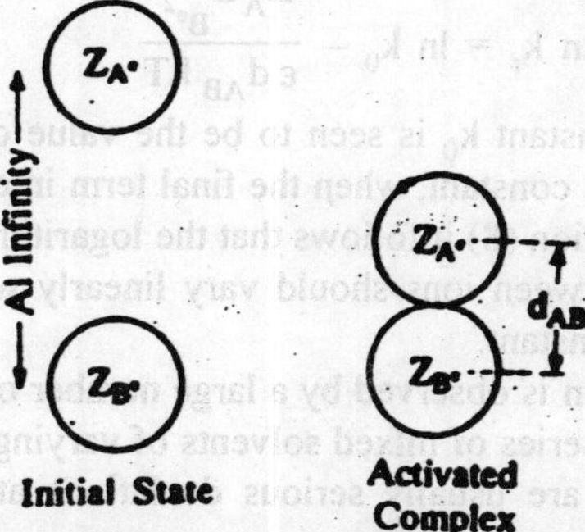

Fig. 3.12 : A simple model for a reaction between two ions of charges Z_Ae, Z_Be, in a medium of dielectric constant e. This is known as 'double shpere' model.

$$G = \frac{Z_A Z_B e^2}{\varepsilon x^2} \qquad ...(3)$$

The work that must be done in together a distance *dx* is

$$dw = -\frac{Z_A Z_B e^2}{\varepsilon x^2} dx^2 \quad \text{...(3)}$$

The work that must be performed in moving the ions from an initial state of infinity to final distance of d_{AB} is therefore

$$w = -\int_{\infty}^{d_{AB}} \frac{Z_A Z_B e^2}{\varepsilon x^2} dx = \frac{Z_A Z_B e^2}{\varepsilon d_{AB}} \quad \text{...(4)}$$

If the signs on the ions are the same, this work is positive; otherwise it is negative. In addition to this electrostatic distribution, there is a non-electrostatic term ΔG^*_{nes}. Therefore, the free energy of activation molecule may therefore be put as

$$\frac{\Delta G^{*\bullet}}{N} = \frac{\Delta G^*_{nes}}{N} + \frac{Z_A Z_B e^2}{\varepsilon\, d_{AB}} \quad \text{...(5)}$$

From equations (1) and (5), we get

$$k_r = \frac{kT}{h} e^{-\Delta G^*_{nes}/RT} e^{-Z_A Z_B e^2/\varepsilon d_{AB} kT} \quad [\because R = Nk] \quad \text{...(6)}$$

This equation may be written in the logarithmic form

$$\ln k_r = \ln \frac{kT}{h} e^{-\Delta G^*_{nes}/RT} - \frac{Z_A Z_B e^2}{\varepsilon\, d_{AB}\, kT} \quad \text{...(7)}$$

or

$$\ln k_r = \ln k_0 - \frac{Z_A Z_B e^2}{\varepsilon\, d_{AB}\, kT} \quad \text{...(8)}$$

The rate constant k_0 is seen to be the value of k_r in a medium of infinite dielectric constant, when the final term in equation (8) becomes zero. From equation (8) it follows that the logarithm of the rate constant of a reaction between ions should vary linearly with the reciprocal of the dielectric constant.

This equation is observed by a large number of reactions which are carried out in a series of mixed solvents of varying dielectric constants. However, there are usually serious deviations at very low dielectric constants.

Frequency Factors. The above treatment can be extended to get an interpretation of the magnitudes of the frequency factors of reactions of this type. The electrostatic combination to the free energy of activation per mole has been seen to be given by

$$\Delta G^*_{es} = \frac{NZ_A Z_B e^2}{\varepsilon\, d_{AB}} \quad ..(9)$$

We $$S = -\left(\frac{\partial G}{\partial T}\right)_P \quad ...(10)$$

Therefore, the electrostatic contribution to the entropy of activation is

$$\Delta S^*_{nes} = -\left(\frac{\partial \Delta G^*_{nes}}{\partial T}\right)_P \quad ...(11)$$

The only quantity in (9), that is temperature dependent is ε, and it, therefore, follows

$$\Delta S^*_{nes} = \frac{N Z_A Z_B}{\varepsilon^2\, d_{AB}}\left(\frac{\partial \varepsilon}{\partial T}\right)_P = \frac{N Z_A Z_B}{\varepsilon^2\, d_{AB}}\left(\frac{\partial \ln \varepsilon}{\partial T}\right)_P \quad ...(12)$$

In aqueous solution ε is about 80 and $(\partial \ln \varepsilon/\partial T)$ p remains constant at – 0.0046 over a fairly wide temperature range. If d_{AB} is taken as equal to 2×10^{-2} cm it follows from equation (12) that

$$\Delta S^*_{nes} \approx -10\, Z_A Z_B \quad ...(13)$$

According to this the entropy of activation in aqueous solution should decrease by 10 units for each unit of $Z_A Z_B$. Moreover since frequency factor is proportional to $e^{\Delta S^*/R}$ with equal to $10^{\Delta S^*/2.303\,R}$ or $10^{\Delta S^*/4.57}$, it follows that frequency factor should decrease by a factor of $10^{10/4.57}$, *i.e.,* by about one hundred fold, for each unit of $Z_A Z_B$. This has been found to be true in a large number of cases.

Salt Effect : Since ions exert considerable electrostatic forces on each other, the kinetics of reactions between ions deviate from such reactions which involve non-electrolytes. The velocity constants of ionic reactions depend, upon the charges of reacting ions and also the ionic strength of solutions. "Thus, the effect of electrolytes can be divided into two main types :

(a) Primary Salt Effect : *It refers to the effect of ionic strength on the rate constant.*

(b) Secondary Salt Effect : *It refers to the actual change in the concentration of reacting* ions by the addition of electrolytes.

Firstly, we will discuss the primary salt effect which is involved in non-catalytic reactions.

(a) Primary Salt Effect : A very satisfactory treatment of the primary salt effect was made by *Bronsted* (1922). Latter on it was modified by *Bjerrum* which is more simple than the *Bronsted theory*. According to the *Bjerrum* the reacting ions first form an activated complex which is in equilibrium with the reactants and then, the activated complex decomposes to yield the products. In simple words, the rate of reaction is determined by the concentration of activated complex.

$$A^{Z_A} + B^{Z_B} \underset{k_2}{\overset{k_1}{\rightleftharpoons}} [AB^{Z_A + Z_B}]^* \xrightarrow{k_3} \text{Products} \qquad ...(1)$$

where A and B are reacting ions, Z_A and Z_B are the charges on the respective species, and AB is the activated complex, $Z_A + Z_B$ is the charge on the activated complex, and k_1, k_2, and k_3 are the velocity constants. According to the law of mass action, equilibrium constant is given by

$$K^* = \frac{a_{AB}}{a_A a_B} \qquad ...(2)$$

where K^* (k_1/k_2) is the equilibrium constant for the process $A + B \rightleftharpoons AB^*$, a_A, a_B and a_{AB} represent the activities of reactants A, B and activated complex AB respectively.

But activity = actual concentration × activity coefficient.

$$a = C \times \gamma \qquad ...(3)$$

where C is the actual concentration g is the activity coefficient. The value of activity coefficient is less than unity except in infinitely dilute solutions. Replacing activities in eq. (2), we get

$$K^* = \frac{C_{AB}\,\gamma_{AB}}{C_A\,C_B\,\gamma_A\,\gamma_B} \qquad ...(4)$$

where C_A, C_B and C_{AB} represent the actual concentration of A, B and AB and γ_A, γ_B, and γ_{AB} represent the corresponding activity coefficients. Eq. (4) may be put as

$$C_{AB} = K^* \frac{C_{AB}\,\gamma_A\,\gamma_B}{\gamma_{AB}} \qquad ...(5)$$

Now the rate of reaction = $k_3\ C_{AB}$ ...(6)

where k_3 is a rate constant for the reaction

$$[AB] \xrightarrow{k_3} \text{Products}.$$

Combining Eqs. (5) and (6), we get

$$\text{Rate of reaction} = k_3 K^* C_A C_B \frac{\gamma_A \gamma_B}{\gamma_{AB}} = \frac{k_0 C_A C_B \gamma_A \gamma_B}{\gamma_{AB}}$$

where $k_0 = k_3 K^*$ ...(7)

The experimental rate of the reaction is given by $k_r C_A C_B$, *i.e.*,

Rate of reaction = $k_r C_A C_B$. ...(8)

Comparing equation (7) with (8), we get

$$k_r = k_0 \frac{\gamma_A \gamma_B}{\gamma_{AB}} \qquad ...(9)$$

where k_0 is the rate constant at infinite dilution when $\gamma_A = \gamma_B = \gamma_{AB} = 1$. It means that any added substance which alters the value of activity co-efficients of reactants and the activated complex will change the experimental rate of reaction. The logarithm of eq. (9) is

$$\log k_r = \log k_0 + \log \frac{\gamma_A \gamma_B}{\gamma_{AB}} \qquad ...(10)$$

According to Deby-Huckel's theory, the activity co-efficients g of an ion of charge Z is related to the ionic strength m of the dilute solution by the expression

$$-\log g = 0.509\ Z^2 \sqrt{(\mu)} \qquad ...(11)$$

For A, B and AB, equation (11) can be written as

$$-\log \gamma_A = 0.509\ Z_A{}^2 \sqrt{(\mu)} \qquad ...(12)$$

$$-\log \gamma_B = 0.509\ Z_B{}^2 \sqrt{(\mu)} \qquad ...(13)$$

$$-\log \gamma_{AB} = 0.509\ (Z_A + A_B)^2 \sqrt{(\mu)} \qquad ...(14)$$

Substituting eqs. (12) (13) and (14) in eq. (10), we get

$$\log k_r = k_0 - [0.509\ Z_A{}^2 + 0.509\ Z_A{}^2 + 0..509\ Z_B{}^2 - 0.509\ (Z_A + Z_B)^2] \sqrt{(\mu)}$$

$$\log k_0 + 2 \times 0.509\ Z_A\ Z_B \sqrt{(\mu)}$$

$$= \log k_0 + 1.018\ Z_A\ Z_B \sqrt{(\mu)} \qquad ...(15)$$

Equation (5) is known as *Bronsted-Bjerrum equation.* As the variation of specific rate k with ionic strength depends on the charges of reacting species, three-special cases may arise,

I. *First case* arises when $Z_A Z_B = 0$. It means that one of the reactants is non-electrolyte.

(ii) $[Cr(NH_2CCONH_2)^{5+} + 6H_2O \rightarrow [Cr(H_2O)_6]^{3+} + 6NH_2CONH_2$

(iii) $CH_2ICOOH + CNS^- \rightarrow CH_2(CNS)COOH + I^-$

In both cases, one reactant is non-electrolyte. Hence $Z_A Z_B = 0$.

II. *Second case* arise when $Z_A Z_B$ is positive. It means that Z_A and Z_B *are* of the same sign. The rate constant for such reaction would increase with $\sqrt{(\mu)}$. *For example.*

IIa. For the reaction between presulphate and iodide ions, the rate determining step is

$S_2O_8^{2-} + I$ Products.

The value for $Z_A Z_B$ is + 2 although both ions have negative valency.

IIb. In the reaction of mercury ions with cobalt-ammonium bromide ions, the value of $Z_A Z_B$ is 2 × 2, *i.e.*, 4. Hence the salt effect is positive.

$Co[(NH_3)_5Br]^{2+} + Hg_2+ \rightarrow$ Products.

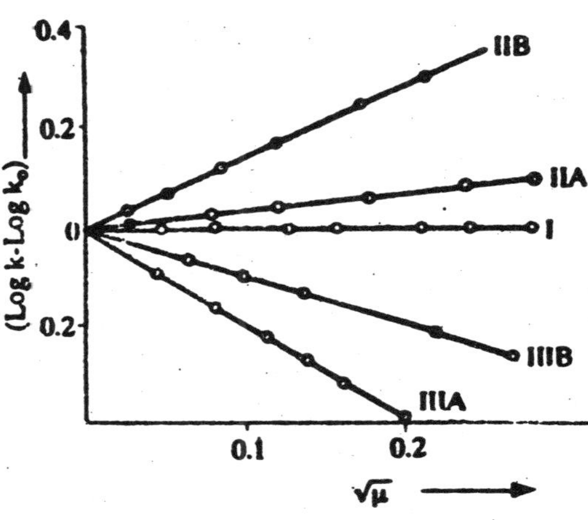

Fig. 3.13 : Effect of ionic strengths on reaction rates.

III. Third case arises when $Z_A Z_B = -1$. It means that Z_A and Z_B are of opposite sign. For such reactions the rate constant would decrease with $\sqrt{(\mu)}$.

IIIa. Example is $H_2O_2 + H^+ + Br^-$ Products. For this reaction $Z_A Z_B = -1$.

IIIb. In the reaction between cobalt ammonium bromide and hycroxyl ion, the value of $Z_A Z_B$ is -2. Hence, the salt effect is negative.

$$[Co(NH_3)_5Br]^{2+} + OH^- \rightarrow [Co(NH_3)_5OH]^{2+} + Br^-$$

Secondary Salt Effect : As the secondary salt effect is involved in catalytic reactions, it can be studied by changing the actual concentrations of the catalytically active ions. If these are produced by a strong acid or a base, the secondary salt effect is negligible. It, however, becomes important when ions are produced by the *dissociation of a weak electrolyte.* Consider the catalytic effect of hydrogen ions which are produced by a mixture of a weak acid and 3 its salt at a definite concentration. The equilibrium constant K_A for the reaction

$$H_A + H_2O \rightarrow H_3O^+ + A^-$$

is given

$$K_A = \frac{(a_{H_3O})(a_{A^-})}{a_{HA}} \quad ...(16)$$

$$\therefore \quad K_A = \frac{C_{H_3O^+} C_{A^-}}{C_{HA}} \times \frac{\gamma_{H_3O^+} \gamma_{A^-}}{\gamma_{HA}} \quad (a = Cg) \quad ...(17)$$

As we know that the pH of a mixture of a weak acid and its salt is of constant composition. It follows that

$$K_A \times \left[\frac{C_{HA}}{C_A}\right] = \text{constant} = K \text{ (say)}$$

Therefore, equation (17) can be put as

$$C_{H_2O^+} = K \frac{\gamma_{HA}}{\gamma_{H_3O^+} \gamma_{A^-}} \quad ...(18)$$

As the activity term varies with the ionic strength of the medium, it follows that hydrogen ion concentration and its catalytic activity will also vary with ionic strength of the medium. This is the secondary salt effect.

As the velocity of a reaction involving hydrogen ions as catalyst is proportional to the concentration of ions, it can be written as

$$k_r = k_0 \frac{\gamma_{HA}}{\gamma_{A^+} \gamma_{H_3O^+}} \qquad ...(19)$$

where k_0 is a constant and its value becomes equal to experimental rate constant provided the values of γ_{HA}, γ^{A^-}~ and γH_3O^+ αρε unity. By using the Debye-Huckel limiting law, equation (19) can be written as

$$\log k_r = \log k_0 + 1.02 \sqrt{(\mu)} .$$

Thus, *the rate constant will increase with increasing concentration of the electrolyte.* The existence of this secondary salt effect has been proved in a number of cases.

KINETICS OF FAST REACTIONS

Introduction : Conventional methods for the study of reactions kinetics can not be applied to reactions which go to equilibrium in a few seconds or less. The reasons why conventional techniques lead to difficulties for very rapid reactions are as follows.

1. The time that it takes to mix reactants or to bring them to a specified temperature may be significant in comparison to the half-life of the reaction. An appreciable error therefore will be made because the initial time cannot be determined accurately.
2. The time it takes to make a measurement of concentration may be significant compared to the half-life.

At one time such reactions were too fast for kinetic study, but this is no longer true as various methods have been developed to study rapid reactions.

Special analytical methods and experimental techniques are used for studying fast reactions. A few of them are described as follows.

SOLVED EXAMPLES

Example 1:

The value of specific rate constant for the decompositions of nitrogen pentoxide.

$$N_2O_5 \rightarrow N_2O_4 + 1/2\ O_2$$

is 3.46×10^{-5} at 25° *and* 4.87×10^{-3} at 65°C. *Calculate the energy of activation for the reaction.*

Solution:

$$k_1 = 3.46 \times 10^{-5},\ T_1 = 273 \times 25 = 298°K$$

$$k_2 = 4.87 \times 10^{-3},\ T_2 = 273 + 65 = 338°K$$

Equation (17) is

$$\log \frac{k_2}{k_1} = \frac{E}{2.303\ R} \frac{(T_2 - T_1)}{T_1 T_2}$$

$$\log \frac{4.87 \times 10^{-2}}{3.46 \times 10^{-5}} = \frac{E}{2.303 \times 1.987} \times \left(\frac{338 - 298}{338 \times 298}\right)$$

or $$E = 24800 \text{ calories.}$$

Example 2:

The following values have been obtained for the velocity constant in decomposition of N_2O_5.

$t°(c)$	0	25	45	65
$k\ min^{-1}$	4.7×10^{-5}	2.0×10^{-3}	3.0×10^{-2}	3.0×10^{-1}.

What information can you derive from these data ?

Solution:

$$k = Ae^{-E/RT} \qquad \text{...(1)}$$

The logarithmic form of equation (1) is

$$\log k = \log A - \frac{E}{2.303\ RT}.$$

If log k is plotted against 1/T a straight line is obtained. The slope being – E/2.303 R and the intercept on the log k axis gives the value of log A. The value of log k and 1/T, shown in the table, are plotted in Fig. 3.14.

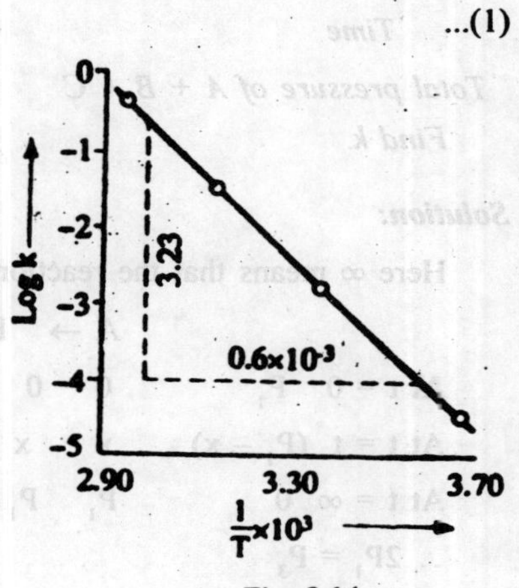

Fig. 3.14

log k	– 4.328	– 2.699	– 1.523	– 0.523
T°K	273	298	318	338
1/T	3.66×10^{-8}	3.36×10^{-3}	3.15×10^{-3}	2.96×10^{-3}.

Inspection of the Fig. (3.14) shows that the slope

$$\frac{-E}{2.303\,R} = -\frac{3.23}{0.6 \times 10^{-3}}$$

$$\therefore \quad E = \frac{2.303 \times 2 \times 3.23}{0.6 \times 10^{-3}} \text{ cal/mole}$$

$$= 24790 \text{ cal/mole.}$$

Taking the data for 0°C and substituting in equation (1),

$$-4.328 = \log A - \frac{24790}{2.303 \times 2 \times 273}$$

$$\text{or} \quad \log A = \frac{24790}{4.606 \times 273} - 4.328$$

$$= 15.382.$$

Example 3:

In this case we have

A → B + C

→ B + C

Time	*t*	*∞*
Total pressure of A + B + C	P_2	P_3

Find k.

Solution:

Here ∞ means that the reaction is complete. Now we have

	A →	B +	C
At t = 0	P_1	0	0
At t = t	$(P_1 - x)$	x	x
At t = ∞	0	P_1	P_1

$\therefore 2P_1 = P_3$

$$\Rightarrow \qquad P_1 = \frac{P_3}{2}$$

At time t, $P_1 + x = P_2$

$$\Rightarrow \qquad \frac{P_3}{2} + x = P_2$$

$$\Rightarrow \qquad x = P_2 - \frac{P_3}{2}$$

$$\Rightarrow \qquad P_1 - x = \frac{P_3}{2} - \left(P_2 - \frac{P_3}{2}\right) = P_3 - P_2$$

$$\therefore \qquad k = \frac{1}{t}\ln\frac{[A]_0}{[A]_t} = \frac{1}{t}\ln\frac{P_3/2}{(P_3 - P_2)} = \frac{1}{t}\ln\frac{P_3}{2(P_3 - P_2)}.$$

Example 4:

A → B + C

Time	*t*	*∞*
Partial pressure of B	P_2	P_3

Find k.

Solution:

Taking clue from the previous illustrations

$P_1 = P_2$ and $x = P_2$

$\therefore P_1 - x = P_3 - P_1$

$$\therefore \quad k = \frac{1}{t}\ln\frac{[A]_0}{[A]_t} = \frac{1}{t}\ln\frac{P_3}{(P_3 - P_2)}.$$

Example 5:

$$A \xrightarrow{D} B + C$$

Time	*0*	*t*	*∞*
Volume of reagent	V_1	V_2	V_3

The reagent reacts with only B, C and D. Find k.

Solution :

$V_1 = V_D$

$$V_2 = 2x + V_D$$

$$V_3 = 2V_A + V_D$$

$$\therefore \quad \frac{V_3 - V_1}{2} = V_A$$

$$\frac{V_3 - V_1}{2} = V_A - x$$

$$\therefore k = \frac{1}{t}\ln\frac{[A]_0}{[A]_t} = \frac{1}{t}\ln\frac{(V_3 - V_1)}{(V_3 - V_2)}.$$

Example 6:

$$A \xrightarrow{D} B + C$$

Time	*t*	*∞*
Volume of reagent	V_2	V_3

Reagent reacts with all A, B and C and have 'n' factors in the ratio of 1:2:3 with the reagent. Find k.

Solution :

	A →	B +	C
At t = 0	V_1	0	0
At t = t	$V_1 - x$	2x	3x
At t = ∞	0	$2V_1$	$3V_1$

$$\therefore 5V_1 = V_3$$

$$V_1 = V_3/5$$

$$V_1 + 4x = V_2$$

$$x = \frac{V_2 - V_1}{4}$$

$$\therefore V_{1-x} = \frac{V_3}{5} - \frac{V_2}{4} + \frac{V_3}{20} = \frac{5V_3 - 5V_2}{20}$$

$$\therefore k = \frac{1}{t}\ln\frac{[A]_0}{[A]_t} = \frac{1}{t}\ln\frac{V_3/5}{5V_3 - 5V_2/20}$$

$$k = \frac{1}{t}\ln\frac{4V_3}{5(V_3 - V_2)}.$$

Example 7:

$A \rightarrow B + C$

Time	*0*	*t*
Volume of reagent	V_1	V_2

Reagent reacts with all of them (A, B & C), A, B and C have 'n' factors in the ratio of 1:2:3 with the reagent (assuming its n factor remains same with all of them). Find k.

Solution:

$$A \rightarrow B + C$$

$$\text{At } t = 0 \quad V_1 \qquad 0 \qquad 0$$

$$\text{At } t = t \quad V_1 - x \qquad 2x \qquad 3x$$

$$\therefore V_1 + 4x = V_2$$

$$\Rightarrow x = \frac{V_2 - V_1}{4}$$

$$\Rightarrow V_1 - x = V_1 - V_1 - \frac{V_2}{4} + \frac{V_1}{4} = \frac{5V_1 - V_2}{4}$$

$$\therefore k = \frac{1}{t}\ln\frac{[A]_0}{[A]_t} = \frac{1}{t}\ln\frac{4V_1}{(5V_1 - V_2)}.$$

Example 8:

Now, we will consider a reaction A $A \xrightarrow{D} B + C$ *which is catalysed by D. We will assume in this problem that the concentration of the catalyst remains constant throughout. The data given to us is*

Time	*0*	*t*	∞
Volume of reagent	V_1	V_2	V_3

The reagent reacts with all (A, B, C and D). Assume that 'n' factor of A, B and C are in the ratio of 1:1:1 and that D is not known. Find k.

Solution:

Let V_A be the volume of the reagent required by A initially and V_D be the volume required by D.

$$\therefore V_1 = V_A + V_D$$

$V_2 = (V_A - x) + x + x + V_D$ (x is the volume of the reagent required for those moles of A that have reacted to give B and C).

$$V_3 = 2V_A + V_D$$

We can see that, $V_3 - V_1 = V_A$

And $\quad V_3 - V_2 = V_A - x$

$$\therefore k = \frac{1}{t}\ln\frac{[A]_0}{[A]_t} = \frac{1}{t}\ln\frac{(V_3 - V_1)}{(V_3 - V_2)}.$$

Example 9:

The rate law corresponding to the below data is :

The data for the reaction

A + B → C are.

	[A]	*[B]*	*Rate*
I.	*0.012*	*0.035*	*0.10*
II.	*0.024*	*0.070*	*0.80*
III.	*0.024*	*0.035*	*0.10*
IV.	*0.012*	*0.070*	*0.80*

(a) $R = k\,[B]^3$

(b) $R = k\,[B]^4$

(c) $R = k\,[A]\,[B]^3$

(d) $R = k\,[A]^2\,[B]^2$.

Solution :

Order of reaction with respect to A from I and III

$$\frac{R_2}{R_1} = \left(\frac{C_2}{C_1}\right)^n \text{ or } \frac{0.10}{0.10;} = \left(\frac{0.024}{0.012}\right)^n$$

Order of reaction with respect to B from II & III

$$\frac{R_2}{R_1} = \left(\frac{C_2}{C_1}\right)^n \text{ or } \frac{0.80}{0.1} = \left(\frac{0.070}{0.035}\right)^n$$

n = 3

The rate law is $R = k\,[B]^3$.

Example 10:

Now let us assume that the first order reaction is $A \rightarrow B + C$ and this solution is taken in a tube. We radiate this solution with a monochromatic light of wavelength λ and we will assume that this radiation is absorbed only by A. Let the intensity of incident light be I_0 and that of the transmitted light be I_t $\log \frac{I_t}{I_0}$ is called transmittance and it is inversely proportional to the concentration of A.

That is $\log \frac{I_t}{I_0}$ = *transmittance* $\propto \frac{1}{[A]}$

The data given is

Time	*0*	*t*
$\log I/I_0$	x	y

Find k.

Solution:

$$x \propto \frac{1}{[A]_0} \text{ and } y \propto \frac{1}{[A]_t}$$

$$\therefore k = \frac{1}{t}\ln\frac{[A]_0}{[A]_t} = \frac{1}{t}\ln\frac{y}{x}.$$

Example 11:

$S \rightarrow G + F$

Time	*0*	*t*	∞
Rotation	r_0	r_t	R_∞

The rotation is due to all (total rotation). Find k.

Solution:

	S →	G +	F
At t = 0	a	0	0
At t = t	a – x	x	x
At t = ∞	0	a	a

$\therefore \quad r_0 = a\ r_1^o$ ...(1)

$r_t = (a - x)\ r_1^o + x\ (r_1^o - r_3^o$ ...(2)

$r_\infty = a\left(r_2^o - r_3^o\right)$ ...(3)

Simplifying equation (2)

$$r_t = ar_1^o - x\left(r_1^o - r_2^o + r_3^o\right)$$

Since $a\,r_1^o$, is r_0

$$\therefore \quad x = \frac{r_0 - r_t}{\left(r_1^o - r_2^o + r_3^o\right)}$$

Now, we can write 'a' at $\frac{r_o}{r_1^o}$ or $\frac{r_\infty}{r_2^o - r_3^o}$. But the problem in writing either of them is that the constants of a and x are different. This implies that the constants of a – x would also be different from that of 'a'.

So we write, $r_o - r_\infty = a\left(r_o^1 - r_2^o + r_3^o\right)$

$$\therefore \quad a = \frac{r_o - r_\infty}{\left(r_1^o - r_2^o + r_3^o\right)}$$

$$\Rightarrow a - x = \frac{r_t - r_\infty}{\left(r_1^o - r_2^o + r_3^o\right)}$$

$$\therefore k = \frac{1}{t}\ln\frac{a}{a-x} = \frac{1}{t}\ln\frac{\left(r_o - r_\infty\right)}{r_t - r_\infty}$$

4

KINETICS OF FAST REACTIONS

Reactions which go to equilibrium in a few seconds or even in less time cannot be kinetically studied by conventional methods. The reason for the inability are as follows :

(a) The time it takes to make a measurement of concentration may be significant as compared to its half-life.

(b) The time that it takes to mix reactants or to bring them to a desired temperature may be significant as compared to the half-life of the reaction. An appreciable error creeps in because the initial time cannot be accurately determined.

So special methods have to be used to study the kinetics of such fast reactions. Some of the methods are as follows.

NUCLEAR MAGNETIC RESONANCE

This method makes use of the Zeeman splitting of nuclear energy levels. The theory of NMR spectra is closely related to the theory of EPR spectra but the resonance frequency is different appreciably from the EPR frequencies. The resonance frequency is in the radio frequency range.

NMR is shown by any isotope of an element which has a non-zero nuclear spin, *i.e.*, C_{12} and O_{16} have 1 = 0 and are not NMR active. It is possible to calculate the rate of reaction determining the space and width of absorption lines in the resonance spectrum.

The charge caused by the conditions such as temperature and amount of reference added should also be determined.

Kinetic data on reactions in solution are obtained by nuclear magnetic resonance methods. This method is based on the fact that when two compounds with different NMR chemical shifts change rapidly from one to another, their two NMR peaks merge into one. This type of behaviour found is shown in figure In the figure, it is assumed that there is not splitting of the NMR lines, that the equilibrium constant for the reaction A $\rightleftharpoons$ B has the value I and all temperatures, and, therefore, the rate constants and relaxation times for the forward and back ward reaction would be equal to one another at all temperature.

The NMR spectra of a zicronium complex (II – cyclo-pentadiene) – Zr – (acetonylacetone)$_2$ Cl in benzene was studied. The complex, which exists in two stereoisomers, showed a pair of clearly defined peaks, one from each isomer. The peaks are separated by about 3Hz. These remained sharp from room temperature to 65°C, then went through a transition very similar to that of figure, merging at 87°C and again showing a sharp single peak above 100°C. Calculated relaxation times varied from 0.26 sec, at 76°C to 0.026 sec at 1000°C. The small $\Delta\delta$ value may be due to the small difference in environment of the proton caused by the chemical change.

The NMR method is a relaxation method in which the system studied are usually close to equilibrium, and those molecules which are excited by the absorption of radiation lose their excess energy rather quickly due to collisions,. With relaxation methods, the rate constant of second order reaction with numerical values appreciably above $\Delta\delta$ may be studied when the concentrations of reactants are substantially below IM. A schematic diagram of NMR spectra photometer is given in Fig.

The apparatus consists of a simple tube S which is kept between the poles of a huge magnet M of adjustable field. The sample is surrounded by two coil A and B. Coil A emits a definite radio-frequency from R.F. transmitter. Coil B picks up any radiation and carries it to and receiver then to the cathode ray oscillograph or recorder. When the field strength is increased, the radiation will only get absorbed when the nuclear energy level spacing matches the quanta of radiation of given frequency. The absorption occurs only when this condition is obtained. These absorptions are detected by a recorder. The pyrolysis of a hydrocarbon is studied by this method.

ELECTRON PARAMAGNETIC RESONANCE METHOD

As free radicals have unpaired electrons they are paramagnetic and their interaction of the electron spin with an applied magnetic field results in splitting of energy levels. This has been detected by resonance absorption of radiation in the micro wave region. As the atoms and radicals have characteristic electron spin resonance spectra (some times called electron paramagnetic resonance spectra) the measured absorption intensities give the relative atom and radical concentrations in the system. How every by direct calibration, using known pressures of oxygen or nitric oxide, these can be concerted to absolute concentrations.

EPR was previously used in direct studies of elementary radical reactions. How ever, its sensitivity of around 10" species cm^3 combined with reliability; has resulted in many used.

From the study of the conditions of disappearance of the fine structure, it is possible to determine the average life time of radicals which gets associated with the width of the absorption line in the EPR spectrum. The width of the absorption line in the spectrum may have the following relationship with kinetics. If no reactions occurs, a narrower sharp line is observed. The absorption line is broadened when the reaction occur because the average life time of the radical gets shortened.

EPR is useful for studying the reactions of short lived paramagnetic intermediates like free radicals presents at low concentrations. The rate constants that can be measured are as large as 5×10^{-1} 5^{-1} EPR finds use after rapidly freezing the reaction mixture and in combination with flow or stopped flow techniques. The slow step in such combinations is freezing or mixing, which needs time of the order of millisecond per forms a short coming that it has a relatively low sensitivity.

GAS LIQUID CHROMATOGRAPHY

Principle

The principle consist of injecting into a flowing gas which is then allowed to pass through a chromatography column. The column is packed with an absorbent solid or an insert solid coated with a liquid solvent. The various components in the sample differ in their solubility in the solvent and so passes into a detector at different rates.

So the components, emerging at different rates from the column, enter a detector at different times. The detector seconds the emergence of each component, and it measured their concentration after suitable calibration. It is also possible to identify a component from the characteristic time at which it emerges from the column.

A gas liquid chromatography can be easily coupled with a reaction vessel by a gas sampling values, and the sample taken may be so small that the concentration in the reaction vessel are not disturbed.

Advantages : The advantages of this method are that (i) It requires small samples for analysis. (ii) Analysis time for each sample can be as small as a few seconds. So, the measurements of concentration can be reported at very short intervals.

MASS SPECTROMETRY

Principle : The principle consist of permitting a sample of reaction mixtures to pass into the ion source region of the instrument where the molecules of the reaction mixture are ionised by bombardment with electrons of controlled energy. The ions so produced in the ion source region, which are characteristic of their parent molecules of the reaction mixture, are allowed to enter the analysing region. In this the ions with different charge/mass rations are separated and detected. The ions can be separated in two ways.

(i) Ions are deflected in electrical and magnetic fields. The amount of deflection depends on the charge/mass ration.

(ii) Ions take different times to traverse a flight tube before reaching the detector.

Relative concentration at different charge/mass rations are measured from the ion currents at the detector. The mass spectrum, the pattern of ions derived from a given parent molecule, are characteristic and so helps us to identify the components in the given mixture.

In the mass spectrometry, very small samples are required which can be leaked directly from a reaction vessel. The mass spectrometry is isotopic analysis.

(i) Mass spectrometry is useful for identifying neutral molecules, radicals and ions.

(ii) The method is sufficiently sensitive to detect approximately $10^{''}$ molecules cm^{-3}

However, ass spectrometry is not suitable for chemical analysis of complex mixtures. The mass spectrum of a complex mixture may be complicated, with many different fragment ions resulting in the different patterns which may coincide making results difficult to interpret.

LARGE PERTURBATION METHODS

The principles used by these methods is to apply a large perturbation to a system, there by generating one of the reactants, and to follow the disappearance of this reactive species. In all the methods to be considered, flash photolysis, pulse radiolysis and shock tubes, it is necessary to use some form of rapid-detection apparatus.

(a) Flash Photolysis method. Porter (1950) first used flash photolysis method to study gaseous reactions. It is now widely used in investigating reaction in solution.

If iodine vapour is illuminated with light of suitable wavelength, the dissociation reaction occurs.

$$I_2 + h\nu \longrightarrow 2I$$

The absorbed photon is written by hν where h is Planck's constant (6.625×10^{-27} erg sec) and ν is the frequency of the light (sec^{-1}). Rapid recombination of iodine atoms occurs in a non-photochemical reaction. The reaction uses the iodine produced by the photochemical dissociated reaction. Under steady illumination with light of moderate intensity, steady state is reached with only a very small fraction of iodine dissociated into atoms ($2I \rightarrow I_2$). This steady state is obtained when the rates of photochemical dissociation of iodine molecules and non-photochemical recombination of iodine atoms equal each other.

If illumination occurs as an intense flash for a very short time an appreciable number of iodine atoms can be produced, their recombination can be directly observed after the flash. In actual experiments, the peak intensity of a flash may build up in 10^{-5} sec and their decay over a period of 4×10^{-4} sec or even less. After the decay of light intensity from the flash, light of low intensity can be used in a spectrophotometric arrangement to follows the increase in concentration of molecular iodine. An oscilloscope trace in which the concentration increase of molecular iodine following the flash is indicated.

This method called *flash photolysis* has contributed valuable information regarding rates of very fast reactions. This method has been used for the study of reactions in solution as well as those in gas phase.

The exact from of apparatus used in flash photolysis experiment depends to a large extent on the nature of the problems.

In the experiments the reactants are put in a cylindrical quartz vessel 50 cm long and 2 cm in diameter which is mounted next to the photolytic flash tube. The latter is a quartz tube of about the same length as the reactions vessel which is filled with a rare gas, *e.g.*, kryton. The light emitted by the flash tube is a continum extending over the whole of the visible as well as into the infrared and ultraviolet regions on which are supermised the broadened atomic spectral lines of the rare gas. The reaction vessel and flash lamp are mounted on a follow cylinder whose inner surface is coated with magnesium oxide, which has a high reflectivity towards light of most wavelengths. Condensers are connected to the electrodes and charged up to the required voltage and then discharged by applying a triggering pulse to a small central electrode or to a spark gap in series with flash tube. The duration time of the photolysis flash depends on the energy to be dissipated. The flash has an energy of a few hundred joules spread over a few tens of micro seconds.

The main difference between flash photolysis and the relaxation methods is that the amount of energy put in to the gas is relatively large, so that the temperature of the gas is increased by 100° or more in a few micro seconds. In case the energy is absorbed into particular degree of freedom in the molecules, highly excited species or dissociated species are produced. By this method chlorine- oxygen mixtures were flashed. The radiation dissociated a large fraction of chlorine molecules into chlorine atoms.

The other reactions which occur are as follows.

$$Cl_2 \longrightarrow 2Cl$$

$$2Cl + O_2 \longrightarrow 2ClO$$

$$2Cl + O_2 \longrightarrow Cl_2 + O_2$$

$$2ClO \longrightarrow Cl_2 + O_2.$$

The ClO was produced very quickly, and its presence and rate of disappearance were followed by taking a series of absorption spectra cone per experiment) and times up to 10 millisecond after the initial flash. The translation and rational temperatures of the molecules remained close to ambient and the energy is specifically absorbed in the Cl– Cl bond. Later, using a faster flash, Nicholas and Norrish were able to measure the rate constant for the addition of chlorine atom to O_2, finding it to be of third order, with a rate constants of 6×10^{14} mole^{-2} L^2 sec^{-1}, where N_2 is the principle third body.

(b) **Pulse Radiolysis :** Pulse radiolysis is in many ways a similar technique and may be regarded as the radiation equivalent of flash photolysis with a pulse of ionizing radiation replacing the photolytic light flash. It has been applied primarily to reactions in solution and has been used with great success for the investigation of reaction of the hydrated electron.

A Typical Pulse : Radiolysis apparatus is shown in figure (12.23). A pulse of electrons from a linear accelerator is passed through the solution and the change produced are followed spectrophotometrically, either at a known time after the electron pulse with a spectroscopic flash and a spectrograph or continuously at a fixed wave length with a steady light source and a photon multiplier Typically, the pulse which may be of X-rays rather than electron, lasts for one or two micro seconds. With this methods it is important to remove the last traces of impurity from the solution since these interfere with the reaction under study. This fan be done when reactions of the hydrated electron are studied. The aqueous electrons produced on is adiation are found to disappear more slowly after successive pulse and it reams that they react with the residual impurities to give a non-reactive product. The solution are, therefore, pulsed until the electron decays at a reproducible rate, by which time it is assumed that all the reactive impurities have been removed. This generally happens after about twenty pulses are so. The method is used to follows reaction of the secondary or tertiary products of the radiolysis. For example, if an aqueous solution of Fe^{2+} is radiolysis some of the oxidizing fragments of the water radiolysis, such as OH^- radicals, are capable of extracting an electron from the metal ion to produce, Fe^{3+}. The subsequent complex formation of this ion with ligands in the solution, such as SO_4^{2-}, can then be followed.

(c) **Shock Tubes :** The technique has been used for studying fast reaction in the milliseconds or microsecond range at high temperatures. The reactants are heated by a shock wave travelling at supersonic speed, the high temperature stays nearly constant for hundred of a microseconds or a few milliseconds in some cases. Throughout this period, the concentration of radicals is followed by one of the standard methods of detection. The essential features of the apparatus are shown in Fig.

The shock tube consists of a metal or glass pipe several meters long and several centimeters in diameter. A thin metal or plastic diaphragm divides the shock tube in to two reactions. One reaction contains the driver gas (usually hydrogen or helium) at pressure upto ten times atmospheric pressure

The other reaction contains a reactants gas at a few, millimeter pressure. At the end of the tube there is an observation point for measurement of radical concentrations and also detector for measuring the velocity of shock front.

As soon as the diaphragm is mechanically ruptured, the driver gas bursts out and forms a shock wave with a self-sustaining sharp shock front at a very high speed. The front travels with a supersonic speed compressing and heating the reactants gas as it proceeds. As room as the shock front reaches the observation point, the reactant gas is suddenly compressed and heater. The initial reading is recorded. After the reacting gases, the driver gas reaches the observation point. The time interval between the arrival of the showed front and arrival of the driver gas represents the reaction time which is nearly a millisecond.

1. Nuclear Magnetic Resonance : This method makes use of the Zeeman splitting of nuclear energy levels. The theory of NMR spectra is closely related to the theory of EPR spectra but the resonance frequency is different considerably from the EPR frequencies. The resonance frequency is in the radio frequency range.

NMR is exhibited by any isotope of any element having a non-zero nuclear spin, *i.e.*, C^{12} and O^{16} have $I = 0$ and are not NMR active. It is possible to calculate the rate of reaction determining the space and width of absorption lines in the resonance spectrum and also their charge caused by the conditions such as temperature and amount of reference added.

Kinetic data on reactions in solution are obtained by nuclear magnetic resonance (NMR) methods. The basis of these methods is the fact that when two compounds with different NMR chemical shifts convert rapidly from one to the other, their two NMR peaks merge into one. The type of behaviour found is shown in Fig. 4.1. In this figure, it is assumed that there is no splitting of the NMR lines, that the equilibrium constant for the reaction $A \rightleftharpoons B$ has the value 1 at all temperatures, and that therefore the rate constants and relaxation times for the forward and reverse reactions would be equal to one another at all temperatures.

Pinnavaia and Lott studied the NMR spectra of (among other compounds) a zirconium complex, (s-cyclo-pentadiene)-Zr (acetonylacetone)$_2$ Cl in benzene. The complex, which exists in two stereoisomers, showed a pair of clearly defined peaks, one from each isomer, separated by about 3 H_2. These remained sharp from room temperature to 65°C, then went through a transition very similar to that of Fig. 4.1, merging at 87°C and becoming a sharp single peak above 100°C. Calculated relaxation times varied from 0.26 sec at 76°C to 0.026 sec at 100°C. The small Δ^{δ} in this case has been attributed to the small difference in environment of the proton caused by the chemical change.

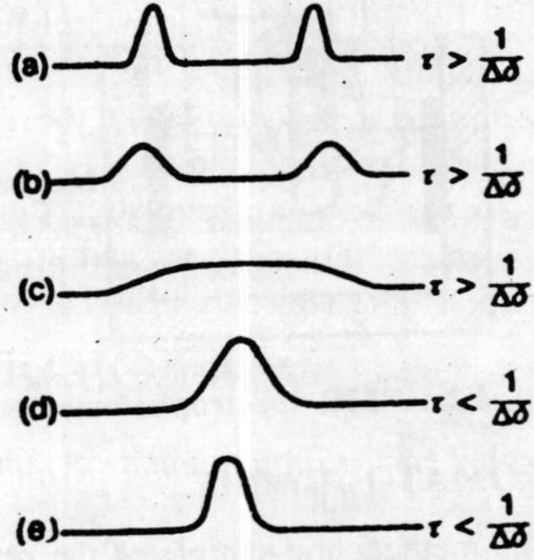

Fig. 4.1 : Combination of two NMR peaks as the rate constant for chemical conversion increases.

The NMR method is somewhat a relaxation one, in that the systems studied are usually close to equilibrium, those molecules which are excited by the absorption of radiation losing their excess energy rather quickly in collisions. As with the relaxation methods, rate constants of second order reactions with numerical values substantially above Δ^{δ} may be studied when the concentrations of reactants are substantially below 1 M.

A schematic diagram of NMR spectrophotometer is given in Fig. 4.2.

The apparatus is having a simple push S which is kept between the poles of a hugemagnet M of adjustable field. Coils A and B surround the sample. Coil A emits a definite radio frequency from R. F. transmitter. Coil B picks up any radiation and carries it to a receiver than to the cathode ray oscillograph or recorder. On increasing the field strength, the radiation will only get absorbed when the nuclear energy level spacing matches the quanta of radiation of given v. The absorption takes place only when this condition is reached. These absorptions are detected by the recorder.

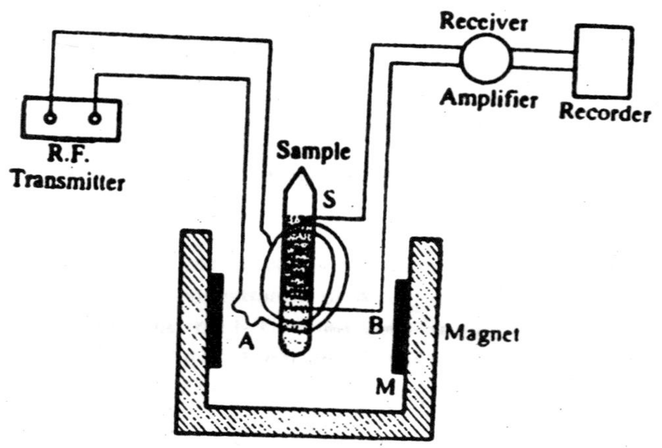

Fig. 4.2 : NMR spectrophotometer.

GAS LIQUID CHROMATOGRAPHY

Principle : In this method, the sample of the reactants is injected into a flowing gas which is then allowed to pass through a chromatographic column. The column is packed with an adsorbent solid or an inert solid coated with a liquid solvent. The various components in the sample differ in their solubility in the solvent, or the strength of adsorption on the solid, thus, passes into a detector at different rates. Hence the components, emerging at different rates from the column, pass into a detector at different times. The detector records the emergence of each component, and when suitably calibrated, it measures their concentration. It is also possible to identify a component from the characteristic time at which it emerges from the column.

A gas-liquid chromatography can be easily coupled with a reaction vessel by a gas sampling value, and the sample taken may be so small that the concentrations in the reaction vessel are not disturbed.

Advantages : (i) This requires small samples for analysis, (ii) Analysis time for each sample can be as small as a few seconds, so the measurements of concentrations can be reported at quite short intervals.

Example:

An example of gas-liquid chromatography is the pyrolysis of hydrocarbons.

MASS SPECTROMETRY

Principle : In this method, a sample of reaction mixture is permitted to pass into the *ion source region* of the instrument where molecules of the reaction mixture are ionised by bombardment with electrons of controlled energy. The ions produced in the ion-source region, which are characteristics of their parent molecules of the reaction mixture, are allowed to enter the *analysing region* where the ions with different charge/mass ratios are separated and detected. The ions are separated in two basic ways :

(i) Ions are deflected in electrical and magnetic fields. The amount of deflection depends on charge/mass ratio.

(ii) Ions take different times to traverse a flight tube before reaching the detector.

Relative concentrations at different charge/mass ratios are measured from the ion currents at the detector. The mass spectrum, the pattern of ions derived from a given parent molecule, is characteristic and it is often possible to identify components in a mixture.

In mass spectrometry, very small samples are required which can be leaked directly from a reaction vessel. The mass spectrometry may also be used as a detector in conjunction with glass-liquid chromatography separation.

Advantages

(i) The area in which mass spectrometry is more successful than gas liquid chromatography is isotopic analysis,

(ii) Mass spectrometry is .useful for identifying neutral molecules, radicals and ions,

(iii) The method is sufficiently sensitive to detect approximately 10^{11} molecules cm^{-5}.

Disadvantages : Mass spectrometry is not suitable for chemical analysis of complex mixtures. The mass spectrum of a complex mixture may becomplicated, with many different fragment ions resulting in the differentpatterns which may coincide making results difficult to interpret.

Uses : Mass spectrometry is mainly employed for identifying radicals in reaction mixtures. A sample leaked directly from the reaction vessel enters into the ion-source region suddenly at a pressure of around

10^{-4} Torr which is reduced further to about 10^{-8} Torr in the analysis region. At such low pressures radicals do not undergo many collisions with reactant molecule, and their life times are long enough for analysis. The radicals generally yield the same fragment ions as a related molecule, but the ions from radicals appear at lower energies of bombarding electrons, that of stable molecules. An important contribution to combustion kinetics was proving the existence of HO_2 radical.

ELECTRON PARAMAGNETIC RESONANCE METHOD

As free radicals have unpaired electrons they are paramagnetic and their interaction of the electron spin with an' applied magnetic field results in splitting of energy levels. This has been detected by resonance absorption of radiation in the microwave region. As the atoms and radicals owe characteristic electron spin resonance spectra (sometimes called electron paramagnetic resonance spectra), measured absorption intensities yield the relative atom and radical concentrations in the system. However, by direct calibration using known pressures of oxygen or nitric oxide, these can be converted to absolute concentrations.

EPR was previously used in direct studies of elementary radical reactions. However, its sensitivity of around 10^{11} species cm^{-3} combined with reliability, has resulted in many uses. Structural investigations of radicals are also done by trapping them in a rapidly cooled solvent and investigating their detailed EPR spectra in the inert matrix of the solid solvent.

The hyperfine structure of multiplicity of lines obtained in the EPR spectrum, because of the interaction of the spin of the odd electrons with magnetic nuclei of the molecule may help in identifying the radicals.

By studying the conditions of disappearance of the fine structure, it becomes possible to determine the average lifetime of radicals which gets associated with the width of the absorption line in the EPR spectrum. The width of the absorption line in the spectrum may be having the following relationship to kinetics.

If no reaction takes place, a narrower sharp line is observed. The absorption line gets broadened when the reaction occurs because the average lifetime of the radical gets shortened.

EPR is specially useful for studying the reactions of short-lived paramagnetic intermediates like free radicals present at low concentrations.

The rate constants that can be measured are as large as 5×10^{-1} s^{-1}.

Chemical processes like electron transfer from a free radical to a diamagnetic species, such as

Naphthalene$^-$ + Naphthalene → Naphthalene + Naphthalene$^-$

or electron exchange between paramagnetic species, such as

$$W(CN)_8^{3-} + W(CN)_8^{4-} + W(CN)_8^{3-}$$

are made to contribute to the relaxation time. The spins tend to get aligned with the applied field, and thermal motion tends to oppose this alignment by randomizing the spin vectors.

Randomization along the field has been characterized by the spin-lattice relaxation time T_1 and perpendicular to the field by the spin-spin relaxation time T_2. It is possible to obtain these times from pulsed experiments T_2 is also obtained from line width analysis.

EPR also finds use after rapidly freezing the reaction mixture and /or in combination with flow or stopped-flow techniques. The slow step in suchcombinations is freezing or mixing, which needs times of the order ofmilliseconds.

One of the shortcomings involving the use to EPR has been its relatively low sensitivity.

FLOW METHOD

Chronologically, the flow techniques were the first of the special methods for following fast reactions to be developed and even now they are possibly used more than all other techniques combined. The flow techniques may be seen as the logical extension of the classical 'mix and shake' method in which the reactants are now mixed within a fraction of a second.

Flow techniques, first developed by Roughton and Hastridge (1923), allow to measure reaction half-time in the range between 10 sec and 10^{-3} sec. Solutions flowing through two tubes into a mixing chamber of appropriate design can be thoroughly mixed in time as short as 10^{-3} sec. The mixed solutions then flow into an observation tube. With flow of completely mixed solution at a constant rate, reaction occurs to a particular extent at each position along the observation tube. This condition prevails because each position along the observation tube corresponds for a particular flow rate to the lapse of some definite time interval after mixing. In this arrangement, the observations need not involve equipment with fast response, since with a constant flow rate the concentrations at a particular point along the observation tube do not change. It is possible to make observations in a few milliseconds after mixing Fig. 4.3.

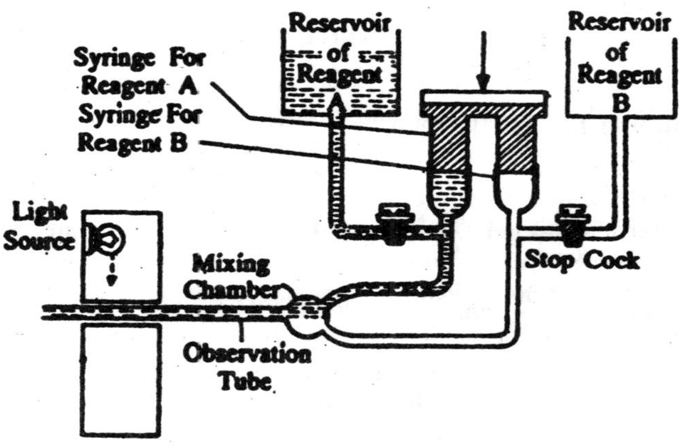

Fig. 4.3

Measurement of light absorption may be used conveniently in this type of experiment to determine the extent of reaction if the absorption spectrum of the product differs from that of the reactant. The spectrophotometric equipment can be set up with the beam of light passing through the solution at a point "downstream" from the point of mixing. If experiments are performed with different rates of flow, values of light absorption at this one particular point provide data from which a plot of extent of reaction *v.s.* time can be prepared. With observation points at several positions, the same data can be accumulated at one particular flow rate.

One of the first reactions studied using this method was the reaction between Fe^{3+} and CNS^- in aqueous solution. Its rate law may be put as follows.

$$\text{rate} = k_2\,[Fe^{3+}]\,[CNS^-]\left[1+\frac{a}{[H^+]}\right]$$

where k_2 denotes the second order rate constant, and *a* an empirical constant which gets related to the dependence of reaction rate on pH. At 25°C the rate constant k_2 has been 127.1 $mol^{-1}\ s^{-1}$.

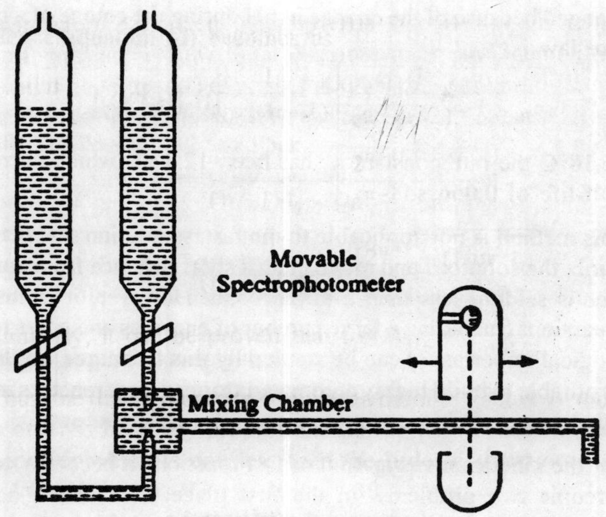

Fig. 4.4 : Diagram of the constant flow apparatus with movable spectrophotometer.

In a variation of the above method, the flow is stopped suddenly, and then the rate at which the system reaches equilibrium can be followed by measuring the concentration change of a reactant or a product as a function of time with a fast response device usually UV-visible spectroscopy. This is the stopped flow method.

In the *quemcide flow method,* after rapid mixing the flowing reactants get discharged into a quenching solution to arrest the reaction. At this instant the concentration as a function of time can be measured using chemical procedures like titration so as to determine the extent

Among the fast reactions studied using flow techniques was the decomposition of carbonic acid.

$$H_2CO_2 \rightarrow H_2O + CO_2$$

Carbonic acid gets formed by mixing a solution of $NaHCO_3$, with a solution of HCl in a flow apparatus.

$$NaHCO_3 + HCl \rightleftharpoons H_2CO_3 + NaCl.$$

The decomposition reaction can be followed by measuring the pH (pH < 8). The initial solutions contain an indicator, the colour of which gets changed because of the change in pH during the course of a reaction. The rate law is

$$-\frac{d[H_2CO_3]}{dt} = k_2\ [H_2[H_2CO_2].$$

At 18°C the rate constant k_2 has been 12.3 s^{-1} which corresponds to a half-life of 0.056 s.

This method is not applicable to the fastest solution reactions as the tine to mix the solutions and move them a short distance from the mixing chamber is seldom less than a millisecond. However, it is found that many reactions (including a large number of enzymes and other reactions of biological importance) can be studied by this technique, so that it has been profitable to build highly automated stopped flow reactors including some that are available commercially.

For the kinetics investigation of fast reactions it becomes necessary to overcome two problems. In the first place the reaction has to be initiated and in the second its time course has to be observed. In flow techniques the classical principle of initiating a reaction by mixing the reactants has been extended, probably to its effective limit. In the remaining techniques novel principles are used for the initiation of the reaction

Modification of Flow Methods : In a modification of the continuous-flow method, where each experiment yields only one point on the reaction profile plot, the reaction is quenched and analysed chemically or physically at leisure. Thus the reagents are mixed as usual and then they are mixed further with a substance which quenches the main reaction and a sample is removed for analysis. A reaction which has been studied in this way is the radiochloride exchange in acetone of organosilicon chlorides with $*Cl^-$. The reaction was quenched by precipitating the free chloride ions as NaCl.

$$R_3SiCl + *Cl^- \rightleftharpoons R_3Si*Cl + Cl^-$$

followed by

$$Na^+ + \begin{Bmatrix} Cl^- \\ *Cl^- \end{Bmatrix} \begin{matrix} NaCl \\ Na*Cl \end{matrix}$$

The fraction of radiochloride had been determined by counting the washed precipitate. In such a study the variables are *v* and *x*. The quenching technique is also applied to reactions of biological importance.

Flow methods are also used for following reactions in the gas phase. One of the difficulties inherent in the study of fast gas reactions is that the absence of large concentrations of "third bodies" means that the temperature equilibration tends to be comparatively slow. This is quite a serious problem in few systems since it is difficult to counteract any thermal effects caused by the heat of reaction. In addition, any change in pressure associated with a mole number change and the pressure drop associated with the flow itself make it rather difficult to define the reaction conditions accurately. Despite this, the flow method has been used to determine the kinetics of many gas-phase reactions, an inert diluent frequency being added.

The general experimental arrangement is similar to that shown in Fig. 4.3. A reaction which has been studied by the flow method is the decay ofnitrogen atoms in the presence of oxygen molecule, $^*N + O_2 \rightarrow NO + O$.

Picosecond and Femtosecond Spectroscopy : With the development of laser system, picosecond (10^{-12} s) and femtosecond spectroscopy (10^{-15}s) have been developed. Laser systems can provide intense picosecond or shorter pulses at various wavelengths, along with the simultaneous measurement of time, wavelength and intensity. The following examples illustrate the importance of this technique.

Example 1:

Excitation of stretching vibration of the C — H bond, in C_2H_5CH takes place at 2900 cm^{-1}. With the help of picosecond spectroscopy, it has been found that during relaxation this vibration gets transformed into two bending vibrations of the C — H bond at 1450 cm^{-1}.

Example 2:

The first stage in this process of vision has been the excitation of rhodopsin. Rhodopsin partially gets deactivated forming an intermediate, prelumirhodopsin or bathorhodopsin. Picosecond spectroscopy reveals that prelumirhodopsin gets formed because of an intramolecular proton transfer – a "jump" of a proton from one position to another.

Large Perturbation Methods : The principle used by these methods is to apply a large perturbation to a system, thereby generating one of the reactants, and to follow the disappearance of this reactive species. In all of the methods to be considered, flash photolysis, pulse radiolysis and shock tube, it is necessary to use some form of rapid-detection device.

(a) Flash Photolysis Method : Though first used by Porter (1950) to study gaseous reactions the flash photolysis method is now widely used in investigating reactions in solution.

If iodine vapour is illuminated with light of appropriate wavelength, the dissociation reaction occurs. The absorbed

$$I_2 + hv \rightarrow 2I$$

photon is represented by hv, where h is Planck's constant, 6.625 × 10^{-27} erg. sec and v is the frequency of the light in second^{-1}. Rapid recombination of iodine atoms occurs in a non-photochemical reaction. The reaction consumes the iodine produced by the photochemical dissociation reaction. Under steady illumination with light of moderate intensity, steady state is established with only a very small fraction of iodine dissociated to atoms.

$$2I \rightarrow I_2$$

This steady state results from equality of rates of photochemical dissociation of iodine molecules and non-photochemical recombination of iodine atoms.

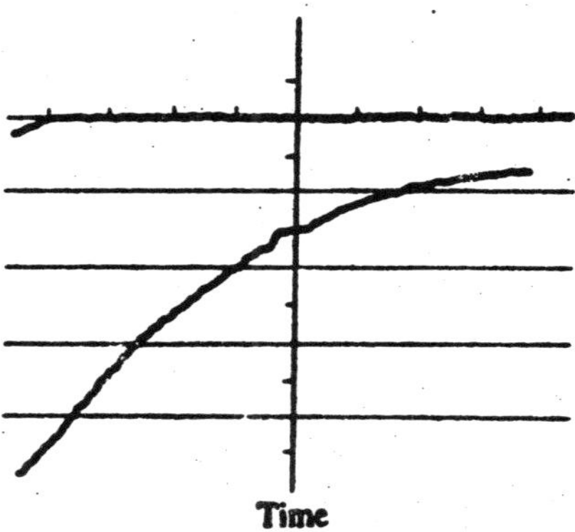

Fig. 4.5

If illumination occurs as a very brief and very intense flash, an appreciable number of iodine atoms can be produced, their recombination can be directly observed after the flash. In actual experiments, the peak intensity of a flash may build up in 10^{-5} sec and their decay over a period of 4 × 10^{-4} sec, or less. After light intensity from the flash has decayed, light of low intensity can be used in a spectrophotometric arrangement to follow the increase in concentration of molecular iodine. Fig. 4.5 is an oscilloscope trace showing the concentration increase of molecular iodine following the flash in an experiment of this type.

This procedure, called flash photolysis, has contributed valuable information regarding rates of very fast reactions such as recombination of iodine atoms. The technique has been used for study of reactions in solution as well as those in gas-phase.

The exact form of apparatus used a flash photolysis experiment depends to a large extent on the nature of the problems, but a typical one is shown in Fig. 4.6.

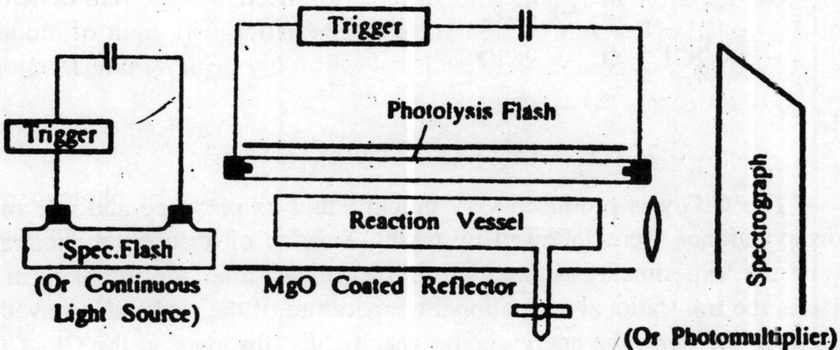

Fig. 4.6 : Typical flash-photolysis apparatus.

The reactants are placed in a cylindrical quartz vessel 50 cm long and 2 cm in diameter which is mounted next to the photolytic flash tube. The latter is a quartz tube of about the same length as the reaction vessel which is filledwith a rare gas, often krypton.

The light emitted by the flash tube is a continuum extending over the whole of the visible and well into the air and u.v. on which are superimposed the broadened atomic spectral lines of the rare gas.

The reaction vessel and flash lamp are mounted in a hollow cylinder whose inner surface is coated wiah magnesium oxide, which has a high reflectivity towards light of most wavelengths. A bank of condensers connected to the electrodes is charged up to the required voltage and then discharged by applying a triggering pulse to a small central electrode or to a spark gap in series with flash tube.

The duration time of the photolysis flash depends on the energy to be dissipated. The two conflicting demands of a short-lived but high-intensity burst have to be balanced, and the flash often has an energy of a few hundred joules spread over a few tens of microseconds.

The principal difference between flash photolysis and the relaxation methods is that the amount of energy put into the gas is relatively large, so that the temperature of the gas is raised by 1000° or more in a few micro seconds or, if the energy is absorbed into particular degree of freedom in the molecules, highly excited species or dissociated species are produced. In one of the first applications of the method, chlorine-oxygen mixtures were flashed, the radiation dissociating a substantial fraction of the Cl_2 to Cl atoms, other reactions following to produce the sequence.

$$Cl_2 \rightarrow 2Cl$$

$$2Cl + O_2 \rightarrow 2ClO$$

$$\rightarrow Cl_2 + O_2$$

$$2ClO \rightarrow Cl_2 + O_2$$

The ClO was produced very quickly, and its presence and rate of disappearance were followed by taking a series of absorption spectra (one per experiment) at times up to 10 millisec after the initial flash. Here, the translational and rational temperature of the molecules stayed close to ambient, the energy being specifically absorbed in the Cl—Cl bond. Later, using a faster flash Nicholas and Norrish were able to measure the rate constant for the addition of Cl atom to O_2, finding it (not surprisingly) to be third order, with a rate constant of 6×10^{14} $mole^{-1}$ cc^2 sec^{-1}, where N_2 is the principal third body.

(b) Shock Tubes : The technique has been used for studying fast reactions in the millisecond microsecond range at high temperature. The reactants are heated by a shock wave travelling at supersonic speed, the high temperature stays approximately constant for hundreds of a microsecond or a few milliseconds in some cases, and throughout this period the concentration of radicals is followed by one of the standard methods of detection. The essential features of the apparatus are shown in Fig. 4.7.

The shock tube is a metal or glass pipe several centimeters in diameterand several meters long. A thin metal or plastic diaphragm divides the shock tube into two sections : one section contains the driver gas (usually hydrogen or helium) at pressure upto ten times atmospheric pressure, and the other section contains a reactant gas at a few torr pressure. At the end of the tube there is an observation point for measurement of radical concentrations and also detector for measuring the velocity of shock front.

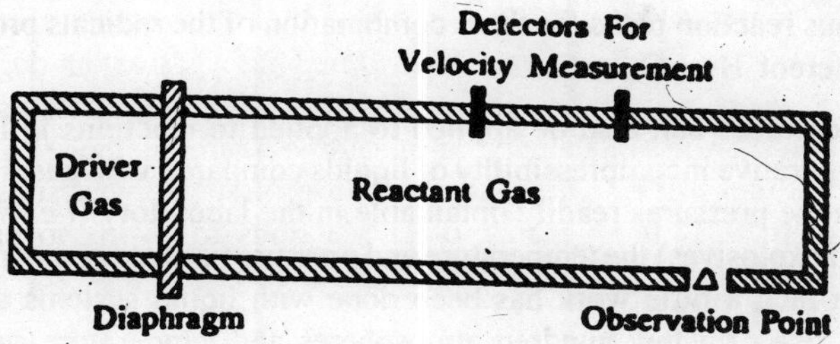

Fig. 4.7

As soon as the diaphragm is mechanically ruptured, the driver gas bursts out and rapidly forms a shock wave with a self's sustaining sharp shock front. The front is travelling with a supersonic speed compressing and heating the reactant gas as it proceeds. As soon as the shock front reaches the observation point, the reactant gas is suddenly compressed and heated. The initial reading is recorded. After the reacting gases, the driver gas will reach the observation point. The time interval between the arrival of the shock front and arrival of the driver gas is the reaction time which is usually about a millisecond.

The temperature in the reaction zone is calculated from the initial conditions in the reactant gas and the velocity of the shock wave. Optical spectroscopy and mass spectroscopy may be employed to follow changes in concentration of the reacting gases.

$$O^{\cdot\cdot} + CO + Ar \longrightarrow CO_2 + Ar$$

$$HCl + M \longrightarrow H. + Cl. + M$$

This method is useful for following gas reactions. It is particularly valuable for studying reactions in which wall effects interfere, such as explosions, since these can be avoided altogether. One reaction which has been studied in shock tubes is that between hydrogen and oxygen. The course of the reaction is followed by measuring the absorption due to OH radicals. These are used because they have comparatively intense absorption lines and are relatively stable. Light is passed through two quartz windows set opposite each other into the side of the tube and is picked up by a photomultiplier. This light is particularly rich in OH spectral lines since it is produced by discharging a flash lamp containing water vapour. Reaction zone temperatures in the range 1150° – 1850°K and pressures of a few atmospheres were achieved. By analysing the change in absorption by the OH radicals as a function of the time behind the -shock front, it was possible to deduce the relative importance of the various reaction paths for the recombination of the radicals produced with different H_2 : O_2 ratios.

Shock tubes can also be applied to applied to reactions in liquids. The comparative incompressibility of liquids compared with gases means that with the pressures readily obtainable in the laboratory (*i.e.*, without the use of explosives) the temperature and pressure rises are comparatively small. In fact, a little work has been done with liquid systems and the pressure rise of a few hundred atmospheres and temperature rise of a few degrees make this an ideal relaxation method." Kinetics of proton-transfer reactions have been studied in this way.

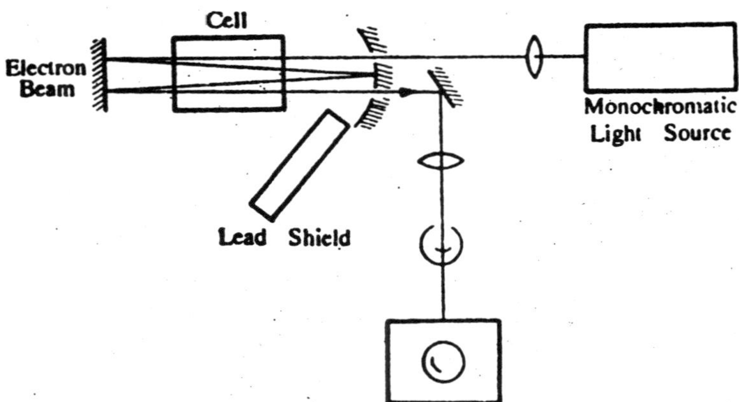

Fig. 4.8 : Schematic diagram of the pulse-radiolysis apparatus.

(c) Pulse Radiolysis : Pulse radiolysis is in many ways a similar technique and may be regarded as the radiation equivalent of flash photolysis with a pulse of ionizing radiation replacing the photolytic light flash. It has been applied primarily to reactions in solution and has been used with great success for the investigation of reaction of the hydrated electron.

A typical pulse-radiolysis apparatus is shown in Fig. 4.6. A pulse of electrons from a linear accelerator is passed through the solution and the changes produced are followed spectrophotometrically, either at a known time after the electron pulse with a spectroscopic flash and a spectrograph, or continuously at a fixed wavelength with a steady light source, and a photomultiplier. The same opposing demands of short duration and high intensity apply for the radiation pulse as for the principal flash in flash photolysis. Typically the pulse, which may be of X-rays rather than electrons, lasts for one or two microseconds. With this technique it is particularly important to remove the last traces of impurity from the solution since these frequently interfere with the reaction under investigation.

This can be done in particularly elegant way when reactions of the hydrated electron are being investigated. The aqueous electrons produced on irradiation are found to disappear more slowly after successive pulse and it seems that they are reacting with the residual impurities to give non-reactive products. The solutions are therefore pulsed until the electron decays at a reproducible rate, by which time it is assumed that all of the reactive impurities have been removed. This generally happens after about twenty pulses.

The method is used to follow reactions of the secondary or tertiary products of the radiolysis. For example, if an aqueous solution of Fe^{2+} *is radiolysed some of the oxidizing fragments of the water radiolysis, such as OH radicals, are capable of extracting an electron from the metal ion to produce* Fe^{3+}*. The subsequent complex formation of this ion with ligands in the solution, such as* SO_4^{2-}*, can then be followed.* Evidently the scope for such application is considerable, although the reactions are frequently so complex that it is not quite certain what is going on.

Small Perturbation : *Chemical Relaxation, Methods.* In all relaxation methods a chemical equilibrium is perturbed by a rapid change in one of several possible external parameters (electric field intensity, temperature or pressure), and the equilibrium process is then followed by spectrophotometric or conductometric techniques. Changes in the solvent structure affected by this same perturbation proceed so much faster than the chemical reactions under investigation that they are unobservable within the time range accessible to the relaxation methods and cause no ambiguity in the experimental results. A broad spectrum of reaction half-lives ranging from a few seconds down to 5×10^{-10} sec have been measured with those methods, although no one of the methods covers this entire time range. A broad range of accessible times is important when the reaction mechanism is unknown and more than one fast reaction step must be detected.

Let us consider a single-step chemical process

$$AB \underset{k_R}{\overset{k_D}{\rightleftharpoons}} A^+ + B^-$$

The rate equation for the above equilibrium is

$$\frac{dx}{dt} = k_D (a - x) - B\ x^2 \qquad ...(1)$$

where *a* is the total concentration of AB and* is the concentration of A^+ or B^-. Now let $\Delta x = x - xe$ where *xe* is the ionic concentration at equilibrium and A* is small. This restriction to small perturbations is important, since we wish for mathematical simplicity to deal only with linear differential equations.

Since x_e is a constant, it necessarily follows that

$$\frac{dx_e}{dt} = 0.$$

Substituting $x = \Delta x + x_e$ in eq. (1), rearranging terms and rearranging the product $k_R \Delta x^2$, we then have

$$\frac{d\Delta x}{dt} = -(k\Delta + 2k_R\ x_e)\ \Delta x + k_D\ (a - x_e) - k_R\ x_e^2 \qquad ...(2)$$

From eq. (1) and the fact that xe is a constant, we see that

$$\frac{dx_e}{dt} = 0\ k_D\ (a - x_e) - k_R\ x_e^2 \qquad ...(3)$$

Eq. (2) reduces to

$$\frac{d\Delta x}{dt} = -(k_D + 2k_R x_e)\ \Delta x \qquad ....(4)$$

Eq. (4) is interesting as the paranthesized quantity on the right side has the dimensions of a reciprocal time.

Thus we may rewrite equation (4) as

$$\frac{d\Delta x}{dt} = -\frac{\Delta x}{\tau} \qquad ...(5)$$

where the "relaxation time" τ is given by

$$\tau = (k_D + 2k_R\ x_e)^{-1} \qquad ...(6)$$

This time is not to be identified with any specific simple, physical (chemical) process taking place in the liquid.

Determination of Relaxation Time : Relaxation time is determined by various methods such as temperature jump and pressure jump methods (used for times larger than 10^{-5} s), high electric/magnetic fields or ultrasonic vibrations (used for times smaller than 10^{-5} s).

In *temperature jump* method, a temperature change of several degrees ($\approx 10°C$) in 10^{-6} sec, is created by a discharge of high voltage condenser (100 KV) through a small quantity of solution. Then the time dependence of concentrations of followed by the absorption spectroscopy or by measuring electrical conductivity as a function of time. Apparatus required in temperature jump method is shown in Fig. 4.7. A temperature range of 1 degree centigrade (near room temperature) has been found to change the equilibrium concentration by about 3 percent.

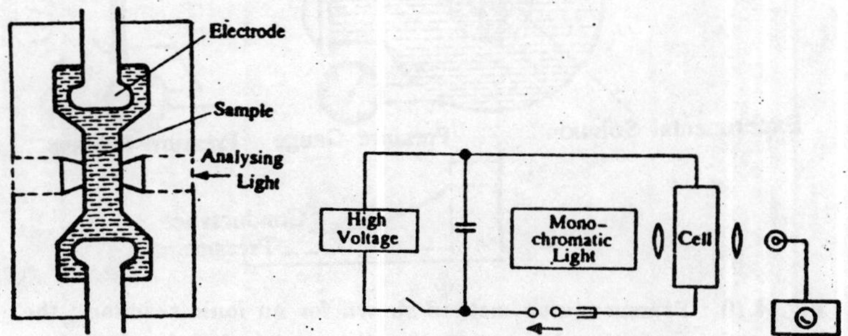

Fig. 4.9 : (a) Temperature-jump cell.

Fig. 4.9 : (b) Schematic diagram of temperature-jump method which uses electrical discharge for heating solution.

An alternative approach to the problem of suddenly raising the temperature is to use a pulse of microwaves. The advantage of this method is that non-conducting solutions may be used, though only a relatively small temperature rise can be obtained, generally less than 1°K. Q-switched laser has also been used as the heat source.

By combining flow and T-jump techniques it is possible to study the reactions of species in quasi-equilibrium.

The *pressure jump* method uses a sudden charge of pressure to displace the equilibrium. The sensitivity of a reaction to pressure depends on the change in volume $\Delta V°$ and is represented quantitatively by the equation

$$(\partial \ln K/\partial P)_T = - \Delta V^*/RT.$$

The sample is placed in a flexible cell which is contained in a pressurevessel filled with an inert liquid such as xylene. A pressure of about 65 atmospheres is set up in the vessel and this is reduced to atmospheric pressure within about 10^{-4} sec by puncturing a thin metal disc set into the wall of the vessel. The attainment of the new equilibrium is followed conduct to-metrically. Pressure-jump method is shown in Fig. 4.10.

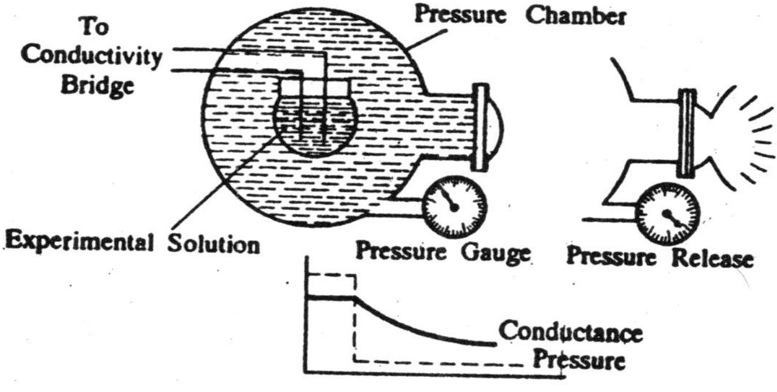

Fig. 4.10 : Pressure-jump method shown for an ionic reactionl; the reaction is followed by measuring the electrical conductance.

MOLECULAR REACTION DYNAMICS AND MOLECULAR BEAM METHOD

Molecular Reaction Dynamics : Both the collision theory and the activated complex theory have been based on the *statistical* treatment of the reacting systems to provide information about the average behaviour of an assemblage of molecules. This subject constitutes what is termed as *macroscopic kinetics*. The disadvantage of macroscopic kinetics is that the reactant molecules are present in a range of energy states and provide only an *indirect* information about the individual molecular events that take place during a chemical reaction. In recent years, another branch called *molecular reaction dynamics* or *microscopic kinetics* has gained importance due to the pioneer work of D.R. Herschbach, Y.T. Lee and J.C. Polanyi. These chemists shared the 1986 Chemistry Nobel prize. Molecular reaction dynamics deals with *the intermolecular and intramolecular motions that take place in the elementary act of chemical change and with the quantum states of the reactant and product molecules.* It is to be remembered that molecular reaction dynamics does not supersede macroscopic kinetics. *It simply aims at gaining a detailed picture of the elementary chemical act.*

The experimental technique of studying molecular reaction dynamics includes the *molecular beam method* in which the dynamics of reactions is investigated by utilizing a *mono-energetic beam of atoms or molecules.* This is described as below.

Molecular Beam Method : Recently many groups of workers have developed techniques for producing monoaminergic beams of molecules. The main aim of this technique is to study chemical reactions under conditions in which there occurs no Maxwellian distribution of velocity. If two molecular beams cross, any species will undergo only one collision in the crossing zone. The result can be elastic, inelastic or reactive scattering away from the direction of the incident beams, but the discussion is limited mainly to the reactive collisions and detecting scattered reaction products. Beam experiments are becoming very sophisticated and, with appropriate additions to beam sources, the relative translational energy on the reactants and perhaps their internal energy states can be defined with some precision.

For each reactant energy, it is possible to detect products which are emerging from the crossing zone at different angles and the reaction cross section can be evaluated from total product yield. Varying the reactant relative translational energy would give the cross section as a function of energy and hence the threshold energy.

Techniques have been developed to measure product velocities at selected angles, which makes the translational and internal energies of the products to get calculated. This gives rise to further deductions such as whether the collisional interaction is short lived or if an intermediate collision complex gets formed and the main features of the potential energy surface for the reaction can sometime be constructed.

The schematic apparatus used in a molecular-beam experiment is shown in Fig. 4.10. The two essential features of this apparatus are :

1. Devices for producing extremely narrow beams of reactants, with provisions for controlling the speeds of the molecules and their rotational, vibrational and electronic states.

2. Movable devices for the detection of the reaction products at various positions.

In the molecular beam experiments performed in the 1930s, the beams got produced by taking the reactants in a small vessel with a very small opening allowing the atoms or molecules to effuse into an evacuated vessel. A rectangular slit served as the orifice and a second small slit, placed close to it made the molecules travel in the required direction. However, these early molecular beam experiments suffered from the fact that the beams produced had a very low intensity and also there was a thermal spread of molecular velocities. Though it was possible to reduce the spread of velocities by suitable means, the intensity of the beams continued to remain low.

In the modern molecular beam experiments, the earlier disadvantages have been overcome by using supersonic *nozzle* sources which can produce beam intensities greater by about 3 orders of magnitude. Fig. 4.11 shows the schematic diagram of a typical nozzle beam source.

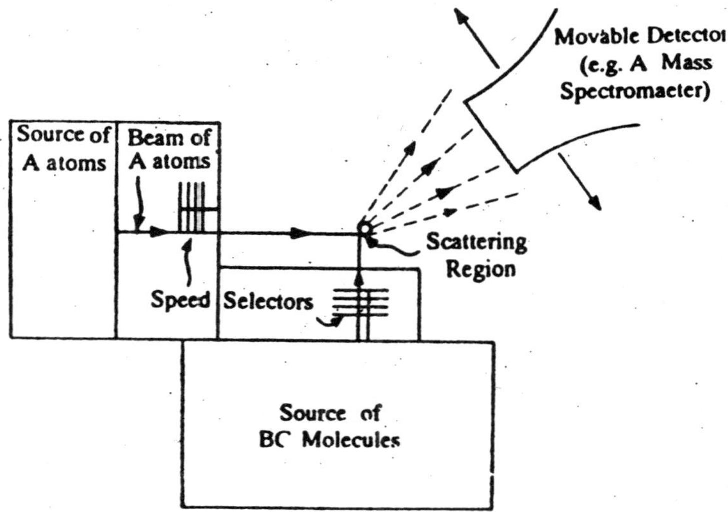

Fig. 4.11 : Schematic diagram for a molecular-beam apparatus for investigating the reaction A + BC → AB + C.

The main components of the apparatus are shown in Fig. 4.12. The apparatus is contained in a chamber at very high vacuum.

The most common sources have been small *ovens,* at pressure of the order 0.1 torr. A series of slits then collimates the molecules into a narrow beam. Beam intensities have been somewhat low but can be greatly improved by employing *supersonic nozzle sources.* The beam (reactant) would leave the source with Maxwellian velocity distribution, but a narrow velocity range can be obtained using a device which rejects species with velocities higher or lower than that needed. This velocity selector, in its simplest form, is having a series of slotted discs which are mounted on a rotating axis which is parallel to the beam direction. The velocity obtained can be known by the speed of rotation of the disc.

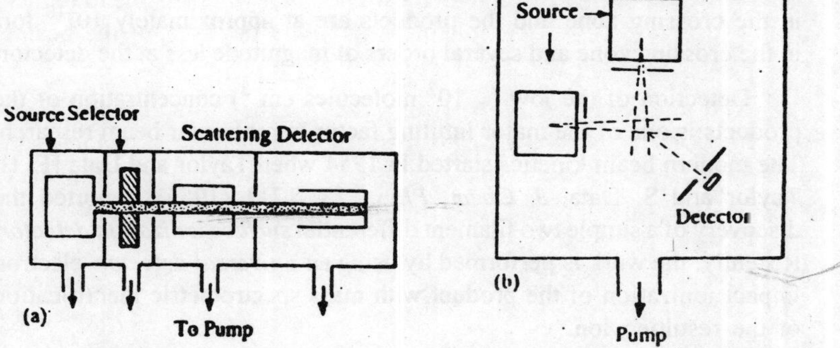

Fig. 4.12 : An expermental set up for a molecular beam method (a) for measuring total scattering cross sections. (b) for measuring the angular variation of scattering cross section.

The translational kinetic energy of the molecules can be varied by varying the oven temperature. A technique known as *seeding* is used for increasing the energy range of heavy molecules. This technique consists in mixing the heavy molecules with an excess of a gas which emerges from the oven at much higher velocities and sweeps the heavier molecules along with it.

The heavier molecules tend to travel along the centre of the beam whereas the lighter molecules are scattered towards the edges of the beam and are trapped out by the skimmer. Techniques have also been devised to put the atoms or molecules in the molecular beams into selected electronic, vibrational and rotational states.

As a result of the collisions of molecules in the crossed molecular beams, the following three processes may occur :

(a) *Elastic scattering* which involves no energy transfer among different degrees of freedom.

(b) *Inelastic scattering* which involves energy transfer among different degrees of freedom but in which no chemical reaction occurs.

(c) *Reactive scattering* in which a chemical change occurs. The chemists are primarily interested in reactive scattering.

The molecular beams are very dilute and background pressure must be minimized for achieving reactant maximum sensitivity for product detection. Typical beams are having reactant pressures of about 10^{-3} torr in the crossing zone and the products are at approximately 10^{-10} torr in the crossing zone and several orders of magnitude less at the detector.

Detection of the low ($\approx 10^2$ molecules cm^{-3}) concentration of the products is one of the major limiting factor in molecular beam research. The modern beam kinetics started in 1954 when Taylor and Data [E. H. Taylor and S. Data. *J. Chem. Phys.*, 23, 1711 (1955)] reported the discovery of a simple two filament differential *surface ionization detector*. Recently, the work is performed by using an *universal detector*–electron impact ionization of the product with mass spectrometric identification of the resulting ion.

The differential surface ionization detector was based on the fact that the alkali metal atoms and their salts could be easily ionized when they came in contact with a heated wire of appropriate work function. This technique was, however, limited to the reactions of alkali metals only.

The universal detectors measure the velocities of the scattered beams in all types of reactions by a time of flight mass spectrometer. A chopper isinserted in the front of the mass spectrometer and is rotated at high speed. After the product molecules manage to pass the chopper, the time required by them to reach the detector depends upon their translational velocities. Thus, the distribution of speeds in the scattered beams can be determined.

With the recent availability of supersonic beam sources and universal mass spectrometric detectors, molecular beam experiments have multiplied in number and application. Some indication of the range of this technique can be obtained from the information given in Table 4.1.

The simplest experimental measurement involves the determination of the flux of primary beam particles that the detector when the scattering chamber is empty and when it is having, scattering gas at a known density, n_2 particles cm^{-3}. Suppose there is a thin layer in the scattering chamber, perpendicular to the beam and with thickness *dl* in the beam direction, and area A in the direction normal to the beam. Then, the number of scattering particles in the layer would be n_2A dl. If the flux of beam particles incident on this layer is I and that leaving is I – dI. The fraction of particle getting scattered would be dI/I. The quantity dI/I then should be equal to the fraction of the area of the thin layer that is opaque to the incident particles. Then, we have

Table 4.1 : Typical reactions studied by molecular beam method.

$M = HX \rightarrow MX + H$	$X + H_2 \rightarrow HX + H$
$M + X_2 \rightarrow MX + X$	$X + HX \rightarrow HX + X$
$M + RH \rightarrow MX + R$	$X + X_2 \rightarrow X_2 + X$
$M + MX \rightarrow MX + M$	$X + C_2H_4 \rightarrow C_2H_3X + H$
	$\rightarrow C_2H_4 + HX$
$M + O_2 \rightarrow MO + O$	
$H + H_2 \rightarrow H_2 + H$	$O + X_2 \rightarrow OX + X$
$H + HX \rightarrow HX + H$	$R + X_2 \rightarrow RX + X$
$H + X_2 \rightarrow HX + X$	
$H + M_2 \rightarrow MH + M$	

*M = alkali or alkaline earth metal atom

H = hydrogen or its isotopes

N = hydrogen atom

R = alkyl radical.

$$\frac{dI}{I} = \frac{-n_2\sigma_{12} A\, dl}{A} = -n_2\,\sigma_{12}\, dl \qquad ...(1)$$

where σ_{12} is termed as the scattering cross section and is equal to the effective target area of the scattering particle. The negative sign appears because dI/I is the fractional *decrease* in flux.

Integration of Eq. (1) between the limits l = 0 and l = L would yield

$$\frac{I}{I_0} = \exp(-n_2 \sigma_{12} L) \qquad ...(12)$$

Eq. 2, is identical to the Lambert-Beer law.

Within the volume element A dl, the total number of particles scattered per second is A dI. The flux of the incident particles gets related to the density n_1 and velocity v by the relationship.

$$I = n_1 v \qquad ...(3)$$

Therefore, using the value of dI from Eq. (1), we get

$$AdI = n_1 n_2 \sigma_{12} (A\ dl) \qquad ...(4)$$

On equating this result with the number scattered, as expressed by second order rate expression, we get

$$k_2 = n_1 n_2 (A\ dl)$$

Therefore, we get

$$k_2 = \sigma_{12} v \text{ cm}^3 \text{ particle}^{-1} \text{ s}^{-1}$$

$$k_3 = 10^{-3} N_A \sigma_{12} \text{ dm}^3 \text{ mol}^{-1} \text{ s}^{-1} \qquad ...(5)$$

$$k_2 = 6.02 \times 10^4 \sigma_{12} v \text{ when } k_2 \text{ is an dm}^3$$

$$\text{mol}^{-1} \sigma_{12} \text{ in}$$

ABSOLUTE RATE THEORY APPLIED TO FAST REACTIONS

Absolute reaction rate theory provides an instructive treatment of reactions slightly displaced from equilibrium. Consider the reaction

$$A + B \underset{k_b}{\overset{k_f}{\rightleftharpoons}} C + D \qquad ...(1)$$

Let (A), (B). (C) and (D) represent the initial equilibrium concentration and (A – x), (B – x), (C + x) and (D + x) the concentrations after time t. Consider as an example what happens if the temperature is jumped from T_1 to $T_1 + \Delta T$. Initially, forward and backward reactions balance. Thus,

$$k_f T_1 (A) (B) - k_b T_1 (C) (D) = 0 \qquad ...(2)$$

If the temperature is jumped to

$$T_2 = T_1 + \Delta T \qquad ...(3)$$

The rate of reaction obeys the equation

$$\frac{dx}{dt} = k_{f\,T_2}\,(A - x)\,(B - x) - k_{b\,T_2}\,(C + x)\,(D + x) \qquad ...(4)$$

Here the T subscripts indicate the temperature of reaction.

$$k_{f\,T_1} = T_1^{\,m}\,Le^{\frac{-\Delta_t \neq}{PT_1}} \qquad ...(5)$$

where L is the temperature independent factor. It will be convenient to consider the very common case where m = 1.

This includes ordinary reactions where the limiting rate is the passage over a single barrier as well as the diffusion where the process involves passage over a succession of barriers of equal height.

In this latter case the temperature independent factors do not affect the analysis of the temperature $e^{\Delta S \neq R}$, but temperature independent factors do not affect the analysis of the temperature jump. Thus consider

$$k_{fT1} = K\frac{kT_1}{h}e^{\frac{-\Delta^G f\neq}{RT_1}} = K\,\frac{kT_1}{h}e^{\frac{-\Delta S_f \neq}{R}}\;e^{\frac{-\Delta H_{f\neq}}{RT_1}} \qquad ...(6)$$

and

$$k_{fT2} = K\,\frac{k\,(T_1 + \Delta T)}{h}e^{-\Delta G_f \neq / RT_1\,(1+\Delta T/T_1)}$$

$$= \frac{kT_1}{h}\left(1 + \frac{\Delta T}{T_1}\right)e^{\Delta S_f \neq / R}\;e^{-\Delta H_f \neq / RT_1\,(1+\Delta T/T_1)} \qquad ...(7)$$

A further expansion of k_{fT_2} gives

$$K_{fT2} \approx \frac{kT_1}{h}\left(1 + \frac{\Delta T}{T_1}\right)e^{\Delta S_f \neq / R}\;e^{-\Delta H_f \neq (1-\Delta T/T_1)\,RT_1}$$

$$= K\,\frac{kT_1}{h}e^{\Delta S_f \neq R}\,e^{-\Delta H_f \neq / RT_1}e^{\Delta H_f \neq \Delta T / RT_1^2}\left(1 + \frac{\Delta T}{T_1}\right)$$

$$= K_{fT1}\,e^{\Delta H_f \neq \Delta T / RT_1^2}\left(1 + \frac{\Delta T}{T_1}\right) \qquad ...(8)$$

Substituting Eq. (8) as well as the analogous result for $k_f T_2$ and also Eq. (2) into Eq. (4) yields

$$\frac{dx}{dt} = k_f\,T_1\,(A)\,(B)\left(1 + \frac{\Delta T}{T_1}\right)\left\{e^{\Delta H_f \neq \Delta T / R_1^2}\left(1 - \frac{(x)}{(A)}\right)\right.$$

$$\left(1-\frac{x}{(B)}\right)-e^{\Delta H_f \neq \Delta T/RT_1^2}\left(1+\frac{x}{(C)}\right)\left(1+\frac{x}{(D)}\right)\}$$

For very fast reactions for which

$$\frac{\Delta G_f \neq \Delta T}{RT_1^2} \leq 1 \qquad ...(10)$$

Eq. (9) may be expanded as follows, neglecting small terms,

$$\frac{dx}{dt} = K_f T_1 \text{ (A) (B)} \left(1+\frac{\Delta T}{T_1}\right)\left\{\left(1+\frac{\Delta H_{f\neq\Delta T}}{RT_1^2}\right)\left(1-\frac{x}{(A)}\right)\right.$$

$$\left.\left(1-\frac{x}{(B)}\right)-\left(1+\frac{\Delta H_{b\neq\Delta T}}{RT_1^2}\right)\left(1+\frac{x}{(C)}\right)\left(1+\frac{x}{(D)}\right)\right\} \qquad ...(11)$$

or $\frac{dx}{dt} = k_{fT1}$ (A) (B)

$$\left(1+\frac{\Delta T}{T_1}\right)\left\{\frac{\Delta H\,\Delta T}{RT_1} - x\,\frac{1}{(A)}+\frac{1}{(B)}+\frac{1}{(C)}+\frac{1}{(D)}\right)\} \qquad ...(12)$$

In deriving Eq. (12) from Eq. (11), we have used the relationship

$$\Delta H = \Delta H_f \neq - \Delta H_b \neq$$

Eq. (12) can be written in the form

$$\frac{dx}{dt} = k\,(a - bx) \qquad ...(13)$$

$$\left(-\frac{1}{b}\right)\frac{-b\,dx}{(a-bx)} = k\,dt \qquad ...(14)$$

$$-\frac{1}{b}\ln(a-bx) + C = kt \qquad ...(15)$$

Evaluating the integration constant C so that t = 0 when x = 0

$$-\frac{1}{b}\ln\left(\frac{a-bx}{a}\right) = kT \qquad ...(16)$$

$$1-\frac{1}{b}\,x = e^{-b\,kt} \qquad ...(17)$$

Thus, Eq. (13) finally brings us to the relationship.

$$x = \frac{a}{b}\left(1 - e^{-bkt}\right) \qquad ...(18)$$

The k_{fT_1} takes the more general form of Eq. 5 with m different from unity, the only effect this has on Eq. 12 is to replace the factor $\left(1 + \frac{\Delta^T}{T_1}\right)$ by $\left(1 + \frac{\Delta^T}{T_1}\right)^n$. Thus we have a useful integrated form for the dependence of concentration on the time for all systems displaced from equilibrium by a temperature jump.

MEASUREMENT OF RATE OF SLOW REACTIONS

This article is mainly concerned with the experimental techniques required to measure reaction rates and rate constant of reactions which may be occurring in solution phases or gaseous phase. In order to formulate the rate equation it is necessary to collect data on the reacting species. This is achieved by determining the concentration of one of the constituents of the reacting system at various times during the course of the reaction. There are two techniques in general use.

(1) Physical Methods : Measurement of physical properties can be used to follow the course of a reaction.- Samples of the reacting mixture can be taken at suitable times, and these samples can be then analysed by measuring some physical property. More conveniently, the measurements can be made from time to time on the reacting system as a whole without disturbing the rate of the reaction. By a proper modification, the latter method can be made to yield continuous record of the values of property and indirectly of the concentrations as a function of time. Following are the indirect methods which are generally used for the study of reaction kinetics.

(a) Pressure, Partial Pressure and Density of Gaseous Systems : Thekinetic of homogeneous gas reactions are very commonly studied bymeasuring the changing pressure of the system at constant temperature andvolume. When the reaction results in a change in the total number of gasmolecules, its progress can be followed by measuring the total pressure.

In some cases, the extent of a reaction can be evaluated by chilling the system and determining the pressure of one or more of the reaction components at a known, low temperature, at which the other reactants and products are frozen out and exert negligible vapour pressure. For example, the synthesis of hydrogen chloride from the elements can be studied by immersing the reaction vessel (after allowing it to react for a known time) in liquid oxygen and measuring the pressure, which is due to hydrogen only.

Alternatively, samples of the reacting mixture may be withdrawn andtheir pressures are determined before and after exposing them in a non-volatile reagent which removes one (or more) of the reaction components. When one of the products is quantitatively absorbed by solid reagent which does not otherwise affect the reaction, this reagent may be introduced into the reduction vessel and its pressure causes the reaction to produce a reduction in the number of gas molecules.

In some instances the gas density changes although there is no accompanying changes in the number of molecules of gas. For example, an exchange reaction may result in the substitution of HD or H, molecules. Such processes can be measured conveniently with a gas density balance. This is an apparatus by which a sealed, thin-walled glass or silica sphere can be weighed in a gas or a vacuum. From the difference in weight and known volume of the sphere, the density of the gas can be calculated. Commonly, the balance is used as a null instrument. The pressure of an unknown gas is adjusted until its density of the gas is the same as that of a temperature. The composition of an unknown mixture of the known compounds can be obtained from such a measurement.

(b) Density Method or Dilatometer Method : When the density measurements are restricted to dilute solutions, the change in density (or in volume of the sample), is a linear function of the percentage completion of the reaction. However, empirical calibration must be employed when the density method is applied to reactions in concentrated solutions or in the absence of a solvent.

The most common use of the density technique in kinetics studies involves the measurement of the change in volume (of the total volume) that is concurrent with the progress of the reaction. The change in volume can be measured conveniently with a dilatometer. A dilatometer is a bulb provided with a side tube through which it can be filled and with an open end capillary. After the bulb is filled with the reacting mixture, the side tube is closed by means of stop cock or a mercury cut off or (for very show reactions) by sealing off the tube of thin walled constriction. Since the volume of the capillary can be made very much smaller than that of the bulb, a small percentage in the volume of the liquid results in an easily detectable motion of the meniscus in the capillary. The position of the meniscus can be compared directly to a scale, fastened to or engraved upon the capillary or it may be read with cathelometer.

Advantages: (i) The dilatometric technique is applicable to most reactions which take place in liquid solution. It is especially useful for slow and moderately rapid first order reactions of any order.

(ii) If suitable precautions are observed, the measurements can be made as precise as is required for any kinetic studies.

(iii) Readings can be taken visually in one or two seconds and these do not disturb the course of the reaction. Automatic records can be obtained with ordinary photographic equipment.

(c) Colorimetry : Colorimetry may be adopted to the study of reactions in which the disappearance of a coloured reactant results in the formation of a differently coloured product. For this purpose a colorimeter is used in which one beam of light passes through a fixed thickness of the unknown solution and the other beam passes successively through the standard solutions of the reactant and product. As the Colorimetry is adjusted, the total amount of coloured substances in the light path is kept constant, but the ratio of the amounts of reactants and products is varied.

Advantages

(i) If the necessary apparatus is available and the experiments are planned to use this apparatus to best advantage, the colorimetry is well suited to kinetic studies.

(ii) The measurements can be made precisely and rapidly.

(iii) It is relatively simple to obtain continuous record of measurements made with a photocell, etc.

(iv) The method can be used in all cases where a marked change in the absorption of the solution occurs at a wave length for which the absorption of the solvent is relatively small.

(d) Optical Rotation : The course of a reaction involving optically active compounds may be followed by polarimetric measurement of the degree of rotation of plane-polarised light passing through the solutions. The amount of plane-polarised light depends among other things, on the concentration of the optically active components. Wilhelmey applied this technique in 1950 to study the catalysed inversion of sugar.

Analyses with the precision required for good kinetic measurements can be obtained with the aid of a well constructed polarimeter.

Advantage : Like colorimetry, polarimetry has the advantage that the resulting solution need not be thermostated more closely than is required to ensure the sensible constancy of the specific reaction rate.

Disadvantages

(i) Polarimeter readings are nearly always made visually and seconds are required for the completion of a single reading. Accordingly, the method is not well adapted to producing a continuous automatic record of the course of a reaction.

(ii) It requires expensive and specialised equipment,

(e) Galvanic Cells : These cells can be so designed that their electromotive force is a direct measure of the concentration of a wide variety of ions and molecules. However, it appears that the method has been seldom used. It has some real disadvantages:

(i) it is limited to ionising solvents, practically to water;

(ii) the electrodes require some time to adjust the changes in concentration; and

(iii) a more serious difficulty is that two electrodes, or an electrode and a salt bridge, must be in contact with the reacting solution,

(f) Polarography : It is an interesting and potentially useful technique for the study of reaction kinetics. The polarographic apparatus can be readily modified to yield continuous automatic record of the course of a reaction. In the present state polarography appears to be incapable of giving results as precise as are required for refined kinetic measurements. Its kinetic usefulness to data has been in the study of biological processes involving the appearance or disappearance of oxygen.

The application of polarography is further limited, since many organic reactions are catalysed by oxygen.

(g) Electrical Conductivity : The rates of ionic reaction occurring in ionising solvents such as water can be determined by measuring the conductivity of reacting mixture as a function of time *t*. In dilute aqueous solutions, where the ionic conductances are almost constant, the conductivity can be regarded as a linear function of the concentrations of the several ions and, therefore, to the extent of the reaction. For moderately dilute solution, after measuring conductivity the Onsager equation is used to know the real conductivity of solutions.

This method has been used extensively to determine the rates of hydrolysis of acid anhydrides, the saponification of esters and similar reactions.

$$CH_3COOC_2H_5 + OH^- \rightarrow CH_3COO^- + C_2H_5OH$$

Since the mobility of the CH3COO$^-$ produced is much less than that of the reactant OH^-, the specific conductance would decrease with time.

Limitations : Two main limitations are :

(i) This method is not applicable if the electrodes of conductivity cells (usually of platinised platinum) catalyse the reaction

(ii) Conductivity measurements require manual adjustments of a variable resistance and take 5 or 10 seconds for their completion. They are, therefore, not directly adaptable for measuring fast reactions.

(h) Nuclear Magnetic Resonance (N.M.R.) : The existence of rapid equilibria can often be established by N.M.R. For example, the proton spectrum of ammonical nitrate solution has a single line. On acidifying this solution with nitric acid, the protons on the ammonium ion show three-fold splitting under the influence of the central nitrogen atom, which has unit nuclear spin. This result shows that rapid exchange of protons occurs under alkaline conditions, so that all the protons in the solution are equivalent. This exchange does not occur in acid solution, and so splitting occurs.

(i) Gas Liquid Chromatography : *Principle.* In this method, the sample of the reactants is injected into a flowing gas which is then allowed to passthrough a chromatographic column. The column is packed with an adsorbent solid or an inert solid coated with a liquid solvent. The various components in the sample differ in their solubility in the solvent, or the strength of adsorption on the solid, and, thus, pass into a detector at different rates. Hence the components, emerging at different rates from the column, pass into a detector at different times. The detector records the emergence of each component, And when suitably calibrated, it measures their concentration. It is also possible to identify a component from the characteristic time at which it emerges from the column.

A gas-liquid chromatograph can be easily ccupled with a reaction vessel by a gas sampling value, and the sample taken may be so small that the concentrations in the reaction vessel are not disturbed.

Advantages : (i) This requires small samples for analysis (ii) Analysis time for each sample can be as small as a few seconds, so the measurements of concentration can be reported at quite short intervals.

Example.

An example of gas-liquid Chromatography is the pyrolysis of hydrocarbons.

(j) Mass Spectrometry : *Principle.* In this method, a sample of reaction mixture is permitted to pass into the *ion source region* of the instrument where molecules of the reaction mixture are ionised by bombardment with electrons of controlled energy. The ions produced in the ion source region, which are characteristics of their parent molecules of the reaction mixture, are allowed to enter the *analysing region* where the ions with different charge/mass ratios are separated and detected. The ions are separated in two basic ways :

(i) Ions are deflected in electrical and magnetic fields. The amount of deflection depends on charge/mass ratio.

(ii) Ions take different times to traverse a flight tube before reaching the detector.

Relative concentrations at different charge/mass ratios are measured from the ion currents at the detector. The mass spectrum, the pattern of ions derived from a given parent molecule, is characteristic and it is often possible to identify components in a mixture.

In mass spectrometry, very small samples are required which can beleaked directly from a reaction vessel. The mass spectrometry may also beused as a detector in conjunction with glass-liquid Chromatography separation.

Advantages : (i) The area in which mass spectrometry is more successful than gas liquid Chromatography is isotopic analysis,

(ii) Mass spectrometry is useful for identifying neutral molecules, radicals and ions. The method is sufficiently sensitive to detect approximately 10^{11} molecules cm^{-3}.

Disadvantages : Mass spectrometry is not suitable for chemical analysis of complex mixtures. The mass spectrum of a complex mixture may be complicated, with many different fragment ions resulting in the different patterns which may coincide, making results difficult to interpret.

Uses : An important contribution to combustion kinetics was proving that HO_2 radical did exist.

(k) Absorption Spectroscopy : Ultraviolet, visible or even infrared spectra can be utilised to measure changes in the concentration of reactants and products if the absorption is very strong and does not overlap the spectrum of another molecule present. Background radiation from a suitable light source is allowed to pass through the reaction vessel into a monochromator which isolates that spectral region at which molecule is absorbing. The intensity of transmitted light can be measured photoelectrically and the photoelectric current can be recorded continuously. Thus, the changes in absorption in the selected wavelength range and so changes in concentration can be recorded.

Examples:

Typical examples involving this technique are hydrolysis of methyl salicylate. The spectrophotometer can be used to study the following reaction.

$$\underset{O^-}{\overset{COOCH_3}{\bigcirc}} + OH^- \longrightarrow \underset{O^-}{\overset{COO^-}{\bigcirc}} + CH_3OH^-$$

The methyl salicylate absorbs strongly in the ultraviolet region at 332 nm but at a higher wavelength than the salicylate ion which absorbs at 305 nm. At the wavelength 38 nm, the methanol possesses negligible absorbance. The reaction can, therefore, be followed by measuring the decrease in absorbance at 332 nm. But at this wave length, the absorbance of the salicylate ion is negligible. Therefore, the system must be analysed spectrophotometrically as a two-component system.

II. Conventional Kinetic Systems : The various kinetic systems are as follows :

(a) **Static System :** In a static system, reaction may take place in a closed reaction vessel, giving in this sense, a static system.

(b) **Flow Systems :** In this system analysis is carried out at the end of a flow tube.

For the study of most of reactions, both static as well as flow systems are employed. We will discuss both these methods one by one.

(a) **Static Systems :** The diagrammatic representation of a static systemis shown in Fig. (4.13). Its various parts are as follows.

(i) R denotes the reaction vessel. This is a cylindrical or spherical vessel, probably a few hundred cubic cms in volume. It is generally made of pyrex glass or quartz. Glass vessel can withstand temperatures upto 600°C whereas quartz vessel may withstand up to 1100°C. The steady high temperatures can be maintained by thermostated furnace F in which the reaction vessel R is kept.

Reactants may be mixed in the vessel or pre-mixed. The time of mixing and temperature equilibration should be as low as one second if we are studying reaction with half-lives greater than ten seconds.

R is connected to a manometer M through a common vacuum lining. R may have another arm for direct sampling to gas-liquid chromatography Fig. 4.13. Both F and R may have suitable windows for spectroscopic measurement.

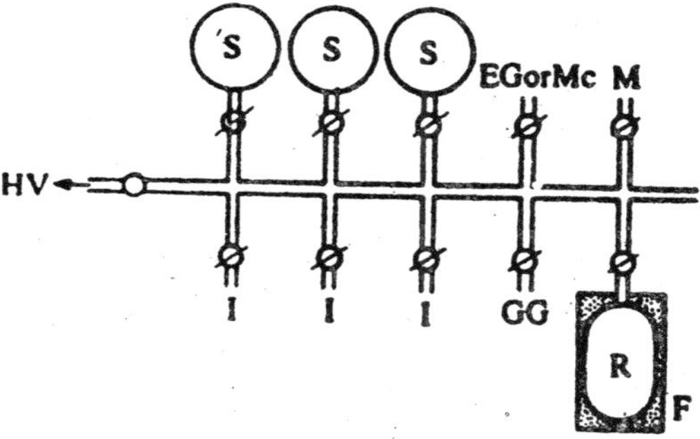

Fig. 4.13

(ii) EG is electrical gauge Me is the Moleod gauge and GG is glass gauge.

(iii) S, S and S are three gas storage vessels, I, I and I are three inlets in storage vessels for gases.

(iv) The end HV is connected to liquid nitrogen traps, diffusion pump and rotatory oil pump.

Working : Before introducing the reaction mixtures in the reaction vessel R, operations are carried out on a high-vacuum line to which all components of the apparatus, are attached. The line is evacuated by a combination of oil rotatory pump and mercury or oil diffusion pump with liquid nitrogen. Generally after evacuation, the pressure remains 10^{-5} Torr which can be measured with electrical gauges or Moleod gauges.

Then, the gases are introduced in the reaction vessel R from storage vessels S. Reactant pressures are best measured with sensitive glass gauges such as the spoon and spiral type.

In order to determine the rate of reaction occurring in a reaction vessel R, there may be continuous or intermittent analysis or even one analysis at a chosen time.

In one analysis at a time, it means the stopping of the reaction at a certain time by feezing the reaction vessel in liquid nitrogen and then analysing the contents. Thus, there will be a separate experiment for each reaction time.Hence, the method is not suitable.

In intermittent measurement, samples are withdrawn at a series of times and then analysed by gas-liquid chromatography. This method gives accurate and almost complete sample analysis at intervals as low as a few seconds.

The continuous method is used for such reactions in which the number of moles changes and this method is followed by the change in total pressure or by its absorption spectra or by mass spectrometry with a continuous small leak from the reaction vessel to the spectrometer.

Uses : This static method has been used to measure reaction rates andso deduce mechanisms and kinetic parameters for the majority of importantcomplex gas reactions like hydrogen-halogen reactions and the pyrolysis-and combustion of hydrocarbons.

(b) Flow Systems : These are mainly of two types.

(i) *Linear flow methods* : These methods can be used for slightly fasterreactions and involve measurements of smaller concentrations than staticmethods. A schematic .representation of a linear flow method is shown in Fig. 4.14. In this figure, M and M are manometers (now shown in figure), F isfurnace, P is a bulb for permanent gases and T is a trap for liquid products (to be immersed in coolant).

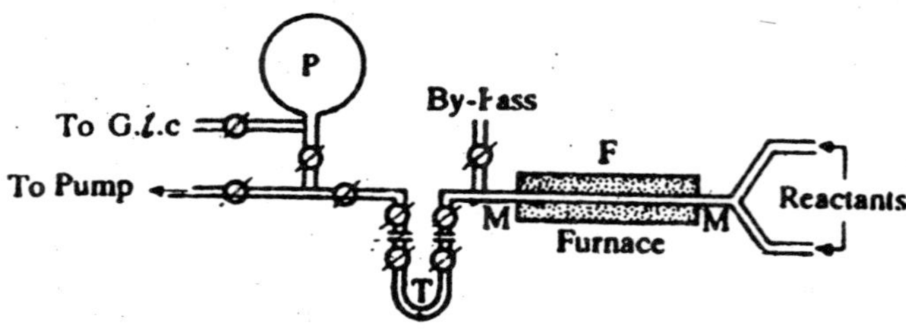

Fig. 4.14

Reactants are mixed and allowed to flow at a constant measured rate through a tube between two manometers M and M and after which the mixture is analysed by the methods discussed in the analytical methods.

The linear flow method is very suitable for reactions with simple kinetics; *i.e.*, for first-order or pseudo-first-order reactions. For example, this method has been used to study the decomposition of toluene which gives hydrogen, methane and benzene where the initial concentration is probably as follows :

$$C_6H_5CH_3 \rightarrow C_6H_5CH_2 + H^-$$

The linear flow methods cannot be applied to complex reactions.

(ii) *Stirred flow*: In this method, reactants are allowed to flow into a vessel which is specially designed to achieve rapid uniform mixing of reactants by ensuring vigorous convection and diffusion of the reactants. When a steady state is reached the reaction mixture flowing from the vessel is collected for analysis.

A typical example of V reaction studied by this method is the decomposition of ditertiary-butyl peroxide, a first-order radical reaction.

LINEAR FREE ENERGY RELATIONSHIPS

We know that the introduction of a group in an organic molecule altersthe reaction rate to an appreciable extent. The substituent is having a definite effect on the distribution of electron density in a molecule. For instance the alkaline hydrolysis of enthyltrichloroacetate (I) takes place about 8 million times faster than the alkaline hydrolysis of trimethylacetate.

$$\underset{\text{(I)}}{\mathrm{Cl_3C{-}C({=}O){-}O{-}C_2H_5}} \qquad \underset{\text{(II)}}{\mathrm{(CH_3)_3C{-}C({=}O){-}O{-}C_2H_5}}$$

Also, the rate of reaction of methyl pyridine with methyl iodide has beenlarger than that of pyridine without a substitute. In 1933, Hammett gave aquantitative relationship between structure and reactivity of substancesand showed that for a number of reactions of methyl esters with NMe_3,

$$RCOOH_3 + N(CH_3)_3 \xrightarrow{k} RCOO– + N^{+}(CH_3)_4 \quad ...(1)$$

the rates of reaction has been linearly related to the ionisation constantsof the corresponding carboxylic acids hi water

$$RCOOH + H_2O \overset{k}{\rightleftharpoons} RCOO^{-} + H_3O^{+} \quad ...(2)$$

A plot of the logarithm of rate constant (– log *k)* for reaction (I) *vs.*equilibrium constant (– log K) for ionisation of the acid would yield areasonably good straight line Fig. 4.15.

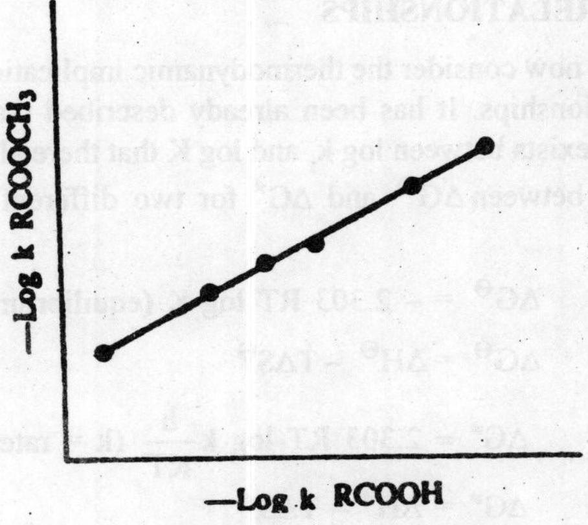

Fig. 4.15 : Linear relationship.

It is possible to characterise the reactivities of substances quantitatively either by free energy change $\Delta G^{\ominus}$ (if equilibrium is taken into consideration) or by free energy of activation $AG^{\neq}$ (if reaction rate is considered). Hence, the relation between equilibrium constant, K, rate constant, *k*, and free energy change may be put as follows.

$$\Delta G^{\ominus} = -\ 2.303\ RT\ \log K$$

$$\Delta G^{\neq} = -\ 2.303\ \log k_r \left(\frac{h}{K'T}\right)$$

where,

K = equilibrium constant

k_r = rate constant

h = Planck's constant

K' = Boltzmann's constant

Thus Fig. 2 has been equivalent to a straight line relationship between $AG^{\neq}$, the energy of activation for the reaction (related to k_r, rate constant) and AG^{-}, the free energy change for the ionisation of acid (related to K). Thus there exists a straight line or linear relationship between free energy terms for these two different reaction series. These types of straight line Hammett plots in Fig. 1 are termed as linear free energy relationships.

THERMODYNAMIC IMPLICATIONS OF LINEAR FREE ENERGY RELATIONSHIPS

We will now consider the thermodynamic implications of linear free energy relationships. It has been already described that a straight line relationship exists between log k_r and log K that there also exists a linear relationship between $\Delta G^{\ominus}$ and $\Delta G^{\neq}$ for two different reaction series. Hence.

$\Delta G^{\ominus} = -\ 2.303\ RT\ \log K$ (equilibrium constant). But

$$\Delta G^{\ominus} = \Delta H^{\ominus} - T\Delta S^{\ominus}$$

and $\quad \Delta G^{\neq} = 2.303\ RT\ \log k\dfrac{h}{KT}$ (k = rate constant)

butt $\quad \Delta G^{\neq} = \Delta H^{\neq} - T\Delta S^{\neq}$

Such a linear relationship between $\Delta G^{\ominus}$ values and $\Delta G^{\neq}$ values would be only expected if, for each series, one or other of the following conditions get satisfied.

(1) DH has been linearly related to DS for the series.

(2) DH has been constant for the series.

(3) DS has been constant for the series.

Hammett tried to apply his correlation to 39 cases and reported good' agreement in 30 cases.

WHY ARE HAMMETT'S RELATIONS CALLED LINEAR FREE ENERGY RELATIONS?

Hammett's relations and similar other relations are also termed as linearfree energy relations. The reason for this name would be evident from thefollowing considerations :

We know that the rate constants and equilibrium constants are related to free energy changes in accordance to the following equations.

$$k = \frac{k_B T}{h} e^{-\Delta G^*/RT} \quad ...(1)$$

and
$$K = e^{-\Delta G^\circ/RT} \quad ...(2)$$

where $AG^{\neq}$ refers to free energy of activation and ΔG° refers to the standard free energy of the reaction.

On taking logs of Eq. (1) and substituting the value of log *k* in equation *(viz.,* $\log k_0 = \log k_0 + \sigma\rho$) for a given set of reactions with a particular value of ρ, we obtain

$$\Delta G^{\neq} = (\Delta G^{\neq})_0 - 2.303\ RT\ \sigma\rho \quad ...(3)$$

where $(\Delta G^{\neq})_0$ represents free energy of activation of the parent compound.

For another set of reactions with a different value of ρ, say p', we have

$$(\Delta G^{\neq})' = (\Delta G^{\neq})'_0 - 2.303\ RT\ \sigma\rho' \quad ...(4)$$

On dividing each of the above equations by the respective value of p and subtracting one from the other assuming that the substituent is the same so that *a* is the same for both, we obtain

$$\frac{\Delta G^{\neq}}{\rho} - \frac{(\Delta G^{\neq})'}{\rho'} = \frac{(\Delta G^{\neq})_0}{\rho} - \frac{(\Delta G^{\neq})'_0}{\rho'} = 0 \quad ...(5)$$

or
$$\Delta G^{\neq} - \rho/\rho¢\ (\Delta G^{\neq})¢ = 0$$

The coefficient p/p' is having the same value for all reactions in a givenset. Eq. (6) is thus a linear relation between free energies of activation of oneset of reactions and those of the corresponding set.

It is possible to write an equation similar to Eq. (4) for the total free energy of reaction, *i.e.*,

$$AG^{\circ} = (AG^{\circ})\ 0^{\ -0.2\text{-}303\ RT\ \sigma\rho} \qquad ...(7)$$

From Eqs. (3) and (7), we have

(8) Eq. (8) gives the *linear relationship* between free energy of activation and free energy of the reaction (the term in brackets is a constant).

The Hammett's and other relationships can thus be restated in terms of linear free energy relationships (6) and (8).

HAMMETT'S RELATIONSHIPS

It is known that the rate of a reaction at a given temperature *depends on the nature of the reactants and the products formed.* Scientists have been all the time making attempt to establish some theoretical correlations with the help of which the reaction rates and mechanisms of reactions could be predicted from a knowledge of the structure and properties of isolated reactants and products. The success in this direction, however, has been almost nil.

Nevertheless, there are few scientists who did succeed in establishing at least *empirical correlations* which could be able to describe the trends in rate constants and equilibrium constants over sets of closely related reactions involving, in particular, aromatic compounds. By using these relations, it became possible to predict the rate constants or equilibrium constants of the reactions of substituted aromatic compounds provided the values for the parent compound are known shown in Fig. 4.16.

Rate constants and equilibrium constants are measured for numerous reactions involving hydrolysis of substituted ethyl benzoates, the substituents (X) being F, Cl, Br, I, NO_2, – NH_2, – OH, etc., present in meta or para positions. Hence

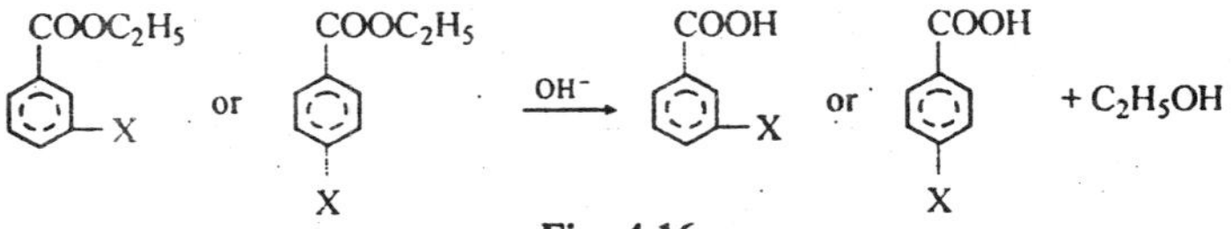

$COOC_2H_5$ (meta-X) or $COOC_2H_5$ (para-X) $\xrightarrow{OH^-}$ COOH (meta-X) or COOH (para-X) + C_2H_5OH

Fig. 4.16

Similar measurements are made for several other sets of reactions of substituted ethyl benzoates. Rate constants and equilibrium constants are now known for hundreds of reactions involving differently substituted aromatic compounds.

L P. Hammett critically analysed the huge mass of data of the type given above and observed that there exists a striking correlation between the rate constants (or equilibrium constants) of substituted aromatic compoundsand the values for the parent compounds. He deduced the Hammett's equation as follows.

DERIVATION OF HAMMETT EQUATION

In Fig. 4.17 there is a plot of log *k* for a series of *m*- or p-substituted aromatic acids v. log K for ionisation of corresponding acid which happens to be astraight line.

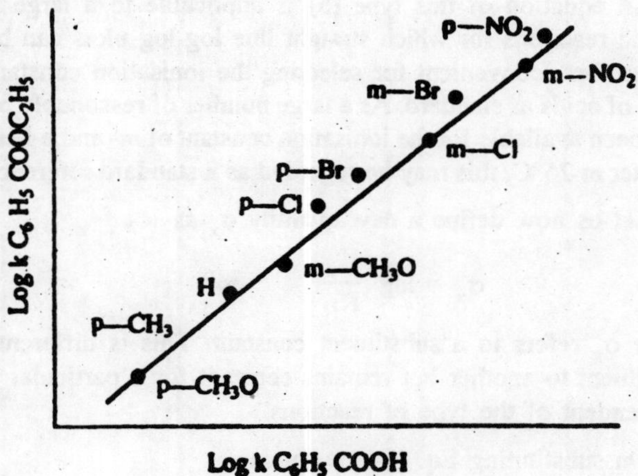

Fig. 4.17 : Plot of log k vs log K.

On applying the general equation for a straight line (y = mx + c) to the straight line curve of the type in Fig. 4.17, it is possible to write

$$\log k_x = \rho \log K_x + c \qquad ...(3)$$

where ρ. = slope of the straight line and is referred to as reaction constant

c = intercept.

x = m- or p-substituent in the benzene ring of the species concerned.

k = rate constant

K = equilibrium constant.

An analogous expression for unsubstituted ester and acid may be put as follows.

$$\log k_H = \rho \log K_H + C \qquad ...(4)$$

Subtraction of Eq. (4) from Eq. (3) yields

$$\log k_x - \log k_H = \rho (\log K'_x - \log K_H)$$

or $$\log \frac{k_x}{k_H} = r \log \frac{K_x}{K_H} \qquad ...(5)$$

An equation of this type (5) is applicable to a large number of organic reactions for which straight line log-log plots can be made. It is, therefore, convenient for selecting the ionisation constants for one series of acids as standard. As a large number of reasonably precise data have been available for the ionisation constant of *m*- and p-benzoic acids in water at 25°C, this may be regarded as a standard reference reaction.

Let us now define a new quantity σ_x as

$$\sigma_x = \log \frac{K_x}{K_H} \qquad ...(6)$$

where σ_x refers to a substituent constant. This is different from one substituent to another but remains constant for a particular substituent independent of the type of reactions.

On substituting Eq. (5) we obtain

$$\log \frac{k_x}{k_H} = \rho \times \sigma_x \qquad ...(7)$$

This correlation is termed as Hammett equation.

If we known the value of the constant σ_x a plot of $\log\left(\frac{k_x}{k_H}\right)$ agains σ_x can be used to find the values of reactions constant ρ. The ρ values for a particular reaction, carried out under specified conditions, constants, no matter what the m-or p-substituent present in the compounds gets involved.

The value of the constants r and s_x have included in the following tables.

Table 4.2 : Some Values of Hammett's Reaction Constants, r (Ar = Substituted Phenyl Group)

Reaction	ρ
$ArCOOH \xrightarrow{H_2O} ArCOO^- + H^+$	1.00
$ArCOOH \xrightarrow{C_2H_5OH} ArCOO^- + H^+$	1.96
$ArCH_2OOH \xrightarrow{H_2O} ArCH_2COO^- + H^+$	0.56
$ArOH \xrightarrow{H_2O} ArO + H^+$	2.26
$ArNOH_3 \xrightarrow{H_2O} ArNH_2^- + H^+$	2.94
$ArCOOH + CH_3OH \xrightarrow{H_2O,H^+} ArCOOCH_3 + H_2O$	– 0.58
$ArCHCl + C_2H_5OH \xrightarrow{H_2H_5OH} Ar_2CHOC_2H_5 + HCl$	– 4.03

Table 4.3 : Hammets Substituent Constants (σ_x)

X	σ_{meta}	σ_{para}	X	σ_{meta}	σ_{para}
NH_3	– 0.16	– 0.66	F	+ 0.34	+ 0.06
OH	+ .12	– 0.37	Cl	+ 0.37	+ 0.23
OCH_3	– 0.12	– 0.27	Cl	+ 0.39	+ 0.23
CH_3	– 0.07	– 0.17	I	+ 0.39	+ 0.23
C_2H_5	– 0.07	– 0.15	NO_2	+ 0.71	+ 0.78
C_6C_5	– 0.06	– 0.01	CN	+ 0.56	+ 0.66

The σ_x, the substituent constant, may be regarded as quantitative measure of the electron-releasing or electro-withdrawing power of the substituent group and may be having a positive or negative value. A positive value of *a* reveals stronger electron-attracting ability relative to hydrogen, while a negative value of this constant reveals that it attracts less strongly than hydrogen. The substituent may also exert a field effect.

The magnitude of the reaction constant, ρ refers to a measure of the sensitivities of the reaction to changes in the polarity of the substituents in the adjacent benzene ring. The larger the absolute value of ρ, the more sensitive would be the reaction in question. The signs of the constant, ρ, characterise the kind of reaction transformation. The negative value of ρ reveals the development of positive charge at the reaction centre in the transition state for the rate-determining step of the overall reaction and whereas a positive value of ρ reveals the pulling back of electron density from the reaction centre in the transition state for the rate-determining step of the overall reaction.

THE TAFT EQUATION

The Hammett equation is not applicable very well to either o-substituted benzoic acid or aliphatic compounds as steric effects play an important role in governing the rate of reaction. This gives rise to non-linear, or even to apparently random plots.

For aliphatic compounds, R, Taft postulated that

$$\log k = \log k_0 + \sigma^* \rho^* \quad \text{...(8)}$$

where k = rate (or equilibrium) constant for a particular representative of the reaction series.

k_0 = rate (or equilibrium) constant for the standard compound.

ρ^* = reaction constant.

σ^* = polar constant of the substituent.

The Taft constant σ^* refers to a measure of the electron-attracting ability of the substituent. It is purely an inductive effect because it gets transmitted through an aliphatic chain.

The σ^* values in Eq. (8) cannot be defined in terms of the dissociation constant. It was determined by Taft by separating the total effect of a substituent into a steric term and a polar term on the suggestion of Ingold. The basis of the method has been as follows. It is found that σ values for the acid-catalysed hydrolysis of *m*- and ρ-substituted benzoic acid have been nearly + 0.03, very nearly to be zero. The lack of effect of substituents is due to the fact that it is a multistep reaction and the effect of substituents for some steps can be cancelled out by others.

In the above reaction, an electron-releasing substituent will tend to favour step I but hinder step II. Due to this balancing effect, the overall ρ for the acid-catalysed reaction would be zero σ values for the base-catalysed hydrolysis has been positive and large (+ 2.2 to + 2.8). Experiments reveal that the rates of base-catalysed hydrolysis are very sensitive to the substituents while in acid hydrolysis the nature of the substituents are having no effect on the rate. It is, therefore, reasonable to regard that steric effect plays an important role in the acid-catalysed hydrolysis shown in Fig. 4.18.

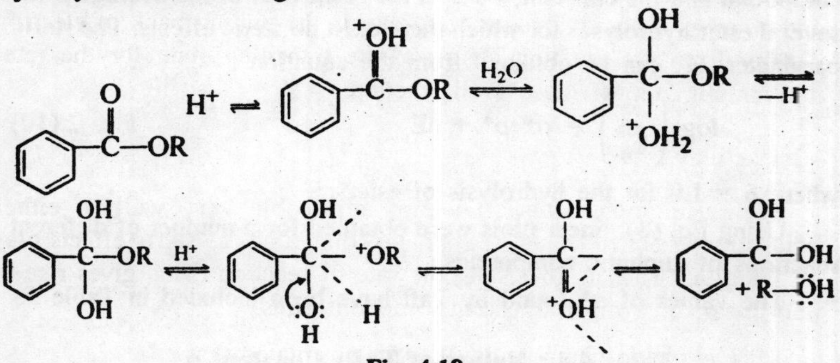

Fig. 4.18

It is evident that close similarity exists between the transition state oftwo reactions and they may be represented by structures I and II in Fig. 4.19. They have been both tetrahedral but differ only by the presence of two additional protons in the acidic case.

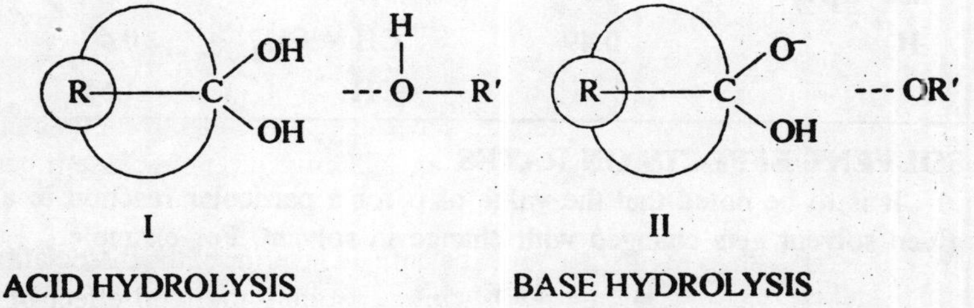

I — ACID HYDROLYSIS

II — BASE HYDROLYSIS

Fig. 4.19

According to Taft the steric effect has been substantially the same inboth acid and base catalysed hydrolysis due to close spatial similarity of the two activated complexes. The ratio k/k_0 in alkaline hydrolysis refers to a measure of the two effects (polar and steric) whereas in acid hydrolysis the relative rate has been proportional to the steric effect of the substitution.

The values of the polar parameter (σ^*) may be obtained from.

$$\sigma^* = \frac{1}{2.5}\left[\log\left(\frac{k}{k_0}\right)_B - \log\left(\frac{k}{k_0}\right)_A\right] \quad ...(9)$$

where A and B refer to the acid- and base-catalysed hydrolysis, *k* denotes the observed rate constant, k_0 denotes the rate constant for the standard compound and the constant, 1/2.5 is the reciprocal of the average ρ for several ester hydrolyses for which there exist no steric effects. The steric parameter, E_s, can be obtained from the equation :

$$\log\left(\frac{k}{k_0}\right) = \sigma^* \rho^* + \delta E_i \qquad ...(10)$$

where δ = 1.0 for the hydrolysis of esters.

Using Eq. (8), linear plots were obtained for a number of different reactions of aliphatic compounds.

The values of σ* found by Taft have been included in Table 3.

Table 4.4 : Values of Tafts Constant σ*

Substituent	σ*	Substituent	σ*
CH_3	0.00	C_6H_5	0.60
C_2H_5	– 0.10	$C_6H_5CH_2$	0.22
iso–C_3H_7	– 0.19	CH_3CO	1.65
H	0.49	$CH=CH_2$	0.40
		OH	1.55

SOLVENT EFFECTS ON RATES

It is to be noted that the value of ρ for a particular reaction in a given solvent gets changed with change in solvent. For example.

Table 4.5

Reaction	σ
$C_6H_5COOH + H_2O \rightleftharpoons C_6H_5COO^- + H_3O^+$	1.00 (by definition)
$C_6H_5COOH + H_2O \rightleftharpoons C_6H_5COO^- + H_2O$ (53% aqueous ethanol)	1.60
$C_6H_5COOH + OH^- \rightleftharpoons C_6H_5COO^- + H_2O$ + (ethanol)	1.96
$C_6H_5COOC_2H_5 + OH^- \rightleftharpoons C_6H_5COO^- + C_2H_5OH$ (70% aqueous ethanol)	1.83
$C_6H_5COOC_2H_5 + OH^- \rightleftharpoons C_6H_5COO^- + C_2H_5OH$ (85% aqueous ethanol)	2.54

A perusal of the values of p reveals that it gets changed on changing the solvent from water to ethanol.

The change in rate of a particular reaction in a range of differing solvents can be correlated with change in dielectric constant. However, this correlation is not proved very useful Grunwald and Winstein tried an attempt to establish reactivity/solvent correlations employing linear free energy equation of the type.

$$\log\left(\frac{k}{k_0}\right) = my \qquad ...(11)$$

where k = rate constant for the solvolysis of a compound in any solvent

k_0 = rate constant for the solvolysis of the same compound in the standard solvent (80%ethanol)

y = measure of the ionising power

m = sensitivity of the substrate to changes in the medium.

m = 1.00 for standard halide

The application of this equation needs that *y* values can be determined with respect to standard. Grunwald and Winstein selected S_N1 solvolysis of t-butylchloride and 80% aqueous ethanol as standard reaction and standard solvent, respectively.

$$R_3C-Cl \xrightarrow[\text{Slow}]{S_N1} \underset{\text{ion-pair intermediate}}{[R_3C^{\delta+}-Cl^{\delta-}]} \xrightarrow[(S = \text{solvent})]{S} R_3C-S$$

On setting up a Hammett-like relation, we get

$$\log k_A - \log k_0 = y_A - y_0$$

where, k_A = rate constant for the solvolysis of tertiary halide in a solvent A

k_0 = rate constant in the standard solvent (80% ethanol)

y_A = empirical solvent parameter in solvent A.

y_0 = empirical solvent parameter in standard solvent.

The y_A value can be found out by setting the value of y_0 zero when k_a in the differing range of solvents are known. Typical values of *ya* are given in Table 4.6.

Table 4.6 : Typical values of y_A

Solvent A	y_A
Water	+ 3.56
50% Ethanol	+ 1.604
80% Ethanol	+ 0.0000
69.5% Methanol	+ 1.023
Formic acid	+ 2.08
50.6% Dioxane	+ 1.292

The use of *m* can be done for diagnostic purpose because it provides some measure of the extent of ion-pair formation in the transition state for the rate-determining step of the solvolysis reaction. Higher values of *m* have been indicative of well advanced ion pair formation in the transition state for S_N1 solvolysis of halide. A lower value of *m* has been characteristic of the S_N2 pathway of solvolysis of halide.

Grunwald-Winstein treatment is having limited applications and is its major defect.

SOLVED EXAMPLES

Example 1:

$S \rightarrow G + F$

	Time	t	∞
Rotation of Glucose & Fructose		r_t	R_∞

Find k.

Solution:

Let the rotation of Glucose be r_2^o and that of fructose be $-r_3^o$ (since it is leavo-rotatory) per mole.

∴	S →	G +	F
At t = 0	a	0	
At t = t	a − x	x	x
At t = ∞	0	a	a

$$\therefore \quad r_1 = x\left(r_2^o - r_3^o\right)$$

$$r_\infty = a\left(r_2^o - r_3^o\right)$$

$$a = \frac{r_\infty}{r_2^o - r_3^o},\ x = \frac{r_t}{r_2^o - r_3^o}$$

$$\therefore \qquad a - x = \frac{r_\infty - r_t}{r_2^o - r_3^o}.$$

Example 2:

Now we shall see how to find the rate constant of a reaction using a very different set of data. There are some organic compounds which have a property of rotating a plane polarized light in a particular direction by a particular value. The compounds are called optically active compounds. One reaction in which an optically active substance converts to some other optically active substance is,

$$\text{Sucrose} \xrightarrow{H^+} \text{Glucose + Fructose}$$

Sucrose, Glucose and Fructose are all optically active and while the first two compounds are dextro rotatory (rotating the plane polarised light in the right hand direction and the last is laevo rotatory (rotating the plane polarized light in the left hand direction). All the three compounds rotate the plane polarised light by different angles and their rotation is directly proportional to concentration.

Now the problem is $S \rightarrow G + F$ and the data is

Time	*0*	*t*
Rotation of sucrose	r_0	r_t

Find k.

Solution:

Let the rotation of Sucrose be r_1^o per mole and the initial moles of Sucrose be a.

$$\therefore \qquad r_0 = ar_1^o$$

Let the moles of Sucrose that is converted to Glucose and Fructose be x.

$$\therefore \qquad r_t = (a - x)\ r_1^o$$

$$\frac{a}{a - x} = \frac{r_0}{r_t}$$

$$\therefore k = \frac{1}{t}\ln\frac{a}{a-x} = \frac{1}{t}\ln\frac{r_0}{r_t}$$

Example 3:

Let $k_1:k_2 = 1:10$. Calculate the ratio, $\frac{[C]}{[A]_t}$ at the end of one hour assuming that $k_1 = x\ hr^{-1}$

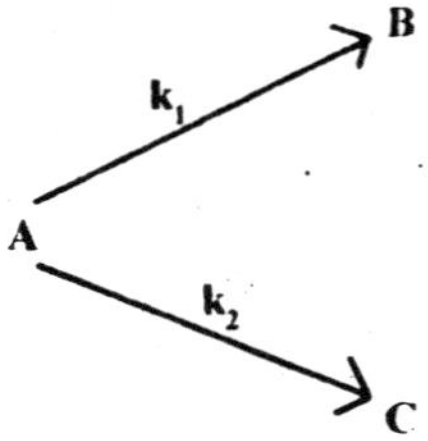

A $\xrightarrow{k_1}$ B

A $\xrightarrow{k_2}$ C

Fig. 4.20

Solution:

$$\frac{-d[A]}{dt} = (k_1 + k_2)[A]$$

$$\therefore \frac{-d[A]}{[A]} = (k_1 + k_2)\ dt$$

Integrating with in the required limits, we get

$$\ln\frac{[A]_0}{[A]_t} = (k_1 + k_2)\ t$$

$$\therefore \ln\frac{[A]_t + [B] + [C]}{[A]_t} = (k_1 + k_2)\ t$$

$$\text{Since } \frac{[B]}{[C]} = \frac{k_1}{k_2} = \frac{1}{10}$$

$$\therefore \ln\frac{[A]_t + \frac{[C]}{10} + [C]}{[A]_t} = 11x$$

$$\frac{[C]}{[A]_t} = \frac{11}{11}\left(e^{11x} - 1\right).$$

Example 4:

5 ml of ethylacetate was added to a flask containing 100 ml of 0.1 N HCl placed in a thermostat maintained at 30°C. 5 ml of the reaction mixture was withdrawn at different intervals of time and after chilling, titrated against a standard alkali. The following data were obtained :

Time (minutes)	*0*	*75*	*119*	*183*	*∞*
Volume of alkali used in ml	*9.62*	*12.10*	*13.10*	*14.75*	*21.05*

Show that hydrolysis of ethyl acetate is a first order reaction.

Solution:

The hydrolysis of ethyl acetate will be a first order reaction if the above data confirm to the equation.

$$k_1 = \frac{2.303}{t}\log\frac{V_\infty - V_0}{V_\infty - V_t}$$

Where V_0, V_t and V_∞ represent the volumes of alkali used in the commencement of the reaction, after time t and at the end of the reaction respectively. Hence

$V_1 - V_0 = 21.05 - 9.62 = 11.43$

Time	$V_\infty - V_t$	k_1
75 min	$21.05 - 12.10 = 8.95$	

$$\frac{2.303}{75}\log\frac{11.43}{8.95} = 0.003259 \text{ min}^{-1}$$

119 min $21.05 - 13.10 = 7.95$

$$\frac{2.303}{119}\log\frac{11.43}{7.95} = 0.003051 \text{ min}^{-1}$$

183 min $21.05 - 14.75 = 6.30$

$$\frac{2.303}{183}\log\frac{11.43}{6.30} = 0.003254 \text{ min}^{-1}$$

A constant value of k shows that hydrolysis of ethyl acetate is a first order reaction.

Example 5:

The reaction given below, involving the gases is observed to be first order with rate constant 7.48×10^{-3} sec^{-1}. Calculate the time required for the total pressure in a system containing A at an initial pressure of 0.1 atm to rise to 0.145 atm and also find the total pressure after 100 sec.

$$2A(g) \rightarrow 4B(g) + C(g)$$

Solution:

$$2A(g) \rightarrow \quad 4B(g) + \quad C(g)$$

initial	P_o	0	0
at time t	$P_o - P'$	2P'	P'/2

$$P_{total} = P_o + P' + 2P' + P'/2$$

$$= P_o + \frac{2P'}{2}$$

$$P' = \frac{2}{3}\ (0.145 - 0.1)\ = 0.03 \text{ atm.}$$

Example 6:

^{227}Ac has a half-life of 22.0 years with respect to radioactive decay. The decay follows two parallel paths, one leading to ^{222}Th and the other to ^{223}Fr. The percentage yields of these two daughter nuclides are 2.0 and 98.0 respectively. What are the decay constants (λ) for each of the separate paths?

Solution:

The rate constant of the decay is

$$k = \frac{0.693}{t_{1/2}} = \frac{0.693}{22}$$

If k_1 and k_2 are the rate constants of the reactions leading to ^{222}Th and ^{223}Fr, respectively we have

$$k_1 + k_2 = \frac{0.693}{22}$$

$$\frac{k_1}{k_2} = \frac{2}{98}$$

On solving for k_1 and k_2, we get

$k_2 = 0.03087\ y^{-1}$

$k_1 = 0.00063\ y^{-1}$.

Example 7:

To 50.00 ml of a solution containing an unknown concentration of Zinc ion was added 0.100 μCi of $^{62}Zn^{2+}$ in 10 ml solution and the total volume was diluted to 100 ml with water. Precipitation of Zinc salt yielded 0.2 g Zinc in the solid phase with an activity of 0.0823 μCi. What was the original concentration of the Zinc ion.

Solution:

$$\% \text{ Zinc recovered} = \%^{62}\text{ Zn recovered}$$

$$= \frac{0.0823}{0.1} \times 100 = 82.3$$

$$\text{Total zinc} = \frac{0.2}{0.823} = 0.243 \text{ g}$$

The mass of the added $^{62}Zn^{+2}$ is negligible, hence 0.243 g is the mass of the Zn^{+2} in the original sample $\frac{0.243 \times 1000}{(65.37)(50)} = 0.0744$ M Zn^{+2}.

Example 8:

A mixture of ^{239}Pu and ^{240}Pu has a specific activity of 6.0×10^9 dis/s. The half lives of isotopes are 2.44×10^4 and 6.58×10^3 years, respectively. Calculate the isotopic composition of this sample.

Solution:

Total activity of a sample is the sum of the individual activities of all its components.

Let the total mass of the sample be 1 gm and the mass of ^{239}Pu be x gm

$$\therefore \frac{x}{239} \times 6.023 \times 10^{23} \times \frac{0.693}{2.44 \times 10^4} \times 6.023 \times 10^{23} \times \frac{0.693}{6.58 \times 10^3}$$

$$= 6 \times 10^9 \times 365 \times 24 \times 60 \times 60$$

On calculating, x = 0.3896

$\therefore$ ^{239}Pu = 38.96% and ^{240}Pu = 61.04%

Example 9:

On analysis a sample of Uranium was found to contain 0.277 g of ${}_{52}Pb^{256}$ and 1.667 g of ${}_{92}U^{238}$. The half the period of ${}_{92}U^{238}$ is 4.51×10^9 years. If all the lead were assumed to have come from decay of ${}_{92}U^{235}$, what is the age of the earth?

Solution:

$${}_{92}U^{238} = 1.667 \text{ g} = \frac{1.667}{238} \text{ mole}$$

$$_{32}Pb^{206} = 0.227\text{ g} = \frac{0.227}{206}\text{ mole}$$

$\because$ All the lead have come from decay of U.

$\therefore$ Moles of Pb formed $= \dfrac{0.277}{206}$

$\therefore$ moles of U decayed $= \dfrac{0.277}{206}$

$\therefore$ Total moles of Uranium present initially

$$= \frac{1.667}{238} + \frac{0.277}{206}\text{, ie } N_0$$

Also N for $U^{238} = \dfrac{1.667}{238}$

$\because$ for U^{238} $t = \dfrac{2.803}{K} \log \dfrac{N_0}{N}$

$$= \frac{2.303 \times 4.51 \times 10^9}{0.693} \log \frac{\frac{1.667}{238} + \frac{0.277}{206}}{\frac{1.667}{238}}$$

$t = 1.143 \times 10^9$ yrs.

Example 10:

The nucleidic ratio of ^3_1H to ^1_1H in a sample of water is 8.0 × 10^{18}: 1. Tritium undergoes decay with half life period of 12.3 years. How many tritium atoms would 10.0 g of such a sample contains 40 years after the original sample in collected.

Solution:

18 g H_2O has H atom in it $= 6.023 \times 10^{23} \times 2$

18 g H_2O has ^3_1H atoms $= 8 \times 10^{-18} \times 6.023 \times 10^{23}$

$\therefore$ 10 g H_2O has ^3_1H atoms

$$= \frac{8 \times 10^{-18} \times 6.023 \times 10^{23} \times 2 \times 10}{18}$$

i.e., No. of $^3_1H = 5.35 \times 10^6$ atoms

$$\text{Now } t = \frac{2.303}{K} \log \frac{N_o}{N}$$

$$40 = \frac{2.303 \times 12.3}{0.693} \log \frac{5.354 \times 10^6}{N} = 5.624 \times 10^5 \text{ atoms.}$$

Example 11:

At 278°C the half life period for the first order thermal decomposition of ethylene oxide is 363 min and the energy of activation of the reaction in 52,00 cal/mole. From these data estimate the time required for ethylene oxide to be 75% decomposed at 450 °C.

Solution:

$$\ln \frac{k_{450}}{k_{278}} = \frac{5200}{2}\left[\frac{1}{551} - \frac{1}{725}\right] = 1.122$$

$$\frac{k_{450}}{k_{278}} = 3.07 = \frac{363}{t_{1/2}\left(\text{at } 450^\circ C\right)} \quad t_{1/2} \text{ (at 450°C)} = 118.24 \text{ min.}$$

$$\text{Now } t_{0.75} = \frac{1}{k} \ln \frac{A_o}{A_o/4} = \frac{1}{k} \ln 4 = \frac{1.386}{k}$$

$$\therefore\ t_{0.75} = \frac{1.386}{0.693} \times 118.24 = 236.48 \text{ min}.$$

5

PHOTOCHEMISTRY

INTRODUCTION

Photochemistry is the branch of chemistry which is mainly concerned with rates and mechanisms of reactions resulting from the exposure of reactants to light radiations.

This board definition of photochemistry would include reactions produced by all radiations of wave lengths ranging from those of radio waves to chose γ rays. But for practical purposes, the light radiations of the visible and ultraviolet regions lying between 2000 to 8000 Å are mainly concerned in bringing about such reactions which are termed as *Photochemical Reactions*. Therefore, the definition of photochemistry may be summarised as follows.

"It is the study of chemical effects produced by light radiations ranging from 2000 to 8000 Å wave-length"

The study of photochemistry helps us to know the changes when a molecule absorbs radiations.

TYPES OF CHEMICAL REACTIONS

A chemical reactions is one in which, *the identity of molecules is changed due to the rupture and formation of chemical bonds.* Chemical reactions are of two types.

Dark of Thermal Reactions : *These are the ordinary chemical reactions which are influenced or induced by temperature, concentration of reactants, presence of a catalyst, etc., except light radiations.*

(i) $N_2 + 3H_2 \rightleftharpoons 2NH_3$

(ii) $H_2 + I_2 \rightleftharpoons 2HI$

(iii) $PCl_5 \rightleftharpoons PCl_3 + Cl_2$

(b) Photochemical Reactions : ***A photochemical reaction may be defined as any reaction which is induced or influenced by the action of light on the system.***

Some of the photochemical reactions are accompanied by increase in free energy unlike dark reactions. Examples of some photochemical reactions are.

(a) *Dissociation.*

$$2HBr \rightarrow H_2 + Br_2$$

(b) *Rearrangement.* Fumaric acid → Maleic acid

(c) *Addition Reaction.*

$$Br_2 + (C_6H_5)_2C = C(C_6H_5) \rightarrow (C_6H_5)_2\,CBr - CBr(C_6H_5)_2$$

(d) *Polymerisation.*

$$nCH_2 \rightarrow (C_2H_5)Nx$$

(e) Photo-catalytic *Reaction*

$$CO_2 + H_2O + \text{chlorophyll} \xrightarrow{\text{light}} \frac{1}{n}(H_2CO)_n + O_2 + \text{Chlorophyll}$$

(f) *Combination.*

$$H_2 + Cl_2 \rightarrow 2HCl$$

(g) *Decomposition.*

$$2O_3 \rightleftharpoons 3O_2$$

(h) *Double decomposition.*

$$C_6H_{12} + Br_2 \rightarrow C_6H_{11}Br + HBr$$

DIFFERENCES BETWEEN DARK AND PHOTOCHEMICAL REACTIONS

Photochemical reactions differ from ordinary dark or thermal reactions in certain aspects which are as follows.

(i) In ordinary thermal reactions the energy of activation is provided by the collisions. On the other hand in photochemical reactions the energy required is gained through the absorption of quanta of visible or ultraviolet light.

(ii) All ordinary chemical reactions are always accompanied by a decrease in free energy. On the other hand, the free energy of the light or photo-chemical reactions increases as some of the light energy is convered into free chemical energy of the products. Some examples of such photochemical reactions are ozonisation of oxygen, polymerisation of anthracene, and photosynthesis occurring in plants.

In these photochemical reactions, the energy of absorbed radiation gets transformed into free chemical energy of the products. However, when the source of radiations is removed, the system tends to return to its original state, though at a very slow rate.

A feature of photochemical activation is its selectivity. The absorbed quanta of light excite and thus activate a separate atom or group of atoms in a given molecule. This is a great advantage of activating molecules with light in comparison with thermal activation.

(iii) The difference between photo excitation and thermal excitation is that while in the former a few molecules succeeding in absorbing photons are strongly excited, in the case of latter a significant rise in temperature increases the average energy of all the molecules by a small amount. Thus, light absorption can bring about reactions at room temperature which might otherwise require a raising of temperature by several tens or hundreds of degrees centigrade.

While the rate of thermal reactions depends upon the temperature, the photochemical reaction rate is independent of temperature. But the reaction rate does depend upon the intensity of the radiation used. In certain photochemical reactions, the rate is seen to vary with the temperature, but this is due to the temperature dependence of thermal reactions which follow the light absorption step.

ABSORPTION OF LIGHT

Introduction : *When light* (monochromatic or heterogeneous) *is incident upon a homogeneous medium, a part of the incident light is reflected, a part is absorbed by the medium and the remainder is allowed to transmit as such.* If I_0 denotes the incident light, I_r the reflected light, I_a the absorbed light and I_t, the transmitted light, then one can write

$$I_0 = I_a + I_t + I_r \quad ...(1)$$

It a comparison cell is used, the value of I_r which is very small (about 4 per cent), can be *eliminated for air-glass interfaces.* Under the condition, equation (1) becomes as

$$I_0 = I_a + I_t \quad ...(2)$$

Bouguer actually investigated the range of absorption of light with the thickness of medium. But the credit was enjoyed by *Lambert* who simply extended the concepts developed by *Bouguer*. Beer later applied Lambert's concept to solution of different concentrations and reported his results just prior to those of *Bernard*. However, the two separate laws governing, absorption are generally known as *Lambert's* Law and *Beer's Law*. We will now discuss these one by one.

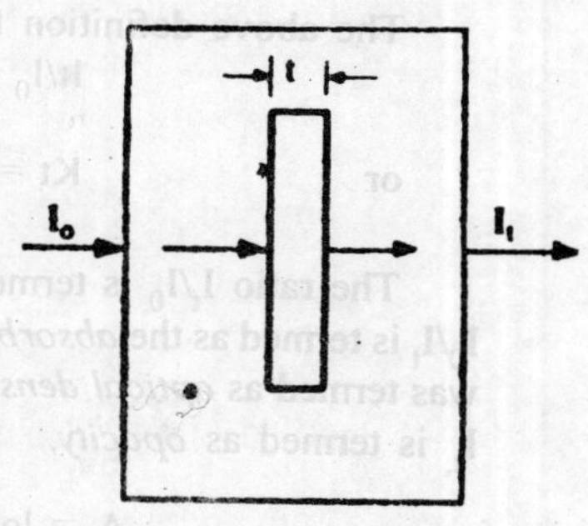

Fig. 5.1

Lambert's Law : This law can be stated as follows. *"When a beam of light is allowed to pass through a transparent medium, the rate of decrease of intensity with the thickness of medium is directly proportional to the intensity of the light."*

Mathematically the Lambert's law may be stated as follows.

$$-\frac{dI}{dt} \propto I \text{ or } -\frac{dI}{dt} = kI \quad ...(3)$$

where I denotes the intensity of incident light of wavelength λ, *t* denotes the thickness of the medium and *k* denotes the proportionality factor. On integrating equation (3) and putting $I = I_0$, when t = 0, we get

$$\ln \frac{I_0}{I_t} = kt \text{ or } I_t = I_0 e^{-kt} \quad ...(4)$$

where I_0 denotes the intensity of the incident light, I_t denotes the intensity of the transmitted light and k is a constant which depends upon the wavelength and absorbing medium used. On changing equation (4) from natural to common logarithms, we get

$$\ln \frac{I_0}{I_t} = kt \text{ or } I_t = I_0 e^{-Kt} \quad ...(5)$$

where $$K = k/2.3026 \qquad ...(6)$$

In equation (6) K is the *absorption coefficient* which is defined as.

"It is the reciprocal of the thickness which is required to reduce the light to 1/10 of its intensity."

The above definition follows from equation (5),

$$It/I_0 = 0.1 = 10^{-Kt}$$

or $$Kt = 1 \quad \text{or} \quad K = K \propto \frac{1}{t}$$

The ratio I_t/I_0 is termed as the *transmittance*, T and the ratio log I_0/I_t is termed as the *absorbance* A, of the medium. Formerly absorbance was termed as *optical density* D or *extinction coefficient* ≤. The ratio I_0/I_t, is termed as *opacity*,

$$A = \log \frac{I_0}{I_t} \qquad ...(7)$$

Lambert's law is very rigid and like Faraday's laws of electrolysis, it has no exception.

Beer's Law : Lambert's law shows that there exists a logarithmic relationship between the transmittance and the length of the optical path through the sample. Beer observed that a similar relationship holds between transmittance and the concentration of a solution, *i.e., the intensity of a beam of monochromatic light decreases exponentially with the increase in concentration of the absorbing substance arithmetically.*

Thus, equation (4) becomes as

$$I_0 = I_0\, e^{-k'0} \qquad ...(8)$$

$$= I_0 .\, 10^{-0.4348k'c} = I_0\, 10^{-K'2} \qquad ...(9)$$

where k′ and K′ are constants and *c* is the concentration of the absorbing substance. On combining equations (5) and (9), we get

$$I_t = I_0 .10^{-act}$$

or $$\log (I_0/I_t) = act \qquad ..(10)$$

where a is the new constant.

Equation (10) is termed as mathematical statement of *Beer-Lambert law*. This is also the fundamental equation of colorimetry and spectrophotometry..

In equation (10), the value of a depends upon the units of concentration. If c is expressed in mole dm^{-3} and t in centimeters, then a is replaced by the symbol ε and is termed as the *molar absorption coefficient* or *molar absorptivity* (formerly the molar extinction coefficient).

It is important to remark here that there exists a relationship between the absorbance A, the transmittance T and the molar absorption coefficient ε, *i.e.*,

$$A = \varepsilon\, ct = \log\frac{I_0}{I_t} = \log\frac{1}{T} = -\log T \qquad ...(11)$$

In spectrophotometers, the scales are calibrated to read directly absorbances. In colorimeters, I_0 is considered to be the light transmitted by the pure solvent whereas I_t is considered to be the light transmitted by the solution.

Equation (11) may be put as

$$\varepsilon = A/ct \qquad ...(12)$$

If c = 1 mole dm^{-3} and T = 1 cm, equation (12) becomes as

$$\varepsilon = A \qquad ...(13)$$

From equation (13) it follows that the molar absorption coefficient is the specific absorption coefficient for a concentration of 1 mole dm^{-3} and a path length of 1 cm.

When a system contains several absorbing substances, each of these contributes to the rate of absorption of light.

$$A = \varepsilon_1 c_1 t + \varepsilon_2 c_2 t + ... = \sum_i \varepsilon_t\ c_i\, t$$

Nature of Molar Absorptivity and Absorbance : Absorbance is an *extensive property* of a substance whereas absorptivity is its *intensive property*. If there is a change in concentration and in the thickness of the container, the value of molar absorptivity will remain constant within the Beer's law range but the absorbance will change significantly.

The value of molar absorptivity will be different at different wavelengths.

This is shown in Fig. 5.2. From this figure, it is alto evident why a monochromatic radiation is used in spectrophotometry and colorimetry. Similarly absorbance also varies with wavelength.

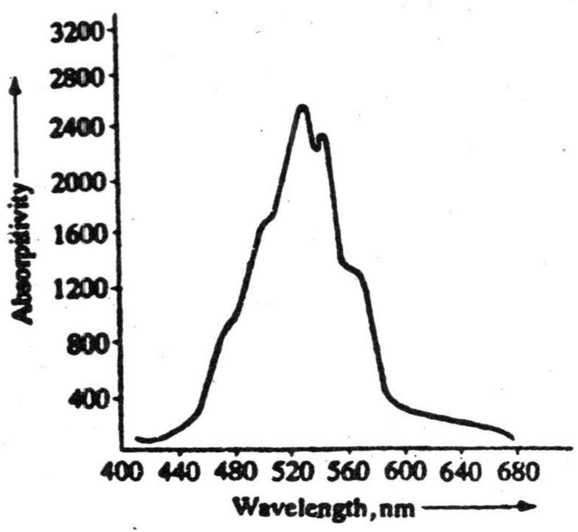

Fig. 5.2 : Molar absorption at different wave lengths.

VALIDITY OF BEER'S LAW

When Beer's law is obeyed by a solution, it means that.

(a) There are no interactions between molecules of the solution.

(b) No new types of molecules will be formed as the concentration is changed.

Deviation from Beer's Law : From Beer's law it follows that if we plot absorbance A against concentration, a straight line passing through the origin should be obtained Fig. 5.3. But there is usually a deviation from a linear relationship between concentration and absorbance and an apparent failure of Beer's law may ensue. Deviations from the law are reported as *positive* or *negative* according to whether the resultant curve is concave upwards or concave downwards.

Deviation from Beer's law can arise due to following factors.

(i) Beer's law will hold over a wide range of concentration provided the structure of the coloured ion or of the coloured non-electrolyte in the dissolved state does not change with concentration. If a coloured solution is having a foreign substance whose ions do not react chemically with the coloured components, its small concentration (foreign substance) does not affect the light absorption whereas its large concentration may affect light absorption and may also alter the value of the extinction coefficient.

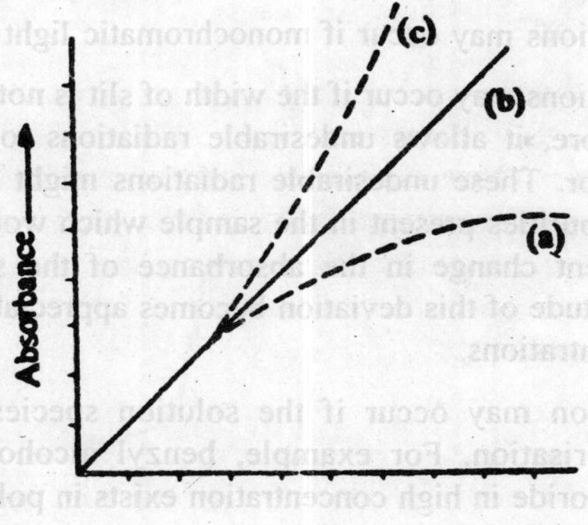

Fig. 5.3

(ii) Deviation may also occur if the coloured solute ionises, dissociates or associates in solution.

For example, benzyl alcohol, in chloroform exists in a polymeric equilibrium.

$$4C_6H_5CH_2OH \rightleftharpoons (C_6H_5CH_2OH)_4$$

Dissociation of the polymer increases with dilution. The monomer absorbs at 2.0 to 2.765 μ whereas the polymer absorbs at 3.000 μ. Hence, absorption at 2.050 μ shows negative deviation whereas at 3.000 μ *positive deviation.*

Another example of deviation is the change of colour of dichromate ion on dilution. The effect can be represented by the equilibrium shown below.

$$\underset{\text{(Orange)}}{Cr_2O_7^{2-}} + H_2O \rightleftharpoons 2HCrO_4^- \rightleftharpoons 2H^+ + \underset{\text{(Yellow)}}{CrO_4^{2-}}$$

It can be seen from above that a solution of dichromate is converted gradually into chromate. The absorption at 450 μ, is positive and at 350 μ is negative.

(iii) Deviations may also occur due to the presence of impurities that fluoresce or absorb at the absorption wavelength. This interference introduces an error in the measurement of absorption of radiation penetrating the sample.

(iv) Deviations may occur if monochromatic light is not used.

(v) Deviations may occur if the width of slit is not proper and, therefore, it allows undesirable radiations to fall on the detector. These undesirable radiations might be absorbed by impurities present in the sample which would cause an apparent change in the absorbance of the sample. The magnitude of this deviation becomes appreciable at higher concentrations.

(vi) Deviation may occur if the solution species undergoes polymerisation. For example, benzyl alcohol in carbon tetrachloride in high concentration exists in polymeric form $[(C_6H_5CH_2OH)_4]$, association of this polymer increases with dilution. Due to this change, absorbance will change.

(vii) Beer's law cannot be applied to suspensions but the latter can be estimated colorimetrically after preparing a reference curve of known concentrations.

Illustration : From the Beer's law it follows that if the absorption remains constant, then the concentration (c) should be related to the thickness of the absorbing layer (t) in a reciprocal manner. This fact was demonstrated by *Halbon* (1922). He observed that a layer of chlorine 20 cm long at a pressure of 0.1 atmosphere would have the same transmission at 10 cm layer at a pressure of 0.2 atmosphere

Applications of Beer's Law : Beer-Lambert's law can be used for the determination of an unknown concentration by comparison with a solution of known concentration by using colorimeter or spectrophotometer, the principle of which is based upon Beer's law.

LAWS OF PHOTOCHEMISTRY

There are three laws which govern the effect of radiation on chemical reactions. These laws are.

I. Grotthuss–Draper Law : This is sometimes referred to as the first law of photochemistry. On the basis of certain theoretical considerations, Grotthuss in 1818 discovered this law. This law was reaffirmed in 1841 by J.W. Draper as a result of his experimental researches on photochemical reaction between hydrogen and chlorine. It may be stated as follows.

"When light falls on any substance, only the fraction of incident light which is absorbed by the substance can bring about a chemical change', reflected and transmitted light do not produce any such effect."

It is important to remark that all light radiations which are absorbed by reacting system are not effective in producing desired chemical reactions. When conditions are not favourable for the molecules to react, some portion or the whole light absorbed by reacting substances is converted into heat in some cases. While in some other cases, the absorbed light is re-emitted as radiations of the same or another frequency.

Grotthuss-Draper law is purely qualitative. It does not give any relationship between the amount of light absorbed by a system and the number of molecules which have reacted,.

II. Law of Photochemical Equivalence : It is one of the most important laws in photochemistry. According to quantum theory of light, energy is absorbed or emitted only in small packets of quanta of magnitude $\varepsilon = h\nu$ where ε is the energy, ν is the frequency W light and h is the Planck's constant. In 1905, Einstein applied quantum theory to photochemical reactions and enunciated the law of Photochemical Equivalence which states that

"When an atom or molecule absorbs light of a given frequency, it absorbs one quantum only,"

The photochemical importance of the above law emphasized in 1909 by Stark and other. In 1913, Einstein considered the work of Stark and restated his law as.

"Each molecule which takes part in a chemical reaction absorbs one quantum of light which induces the reaction " or briefly "one molecule one quantum."

The term one photon (or one quantum) means energy equal to $h\nu$, where h is the Planck's constant and ν is the frequency of radiation

$$AB + h\nu \rightarrow AB^*$$

Thus, in the primary process, the number of molecules that are activated is equal to the number of quantas absorbed. Therefore, amount of energy E, for activation of 1 mole will be N hv where N is the Avogadro's number and is equal to 1 mole, *i.e.*,

$$E = Nh\nu$$

This quantity of energy 'E" *absorbed per mole of the substance is called an Einstein.*

But $N = 6.023 \times 10^{28}$ molecules

$h = 6.624 \times 10^{-27}$ erg sec

$\therefore$ $E = 6.023 \times 10^{28} \times 6.624 \times 10^{-27} \times \nu$ ergs

$= \underline{6.1023 \times 6.624 \times 10^{-27} \times \nu}$ calories ...(1)

$E = 9.53 \times 11-11 \times \nu$ calories

[$\because$ 1 cal = 4.186×10^{7} ergs]

But $$\nu = \frac{c}{\lambda} = \frac{\text{velocity of light in cm/sec}}{\text{wave length in cm}} = \frac{3 \times 10^{10}}{\lambda}$$

Substituting the value of v in equation (1), we obtain

$$E = \frac{9.53 \times 10^{-11} \times 3 \times 10^{10}}{\lambda} \text{ calories}$$

where λ is expressed in cm. But 1 A° = 10^{-8} cm. Therefore,

$$E = \frac{9.53 \times 10^{-11} \times 3 \times 10^{10}}{\lambda \times 10^{-8}} = \frac{2.859}{\lambda} \times 10^{8} \text{ cal}$$

$$= \frac{2.859}{\lambda} \times \frac{10^{8}}{10^{3}} \text{ K cal} \quad [\because 1 \text{ K cal} = 1000 \text{ cal}]$$

$$= \frac{2.859}{\lambda} \times 10^{5} \text{ K cal}$$

Thus, the value of one Einstein of radiation of given frequency or wavelength is given by the relation (2). It is evident from equation (2) that the energy absorbed per mole decreases with increasing wavelength. In other words, the value of E is inversely proportional to the wavelength of light absorbed, *i.e.*,

$$E \propto \frac{1}{\lambda}$$

From the above facts it is evident that shorter the wavelength of light, the greater the energy. It has been verified that.

1. For the wavelength 4000Å (ultraviolet light), the energy is 71 K cal;
2. For the wavelength 7500Å (red light), the corresponding energy is 38 K, cal.

From the above results, it follows that in ultraviolet and violet portions, the radiations will be more active chemically than those of longer wavelengths.

QUANTUM YIELD OR QUANTUM EFFICIENCY

The photochemical equivalence law applies to the primary photochemical process, *i.e.,* as a result of primary absorption of one quantum, only one molecule undergoes dissociation and the products enter no further reaction. In such cases, there will be 1 : 1 relationship between the number of quantas absorbed and the number of reacting molecules.

In most cases, a molecule activated photochemically initiates a series of thermal reactions called secondary reactions. As a result of such reactions many reactant molecules may undergo chemical change by absorbing one quantum only. Under such conditions, there will be no. 1 : 1 relationship between the number of quanta absorbed and the number of reacting molecules.

In some cases, a molecule activated photochemically undergoes deactivation. Thus, less than one molecule may react per quantum

To describe the deviation from 1 : 1 relationship, the idea of quantum yield or efficiency of a process 'ϕ' was introduced. This can be defined as follows :

"It is the number of molecules which undergo chemical transformation per quantum of absorbed energy."

Mathematically, this quantity is defined as follows.

or $$\phi = \frac{\text{No. of molecules reacting in a given time}}{\text{No. of quanta absorbed in the same time}}$$

or $$\phi = \frac{\text{No. of molecules reacting in a given time}}{\text{No. of einsteins absorbed in the same time}}$$

or $$\phi = \frac{\text{Rate of chemical reaction}}{\text{No. of einsteins absorbed}} \quad ...(1)$$

The number of quanta absorbed (n_a) in a unit time i

$$n_a = \frac{Q}{hv} \qquad ...(2)$$

and the number of molecules transformed under the action of light (n_r) is

$$n_r = \frac{Q}{hv} \qquad ...(3)$$

On substituting equation (2) in (1), we get

$$\phi = \frac{n_r}{n_a} = \frac{n_r}{Q/hv} \qquad ...(4)$$

The rate of a chemical reaction; according to equation (4), is as follows.

$$-\frac{dn}{dt} = \frac{dn_r}{dt} = \phi\frac{dn_a}{dt} = \phi\frac{Q}{hv}$$

The concept of quantum yield or quantum efficiency was first introduced by Einstein. Because of the frequent complexity of photochemical reactions, quantum yields as observed vary from a million to a very small fraction of unity.

The concept of quantum yield can be extended to any act, physical or chemical, following light absorption. The quantum yields are dependent on the light intensity.

The concept of quantum yield provides a mode of account keeping for partition of absorbed quanta into various pathways.

If Stark-Einstein law is correct, then value of ϕ should always be unity. However, later on, it was realised that the law of photochemical equivalence is applicable to only primary processes.

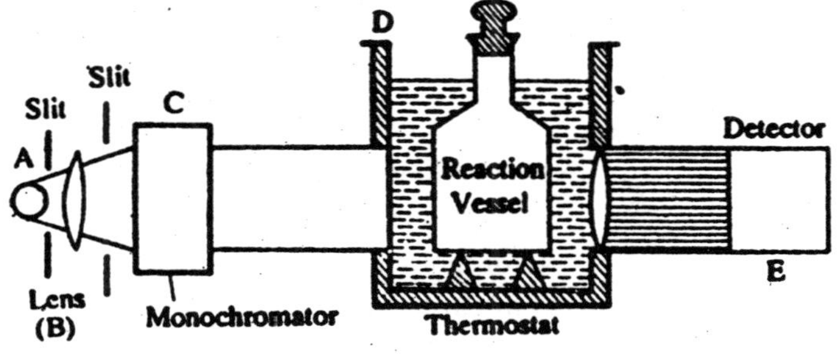

Fig. 5.4

Experimental Determination of Quantum Yields : Knowledge of quantum efficiency of a photochemical reaction offers valuable information about the mechanism of photochemical reactions. In order to determine the value of quantum yield of photochemical reaction, it is essential to know (a) Number of moles reacting, and (b) Number of Einsteins absorbed. For this purpose, an arrangement such as shown in Fig. 5.5 is required.

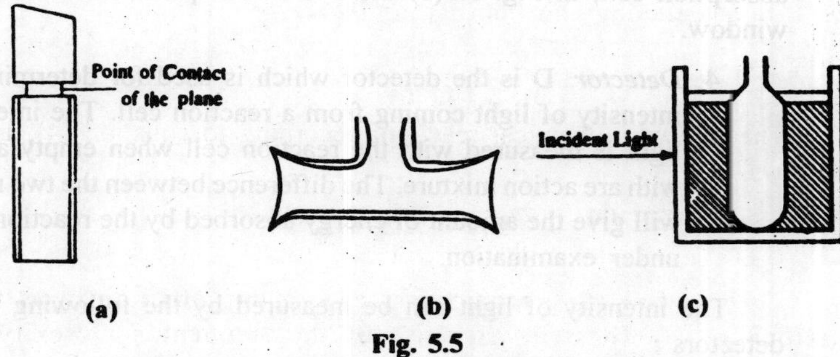

Fig. 5.5

1. *Light source.* 'A' is the source of light which may be sunlight, arc lamp, mercury vapour lamp, discharge tube, etc. For a quantitative type of work, mercury vapour lamp is the best source of ultra-violet radiations because it emits radiation of suitable intensity in the desired spectral range., For visible or ultraviolet regions, the tungsten lamp is often used in place of mercury vapour lamp.

2. *Monochromator*. C is a monochromator or a filter. When light from source, A is passed through the lens B and then allowed to pass through, C_0 the monochromator (C) will absorbed that these colour filters transmit light within certain range of wavelengths. Monochromators are generally made of gelatin or coloured glass or transparent plates with metal films of suitable thickness which absorb undesired wavelengths by interference.

3. *Reaction cell.* The light from a monochromator enters a reaction cell D which contains the reaction mixture. The reaction cell is generally made of glass or quartz with optical plane windows for free exit and entrance of light radiation. For visible spectral range, the reaction cell and other parts of the instrument are generally made of glass. For ultraviolet light, all optical parts

are made of quartz glass. For solutions, the cell is provided for stirring.

The actual design of the cell depends upon the nature of the reaction, but the front and the back should be plane parallel. The types of cells generally used are shown in Figs. 5.5(a), 5.5(b) and 5.5(c). These represent the various types of cells which are generally used. In Fig. 5.5(a), there is sealing plane window absorption cell; in Fig. 5.5(b) there is a tubular absorption cell, in Fig. 5.5(c) there is a absorption cell with in blow window.

4. *Detector.* D is the detector which is used for determining the intensity of light coming from a reaction cell. The intensity of light is measured with the reaction cell when empty and then with are action mixture. The difference between the two readings will give the amount of energy absorbed by the reaction system under examination.

The intensity of light can be measured by the following types of detectors :

(a) *Radiomicrometer:* It consists of a coil (thermocouple) which is suspended between the poles of an electromagnet. When the couple gets heated by the light radiations, a current flows in the circuit and it is deflected in the magnetic field. The deflection can be measured by lamp and scale arrangement. This deflection is thus, proportional to the intensity of light radiations.

(b) *Photo-electric cell.* The principle of photo-electric cell is based upon the photo-electric effect. It is a device which converts light energy into electrical energy, A photo-electric cell consists of lithium or sodium metal which is enclosed in a vessel. The metal is generally connected to gold leaves of an electroscope or to an electrometer. When the light radiations are incident upon lithium metal, it loses electrons and develops a positive charge on itself whereby the gold leaves diverge. This divergence of the gold leaves is proportional to the intensity of light radiations. Photo-electric cells are generally used for measuring very low intensities of light shown in Fig. 5.6.

(c) *Chemical actinometers. When* extreme accuracy is not desired, a chemical actinometer can be used to measure the intensity of light radiation. A chemical actinometer generally consists of gas mixtures or "solutions which are sensitive to light. When radiations Fall upon these substances, a chemical reaction will take place and the extent of which is a direct measure of energy absorbed. Following are the main types of actinometers.

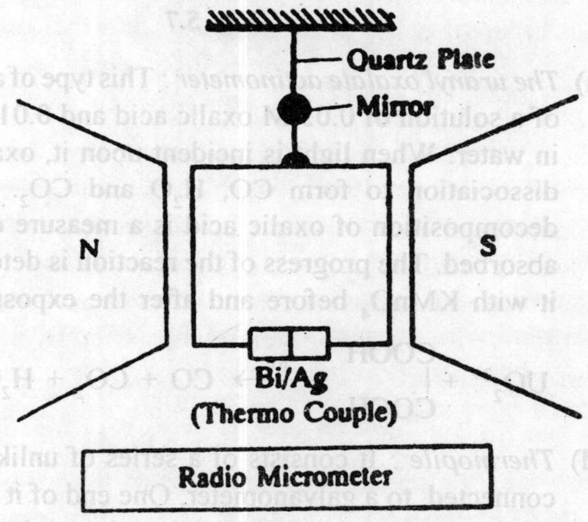

Fig. 5.6

(i) *Bunsen and Roscoe's actinometer.* It is illustrated.

The bulb B′ contains a mixture of hydrogen and chlorine over water. The bulb D′ also contains water. The two bulbs B′ and D′ are connected through a graduated tube C′. When the light is incident on B′, some of the gases in A′ react to form hydrochloric acid.

This is absorbed in water of B' and the water moves from C′ to B′. The motion of the water thread will be proportional to the amount of hydrochloric acid formed and hence to the intensity of light absorbed.

(ii) *Eder's actinometer : J. M. Eder (1979)* utilized the following reaction to determine the absorbed radiations

$$2HgCl_2 + (NH_4)_2C_2O_4 \rightarrow 2NH_4Cl + 2CO_2 + Hg_2Cl_2$$

The system was exposed to radiations and the mercurous chloride formed in the reaction was weighed. From the weight of mercurous chloride or CO_2 evolved, the absorbed radiations may be determined quantitatively.

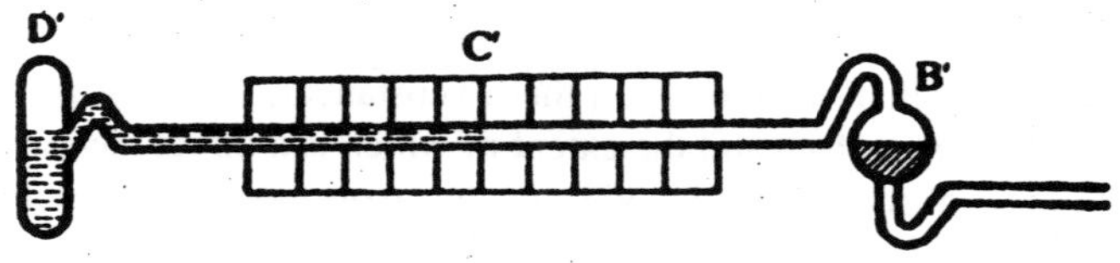

Fig. 5.7

(iii) *The uranyl oxalate actinometer* : This type of actinometer consists of a solution of 0.05 M oxalic acid and 0.01 M uranyl sulphate in water. When light is incident upon it, oxalic acid undergoes dissociation to form CO, H_2O and CO_2. The extent of the decomposition of oxalic acid is a measure of intensity of light absorbed. The progress of the reaction is determined by titrating it with $KMnO_4$ before and after the exposure.

$$UO_2^{2+} + \begin{array}{l} COOH \\ | \\ COCH \end{array} \xrightarrow{hv} CO + CO_2 + H_2O + UO_2^{2+}$$

(d) *Thermopile* : It consists of a series of unlike metals which are connected, to a galvanometer. One end of it is coated with lamp black, or platinum black and the other end "is kept as such, The thermopile is kept in, .glass or quartz vessel.

When the light from a reaction cell is incident upon the black surface, it gets heated up, resulting a difference in temperatures of the black and the other cold end. Thus, a current flows in the circuit. The current produced is proportional to the intensity of light absorbed. Thermopiles are generally calibrated with standard light sources.

(e) *Geiger Muller counter* : It is a modification of the ionisation chamber and is generally used for measuring very low intensities. The apparatus consists of an ionisation chamber with a thin wire of steel or tungsten. When a charged particle enters the chamber the air present in it gets ionised and discharge is initiated in the gas. A current then .flows between the electrodes which can be detected in the usual manner,.

Process For the Determination of Quantum Yield : Set up the apparatus. The choice of the detector depends upon the nature of the experiment. First keep the reaction vessel empty. Allow the light to pass through it. Determine the light intensity as such. In case of solution, it is filled with the solvent.

Now, fill the reaction vessel with reaction mixture. Allow the light to pass through it for a known time. Then note the intensity of light after the reaction is over. The difference between the two readings will give the amount of energy absorbed by the reaction system under examination. Analyse the contents of reaction mixture. Then, determine the no, of moles reacted in given time. Then, apply the following equation to get the value of quantum yield.

$$\text{Quantum yield} = \frac{\text{No. of moles reacting in a given time}}{\text{No. of einsteins absorbed in the same time}}$$

DEVIATIONS IN THE LAW OF PHOTOCHEMICAL EQUIVALENCE

The photochemical equivalence law applies to the primary photochemical process. As a result of primary absorption of one quanta, only one molecule undergoes dissociation and the products enter no further reaction. In such cases, there will be 1 : 1 relationship between the number of quantas absorbed and the number of reaction molecules.

In most cases, a molecule activated photochemically initiates a series of thermal reactions called secondary reactions. As a result of such reactions, many reactant molecules may undergo chemical change by absorbing one quanta only under such conditions.

Class I : High Quantum Yield Reactions : These are the photochemical reactions in which the quantum yield is greater than unity. Examples are :

(i) The quantum yield of the combination of carbon monoxide and chlorine by a light of wavelength ranging from 4000 to 4360 Å is 10^3.

$$CO + Cl_2 \rightarrow COCl_2$$

(ii) Another example is the combination of hydrogen and chlorine to form hydrogen chloride by a wavelength less than 4500°. The quantum yield of this reaction has been found to vary from 10^4 to 10^6, *i.e.*, one quantum brings about the combination of 10^4 to 10^6 molecules.

$$H_2 + Cl_2 \rightarrow 2HCl$$

(iii) Another example is the photochemical decomposition of H_2O_2, by a wavelength of about 3100Å". The quantum yield for this reaction is more than 7.

$$2H_2O_2 \rightarrow 2H_2O + O_2$$

Class II. Low Quantum Yield Reactions : This type includes such photochemical reactions in which the quantum yield is less than unity. Examples are.

(i) Combination of hydrogen and bromine to form HBr is an example of this type. The quantum yield of this reaction is about 0.01.

$$H_2 + Br_2 \rightarrow 2HBr$$

(ii) Photolysis of ammonia by a wavelength 2100Å is another example.

$$2NH_3 \rightarrow N_2 + 3H_2,$$

The quantum yield of this is nearly 0.2.

(iii) The quantum yield for the dissociation of acetone to form carbon monoxide and ethane by a light of wavelength 3000Å is 0.1.

$$CH_3COCH_3 \rightarrow CO + C_2H_6$$

(iv) The quantum yield of the concentration of maleic acid into fumaric acid by a light of wave-length ranging from 2000 to 1800 Å is 0.04.

$$\begin{array}{c} H-C-COOH \\ \| \\ H-C-COOH \end{array} \longrightarrow \begin{array}{c} H-C-COOH \\ \| \\ HOOC-C-H \end{array}$$

Maleic Acid — Fumaric Acid

Class III. Small Integer Quantum Yield Reaction : This type includes the photochemical reaction in which the quantum yield is a small integer like 1, 2, 3, etc. Examples are.

(i) The quantum yield for the dissociation of hydriodic acid by a light of wave length 2070 – 2823Å is 2.

$$2HI \longrightarrow H_2 + I_2$$

(ii) The quantum yield of the following reaction is unity.

$$2Fe^{3+} + I_2 \longrightarrow 2Fe^{2+} + 2I^-$$

This reaction is carried out in liquid phase by a wave-length 5790Å.

(iii) The quantum yield of the combination of SO_2 and chlorine to form SO_2Cl_2 by a light of wavelength 4200 Å is one.

$$SO_2 + Cl_2 \longrightarrow SO_2Cl_2.$$

(iv) The quantum yield of the ozonisation of oxygen by a light of wave-length 1700 – 1900 Å is three.

$$3O_2 \longrightarrow 2O_3$$

REASONS OF HIGH AND LOW QUANTUM YIELD

In order to explain the exceptions to Einstein's Law of photochemical equivalence, *Bodenstein and others* pointed out that photochemical reactions involve two distinct processes I

(a) *Primary process.* It is the process in which each molecule capable of entering into chemical reaction absorbs one quantum of radiation (hv). The absorbed energy may give rise to the formation of excited molecule, *i.e.,*

$$AB + h\nu \rightarrow AB^*$$

or the molecule which absorbs light may undergo dissociation to yield free atoms (some in excited state) or free radicals.

$$AB + h\nu \rightarrow A. + B.$$

(b) *Secondary process.* It is the process which involves the excited atoms or molecules or free radicals produced in the primary process. A secondary process can take place in the dark as well. Due to this process, a different number of molecules may undergo reaction, by absorbing one quanta only. As a result, the quantum efficiency of the reaction, as a whole, will differ from unity.

From the above discussion it follows that the value of quantum efficiency or yield will depend upon secondary processes whereas the primary process remains the same for every reaction. We will now discuss the reasons for different quantum yields for different reactions on the basis of secondary process.

Reasons of High Quantum Yield : These are outlined below.

(a) When the excited atoms or molecules or free radicals produced in the primary state undergo secondary processes, each secondary process gives rise to some further excited particles or radicals; this process continues unless it is checked due to one or the other reason. Thus, by absorbing one quanta only, a large number of reactant molecules undergo reaction. Hence, the quantum efficiency of this type of reaction will be greater than unity.

(b) In some reactions, such as the. Combination of hydrogen and chlorine, a chain is set up and quantum yield becomes as high as 10s. These reactions are known as chain reactions. Some evidences in favour of chain reactions are given below.

(i) The chain is broken when a foreign substance is added in the system to remove H_2 or Cl_2.

(ii) **Marshall and Taylor** found that at least one step in the chain reaction can be carried out experimentally.

Reasons of Low Quantum Yield : The low quantum yield is explained on the basis of following reactions.

(a) If the excited particles formed in the primary process are such that they cannot react due to their deactivation by collisions, by fluorescence or by internal arrangements or by coming in contact with inert molecules the quantum yield will be extremely low.

(b) The excited particles produced in the primary process may recombine to form the reactants so as to give low quantum yield.

To illustrate the above mentioned reasons, a few examples will now be taken.

(i) **Dissociation of HI :** In the primary process, a molecule of hydrogen iodide absorbs light of wavelength less than 4000 Å AND dissociates to form an atom of hydrogen and an atom of iodine. The primary process may be represented as.

$$HI + h\nu \rightarrow H + I$$

The primary process has been established from spectroscopic studies. The possible secondary reactions are..

$$H + HI \rightarrow H_2 + I \quad ...(1)$$

$$I + HI \rightarrow I_2 + H \quad ...(2)$$

$$H + H \rightarrow H_2 \quad ...(3)$$

$$I + I \rightarrow I_2 \quad ...(4)$$

$$I + H \rightarrow HI \quad ...(5)$$

Reaction (2) is endothermic and, thus, it cannot take place at ordinary temperatures. Reactions (3) and (5) are highly exothermic. The heat produced in these reactions is so high that the products formed in these reactions undergo dissociations. Thus, the occurrence of reactions (3) and (5) is highly unlikely. If we eliminate reactions (2), (3) and (5), the reactions (1) and (4) are the only two likely secondary reactions. On this basis, the overall process may be written as.

$$HI + h\nu \rightarrow H + I \qquad \text{(Primary)}$$

$$H + HI \rightarrow H_2 + I \qquad \text{(Secondary)}$$

$$I + I \rightarrow I_2 \qquad \text{(Secondary)}$$

$$2HI + h\nu \rightarrow H_2 + I_2$$

Thus, two molecules of hydrogen iodide can undergo dissociation by absorbing one quanta only. Thus, the quantum efficiency of the reaction is 2 which has been found experimentally.

(ii) Dissociation of HBr : The mechanism of the reaction is analogous to that proposed above for HI.

(iii) Combination of Hydrogen and Bromine : The quantum efficiency of the reaction is about 0.01. The primary process involves the dissociation of bromine molecule into bromine atoms.

$$Br_2 + h\nu \rightarrow 2Br$$

The expected subsequent secondary processes are.

$$Br + H_2 \rightarrow HBr + H \qquad ...(1)$$

$$H + Br_2 \rightarrow HBr + Br \qquad ...(2)$$

$$H + HBr \rightarrow H_2 + Br \qquad ...(3)$$

$$Br + Br \rightarrow Br_2 \qquad ...(4)$$

Reaction (1) is highly endothermic and is, therefore, taking place very slowly at ordinary temperature. At this slow speed, most of the bromine atoms recombine to give the bromine molecules. Hence the reactions (2), (3) and (4) which are due to reaction (1) cannot occur. Therefore, the quantum yield will be extremely low.

(iv) Polymerisation of Anthracene : This reaction can be represented as

$$2C_{14}H_{10} \underset{\text{dark}}{\overset{\text{light}}{\rightleftharpoons}} C_{28}H_{20}$$

The quantum yield of this reaction should be 2 but actually it is only 0.5. It is due to the reason that this reaction involves a back thermal reaction. The low quantum yield of isomeric transformation of maleic acid into fumaric acid can be explained on the basis that a state of equilibrium is reached in this reaction which will lower the quantum yield.

FACTORS AFFECTING QUANTUM YIELD

Except those reactions which obey Einstein's law of photochemical equivalence, it is found that the quantum yield depends upon the following factors.

1. **Temperature :** The change of quantum yield with temperature is given by the relation.

$$\frac{d \log \phi}{dT} = \frac{Q}{RT^2}$$

where f = quantum yield.

Q = amount of heat evolved by the formation of one gm mole of the substance.

Kuhn found that the quantum yield of photochemical decomposition of ammonia increases by about 15% for every 100° temperature increase. It reaches seven fold at 500°.

For some photochemical reactions the values of temperature coefficient and activation energy are given in Table 5.1.

Table 5.1 : Temperature Coefficient and Activation Energy of Some Photochemical Reactions.

Reactions	Temperature coefficient	ΔE (K cal).
$H_2 + Br_2 \rightarrow 2HBr$	1 – 1.02	40.7
$H_2 + Cl_2 \rightarrow 2HCl$	1 – 1.04	44.9
$H_2 + O_2 \rightarrow H_2O_2$	1.09	47.6
$2O_3 \rightarrow 3O_2$	1.27	80.5
$2C1_4H_{10} \rightarrow C_28H_20$	1.04	44.8

2. **Wave-length :** It has been found experimentally that the power of a quantum or photon of light energy to induce photochemical change is greater, the higher the frequency of incident radiation. In other words, the quantum yield will be lower at the higher-wave lengths.

Bonhoeffer and *Harteck* (1930) suggested that the lower quantum yield at longer wavelength is due to a low efficiency of the primary dissociation process in that region. Generally, the quantum yield is always found to rise over a range of frequencies, as in the case of decomposition of nitrogen' dioxide.—

Wavelength	310	365	405	436
Quantum yield	2.97	1.54	0.74	0.009

Henri, Wurmser and Lasareff have obtained a proportionality between the amount of light of different wavelengths absorbed and the photochemical effect produced.

3. **Light intensity :** It was shown by Dhar and his coworkers that decrease in light intensity results in an increase of quantum yield. The light intensity was found to change by varying the iris of the aperture, through which light was allowed to pass into the reacting system. These workers found that.

   (i) The velocity of photochemical reactions is found to be directly proportional to the intensity of light.

   (ii) The quantum yield decreases as the amount of light falling on the reacting substance is increased.

4. **Inert Gases :** It was found by Hartel that in most of the photochemical reactions, the addition of inert gases increases the quantum yield. This may be explained due to the occurrence of an induced pre-dissociation which will increase the number of starting chains as it may retard the diffusion of the atoms to the walls. This latter effect would decrease the rate of chain terminating step thereby increasing the speed of reaction.

LUMINESCENCE

We know that the usual methods of obtaining light are involving the heating of solids, or solid particles to sufficiently high temperatures. For example, by heating a tungsten, platinum wire or a carbon filament electrically or in a flame, bright light can be obtained. These are examples of *incandescence* in which thermal energy is converted into light. If, however, light is produced by processes other than those involving heating, such light is called *luminescence light or cold light.*

The above discussion may be summarised as follows.

"When the emission of visible radiation occurs due to some cause other than temperature, the phenomenon is known as luminescence.

As light is produced at low temperatures, the term luminescence may be regarded as 'light' produced without 'heat' or 'cold light'

But in incandescence and luminescence, emission of light occurs due to the return of electrons from excited outer position to lesser excitation position or ground state. Luminescence is of the following types.

1. **Photoluminescence :** Luminescence caused by light is called photoluminescence. Photoluminescence that ceases immediately after the cause of exciation is cut off is called *fluorescence.* Photoluminescence that persists for an appreciable time after the stimulating process is cut off is called *phosphorescence.*
2. **Chemiluminescence :** Luminescence resulting from chemical reactions is called *cathodoluminescence.*
3. **Cathodoluminescence :** Luminescence caused by bombardment of electrons is called *cathodoluminescence.*

   Let us discuss these one by one.
4. **Electroluminescence :** Luminescence resulting from the application of an electric field to matter is called *electroluminescence.*

Let us discuss these one by one.

FLUORESCENCE AND PHOSPHORESCENCE

Introduction to Fluorescence : *When a beam of light is incident on certain substances, they emit visible light or radiations and they stop emitting light or radiation as soon as the incident light is cut off. This phenomenon is known as fluorescence.*

Such substances which emit radiations during the action of stimulating light are called fluorescent substances.

When a substance absorbs light energy, it will result in the excited state of an atom or molecule. But if the light absorbed is not sufficient energetic to eject an electron, it will cause the electrons to move from inner orbits to outer orbits. When these excited electrons return to the ground state, the light is emitted which may possess different frequency than the incident light. The successive stages may be put as follows.

$$\underset{\text{Normal state}}{X} + \underset{\text{photon}}{h\nu} \longrightarrow \underset{\text{Excited state}}{X^*}$$

$$X \longrightarrow X^{**} + h\nu'$$

$$X^{**} \longrightarrow X + h\nu''$$

where X^{**} represents an energy state between X and X^*. The emitted radiation frequencies ν' and ν'' are different from ν, *i.e.*, the frequency of incident radiations.

Some characteristics of the phenomenon of fluorescence are as follows

(i) This phenomenon is instantaneous and starts immediately after the absorption of light and stops as soon as the incident light is cut off.

Fluorescence is stimulated by light of the visible or ultraviolet regions of the spectrum. Line, band and continuous spectra of emitted light of fluorescence are observed. The character of the spectrum depends essentially on the state of aggregation of the substance.

(ii) It is a general phenomenon and is exhibited by gases, liquids and solids. No fluorescence will be observed in gases, unless the pressure is low.

(iii) Different substances fluoresce with light of different wavelengths. Thus fluorspar fluoresces with blue light, chlorophyll with red light, uranium glass with green light and so on.

(iv) The fluorescent light from solutions is polarised and the degree of polarisation depends in some cases upon the concentration of the solution.

(v) The extent of fluorescence depends upon the nature of the solvent and the presence of certain anions in solution. Thus the thiocyanate, iodide and bromide ions show a marked quenching effect.

(vi) According to *Stoke's law,* during fluorescence light is absorbed at a certain wavelength and should be emitted at a greater wavelength.

The above principle was recognised before the quantum theory was proposed.

(vii) The quantum yield in fluorescence is the ratio of the number of photons of luminescent radiation to the number of photons absorbed from the stimulating light upon a fixed wavelength of the latter. The quantum efficiency of fluorescence increases in proportional to the wavelength l of absorbed radiation. Then, after reaching its maximum value in a certain interval of $\sim \lambda_{max}$ the efficiency drops rapidly to zero upon a further increase in l.

(viii) Fluorescence may be regarded as a secondary effect resulting from the primary process of absorption of a quantum of light by an atom or molecule.

Introduction to Phosphorescence : *When light radiation is incident on certain substances, they emit light continuously even after the incident light is cut off. This type of delayed fluorescence is called phosphorescence and the substances are called phosphorescent substances.*

Some characteristics of the phenomenon of phosphorescence are as follows.

(i) Materials exhibiting fluorescence generally re-emit excess radiation within 10^{-6} to 10^{-4} second of absorption. On the other hand, materials exhibiting phosphorescence re-emit excess radiation within 10^{-4} to 20 seconds or longer. Thus, the life-time of phosphorescence is much longer than fluorescence.

(ii) The phenomenon of phosphorescence is caused chiefly by the ultraviolet and violet parts of the spectrum.

(iii) The phenomenon of phosphorescence is shown mainly by solids.

(iv) The magnetic and dielectric properties of phosphorescent substances are different before and after illumination.

(v) The time for which the light is emitted from phosphorescent substance defends upon the nature of substance and sometimes on the temperature changes.

(vi) Different colours may be obtained by mixing different phosphorescent substances. *Examples of Fluorescent Substances.*

Among the naturally occurring substances, chlorophyll present in green leaves show the phenomenon of fluorescence, *i.e.,* when leaves are strongly illuminated in presence of oxygen, they emit fluorescent light whose intensity changes with time of irradiation. Franck and Herzfeld explained this phenomenon on the basis of complex formation.

Petroleum, vapours of sodium, iodine, acetone and hydrocarbons (paraffins and olefins) have been found to fluoresce in ultraviolet light. For example, acetone absorbs at 2700Å corresponding in > C = C group and emits blue fluorescence. Same thing happens with other aldehydes and ketones.

We know that all molecules are capable of absorption. But fluorescence is not exhibited by a large number of compounds. Fluorescence is generally observed in those organic molecules which have rigid framework and not many loosely coupled substituents through which vibrionic energy can flow out. In analogy with chromophores, following structures are termed as *fluorophores.*

C=C, N=O, —N=N, C=O, —C≡N, —C=S

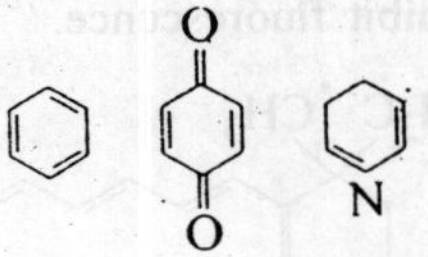

Fig. 5.8

A large number of substances enhance fluorescence. These are known as fluorochromes in the same analogy as *auxochromes.* Generally electron donors act as auxochromes.

–OH
—NH_2
—CH_2

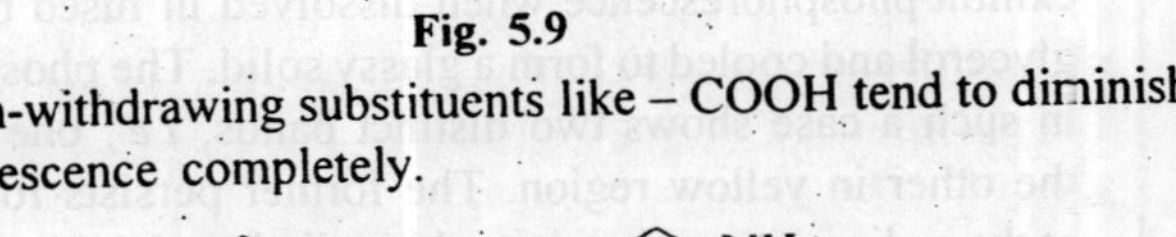

Acridine
(non-fluorescent)

$(CH_3)_2N$ $N(CH_3)_2$
Acridine Derivative
(fluorescent)

Fig. 5.9

Electron-withdrawing substituents like – COOH tend to diminish or inhibit fluorescence completely.

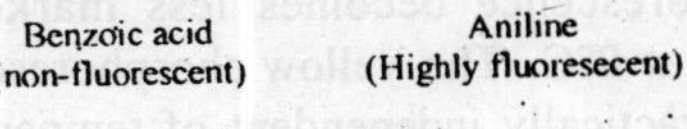

COOH
Benzoic acid
(non-fluorescent)

NH_2
Aniline
(Highly fluoresecent)

Fig. 5.10

Some examples of fluorescent substances are as follows:

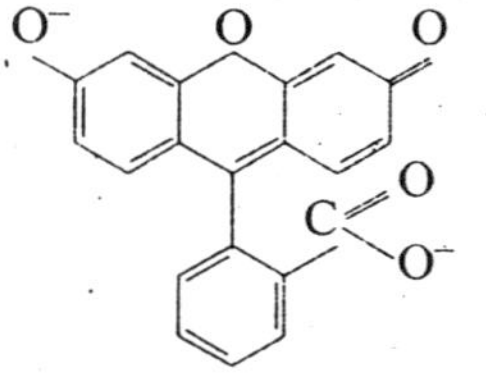

Flourescein
(fluorescent)

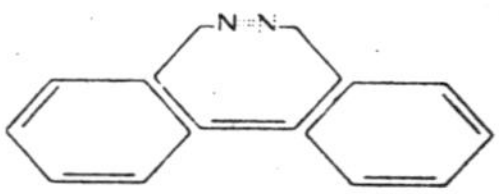

Azophenanthrene
(fluorescent)

Fig. 5.11

Among inorganic substances there are only few examples such as fluorite (the name fluorescence originates from it), uranium compounds and rare earths which exhibit fluorescence.

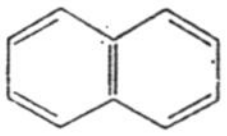

Naphthalene
(fluorescent)

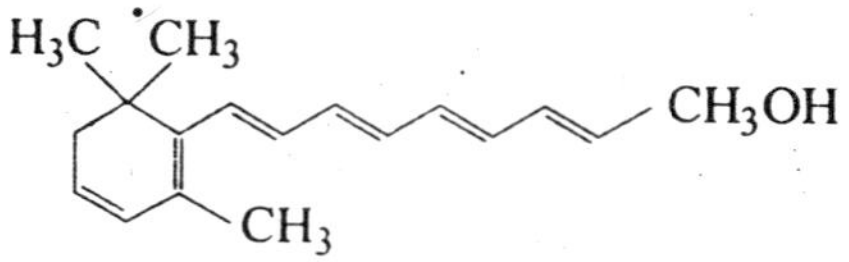

Vitamin A
(fluorescence as that of naphthalen)

Fig. 5.12

Nitrogen peroxide in blue light (4920 to 4550A°) shows intense fluorescence which becomes much weaker in violet light (4550 to 3650Å) and vanishes at 3650Å)

EXAMPLES OF PHOSPHORESCENT SUBSTANCES

(i) Many dyes which fluoresce in ordinary light in aqueous solution exhibit phosphorescence when dissolved in fused boric acid or glycerol and cooled to form a glassy solid. The phosphorescence in such a case shows two distinct bands, *i.e.,* one in blue and the other in yellow region. The former persists for some time at the ordinary temperature and is called α-phosphorescence and is identical with the fluorescence. If the temperature is lowered, α–phosphorescence becomes less marked and disappears altogether at 0°C. The yellow phosphorescence, on the other hand, is practically independent of temperature and has been observed down to – 250°C. The yellow phosphorescence is called β-*phosphorescence.*

(ii) The common substances that exhibit this phosphorescence phenomenon are sulphides of calcium, barium, strontium.

(iii) Many organic substances and certain fungi phosphoresce to emit a faint light This is due to the slow oxidation of organic substances.

(iv) Minerals, *e.g.*, ruby, emarld are the interesting examples of phosphorescent substances.

(v) Many phosphors are prepared by mixing alkaline earth metals with about 2.5 per cent alkali chlorides and a trace of sulphide of some heavy metal.

THEORY OF FLUORESCENCE AND PHOSPHORESCENCE

1. Singlet and Triplet States : In order to understand the theory of fluorescence and phosphorescence, one has to understand the meaning of singlet and triplet states. These terms arise from multiplicity considerations to atomic spectroscopy and simply define the *number of unpaired electrons in the absence of magnetic field.* If there are *n* number of unpaired electrons, it means that (n + 1) -fold degeneracy (equal energy states) will be associated with the electron spin, regardless of the molecular orbital occupied.

Thus, if no unpaired electrons are present (n = 0), where there is only n + 1 or 0 + 1 or 1 spin state. Such a state is a called a *singlet state.* Similarly, systems having 1, 2, 3, 4, ... unpaired electrons refer to doublet, triplet, quartet, etc. respectively.

Most of the molecules in their ground state do not have unpaired electrons *(singlet state).* When such a molecule absorbs ultraviolet or visible radiation of the proper frequency one or more of the paired electrons (generally a re-electron) get raised to an *excited singlet state.* In this excited state, the spin of the electron does not undergo any change and the net spin is still zero.

One more possibility is that one set of electron spins may have undergone unpairing, resulting in two unpaired electrons which make an *excited triplet state.* Fig. 5.13(a) represents a molecule in the ground singlet state, Fig. 5.13(b) exhibits a molecule in the excited singlet state. Fig. 5.13(c) represents a molecule in an excited triplet state.

One should remember that there is very small probability of a direct transition from the singlet ground state to the triplet excited state.

2. Excited-State Processes in Molecules : When molecules are irradiated with light of the appropriate frequency, it will be absorbed in about 10^{-15} second. In the process of absorption the molecules may move from the ground to the first excited singlet electronic state. Although at room temperature molecules may be present in their ground vibrational level, after absorption the excitational molecules can end up in any one of the vibrational levels in the first excited electronic state. From the excited singlet state, one of the following three phenomena will probably occur, depending on the molecules involved and the conditions.

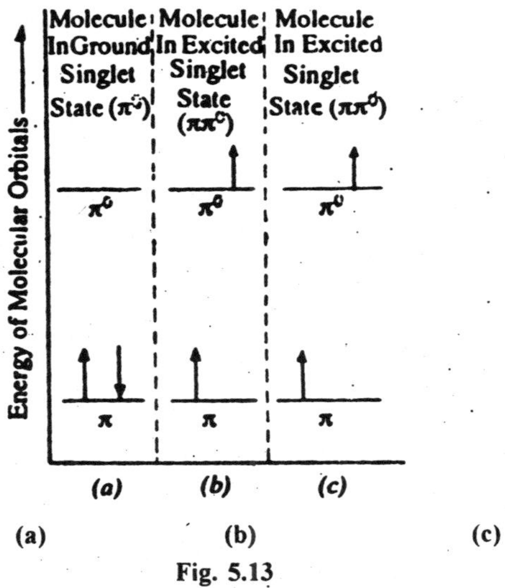

(a) (b) (c)

Fig. 5.13

(a) The *first possibility* is that the excited singlet state is relatively *unstable.* In such a situation, the excited molecule will return to the ground state by *collisional deactivation without emitting any radiation.*

(b) The *second possibility* is that the molecule in the excited singlet state may emit an ultraviolet or visible light photon. This process is known as *fluorescence.*

If we compare the absorption and fluorescence spectrum of the same compound, they do not superimpose on each other as expected but they are mirror images of each other with the fluorescence spectrum shifted to longer wavelengths. The reason for this is that as the time required to execute a vibration is about 10^{-13} second which is much shorter than the decay or mean life of 10^{-9} second, most of the excess vibrational energy will be given to the surroundings and the excited molecules will decay in their ground vibrational levels. This is illustrate in Fig. 5.14.

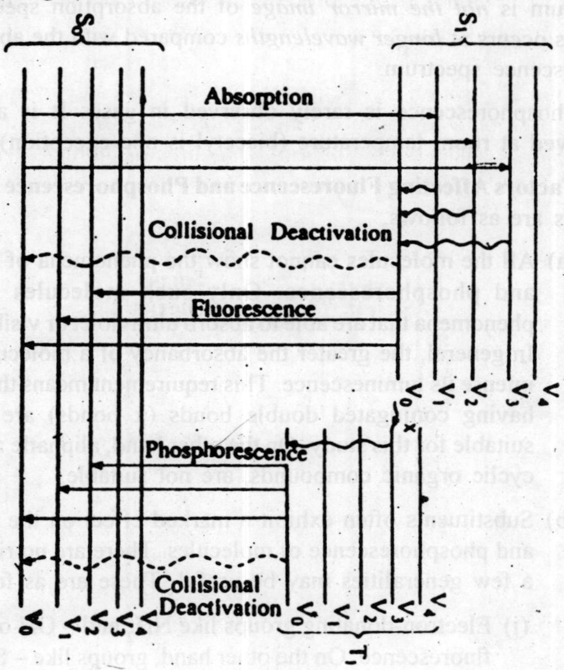

Fig. 5.14

(c) The third possibility is that the molecule with relatively stable *excited singlet state* may undergo transition to a *metastable triplet state* and some time thereafter returns to the ground states, usually by emission of an ultraviolet or visible light photon. This is known as *phosphorescence emission* Fig. 5.14, and the process of crossing from a singlet state (no unpaired electron) to a triplet state (two unpaired electrons) is termed as *intersystem crossing.*

The decay from the triplet to the ground state singlet is *forbidden by spin symmetry* and is therefore *slow.* Thus, the life-time of phosphorescence is much longer than flour-essence.

The above mechanism of phosphorescence involving singlet-triplet decay scheme has been confirmed by the magnetic susceptibility and ESR measurements.

According to Hund's rule, the triplet level always lies *lower* than the corresponding singlet level and for this reason phosphorescence spectrum is *not the mirror image* of the absorption spectrum and it always occurs at *longer wavelengths* compared with the absorption and flourescence spectrum.

Phosphorescence is rarely observed in gases It is almost never observed at room temperature (biacetyl is one exception).

Factors Affecting Fluorescence and Phosphorescence : The various factors are as follows.

(a) All the molecules cannot show the phenomena of fluorescence and phosphorescence. Only such molecules show these phenomena that are able to absorb ultraviolet or visible radiation. In general, the greater the absorbancy of a molecule, the more intense its luminescence. This requirement means that molecules having conjugated double bonds (π bonds) are particularly suitable for this study. On the other hand, aliphatic and saturated cyclic organic compounds, are not suitable.

(b) Substituents often exhibit a marked effect on the fluorescence and phosphorescence of molecules. There are no rigid rules but a few generalities may be useful. These are as follows.

(i) Electron-donating groups like NH_2 and – OH often enhance fluorescence. On the other hand, groups like – SO_2H, – NH_4^+ and alkyl groups do not have much effect on both phosphorescence and fluorescence.

(ii) Electron-withdrawing groups like – COOH, – NO_2. – N = N – and halides decrease or even destroy fluorescence.

(iii) If a high atomic number atom is introduced into a re-electron system, it enhances phosphorescence and decreases fluorescence.

(c) The pH exhibits a marked effect on the fluorescence of compounds. For example, the neutral or alkaline solution of aniline shows fluorescence in the visible region. But if this solution is acidified, the visible fluorescence disappears (aniline shows fluorescence in the ultraviolet regardless of pH).

Relation between Fluorescence Intensity and Concentration : We know that the Beer-Lambert law can be applied to the intensity of radiation transmitted by a substance or a solution. But this cannot be applied to fluorescent radiation directly because it is emitted by a substance. However, the following relation has been developed.

$$F \propto I_D - I$$

or $$F = K (I_0 - I) \quad ...(1)$$

where F is the intensity of fluorescent radiation, K is a proportionality constant. I_0 is the intensity of incident radiation and I is the intensity of transmitted radiation. We know that the Beer-Lambert law is

$$I = I_0 \, 10^{-abc}$$

or $$I_0 - I = I_0 - I_0 \cdot 10^{-abc} \quad ..(2)$$

$$= I \, (1 - 10^{-abc}) \quad ..(3)$$

On substituting equation (3) in (1), we get

$$F = KI_0 \, (1 - 10^{-abc}) \quad ...(4)$$

On rearranging equation (4), we get

$$\log \frac{KI_0}{KI_0 - F} = abc \quad ...(5)$$

In the above equation, K is the fraction of the incident radiation that is adsorbed, *a* is the absorptivity, *b* is the length of cell path and *c* is the concentration of absorbing substance.

For very dilute solutions, equation (4) simplifies to the following equation

$$F = 2.303 \, KI_0 \, abc$$

or $$F = K'c$$

For the above equation it follows that the fluorescent intensity is practically proportional to the concentration of the fluorescent substance. Equation (6) holds good for solutions of few parts per million, *i.e.*, dilute solution. For connected solutions, fluorescence concentration curve will bend towards the concentration axis. Fig. 5.15 shows typical curves at – 196°C, and 35°C.

From Fig. 5.15. it follows that the curve levels off at high concentration at optimum temperature. Factors such as association, dissociation, or solvation which are responsible for deviations in Beer-Lambert law would be expected to show a similar effect in flourescence.

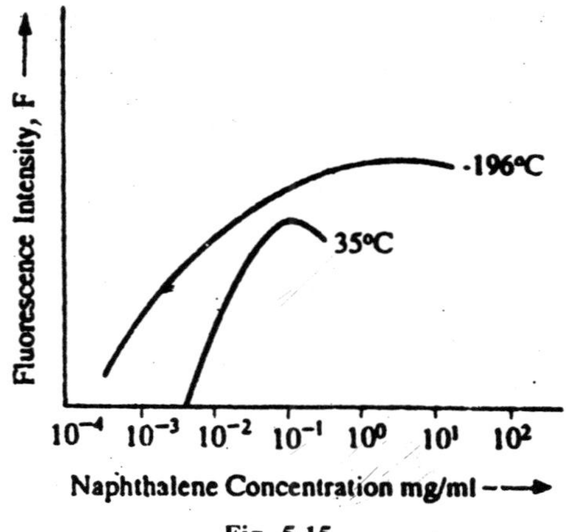

Fig. 5.15

Another important factor in fluorescence is temperature. In most of the instruments, sources of the very high intensity are used. This causes the heating of the sample or sample chamber. An increasing temperature usually decreases flourescence intensity. In order to minimise temperature effects in fluorescence instruments, shutter mechanisms are provided to minimise exposure time and thermally insulating the sample holder.

Relation between Phosphorescence Intensity and Concentration : Exactly the same considerations as discussed for fluorometry are important in relating phosphorescence intensity P to concentration *c*, and thus the final equation is as follows.

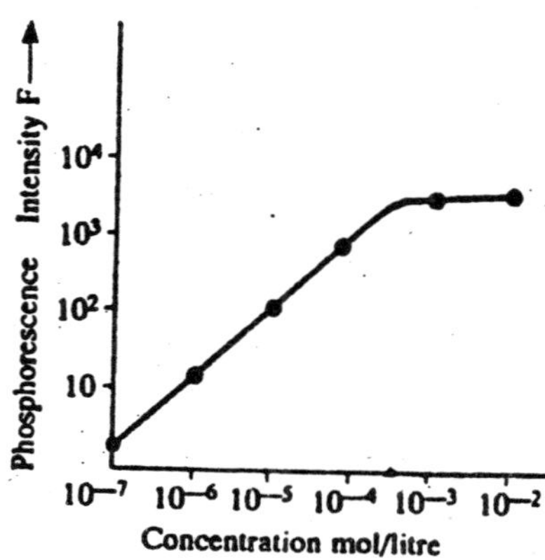

Fig. 5.16

$$\log \frac{K'I_0}{K'I_0 - P} = abc$$

where K′ depends on the instrument used. A typical curve or phosphorimetry is shown in Fig. 5.16.

Applications of Fluorescene : We know that the phenomenon of fluorescence is a well established analytical tool. A large number of applications are known. But we will describe some of these as follows.

(a) An interesting example is the determination of uranium in salts by fluorescence. This is used extensively in the field of nuclear research.

The uranium sample is evaporated with nitric acid to bring about oxidation. Then the sample is fused with sodium fluoride to a melt having fluorides of sodium and uranium. On cooling, this solidifies to a glass which is examined in a specially designed fluorimeter. By this method, one can determine uranium of the order of 5 x 10^{-9} g in a 1 g of solid sample.

(b) In general, inorganic ions do not exhibit fluorescence. However, some of these inorganic ions form fluorescent chelates with non-fluorescent organic molecules. This has provided the basis for very sensitive analysis of many elements including most of the transition elements. Some interesting examples are as follows.

(i) An interesting example is the determination of ruthenium ion in the presence of other platinum metals. With 5-methyl-1, 0-phenanthroline ruthenium forms the complex ion which fluoresces strongly at pH 6. By this method, the determination of ruthenium can be carried out in the range of 0.3 to 2.0 μg/ml in the presence of interfering elements of platinum group which may be present to the extent of at least 30μ/ml.

(ii) Another interesting example is the determination of aluminium (III) in alloys. Aluminium (III) forms a complex with dye Pantachrome Blue-black RM at a pH of 4.8 which fluoresces strongly. One can determine aluminium in the range of 0.2 to 25μg in a volume of 50 ml with a sensitivity of 1 part 10^8.

(iii) Another inorganic fluorimetric determination is the estimation of traces of boron in steel by means of the complex formed with benzoin.

The boron present in the acid solution of steel is first converted into boric acid which is separated from the other constituents by distillation with methyl alcohol. The resulting distillate having boric acid is neutralised with sodium hydroxide. Then, the methyl alcohol is distilled off leaving the sodium salt of boric acid in the distillation flask which on treatment with an alcoholic solution of benzoin yields a fluorescent solution. It has been found that the fluorescent power is linear with the concentration upon 100 *ug* of boron in 50 ml volume but drops off at higher concentrations. This method for its determination is superior to old techniques in speed and sensitivity.

(iv) Cadmium can be estimated by precipitating it with 2-(2- hydroxyphenyl)-benzoxazole in the presence of tartrate. The complex on dissolving in glacial acetic acid yields a solution with an orange tint and a bright blue fluorescence in ultraviolet light. The acetic acid solution forms the basis for the determination of cadmium.

(v) Similarly, calcium can be estimated by fluorimetry with calcein solution.

(c) Fluorescent indicators : The intensity and colour of the fluorescence of many substances depend upon the *pH* of the solution, *i.e.,* their colours depend upon the pH range. These are termed *as flourescent indicators.* These are mainly used in acid-base titrations. These can be employed in the titration of coloured solutions in which the changes in colour of indicators get masked. Some examples of fluorescent indicators are given in Table 5.2.

Table 5.2 : Some Fluorescent Indicators.

Name of Indicators	approx. pH	Colour Change
Eosin	3.4–4.0	Colourless to green
Fluorescein	4.0–6.0	Colourless to green
Quinine sulphate	3.0–5.0	Blue to violet
Acridine	5.2–6.6	Green to violet blue
2 – Naphthaquinone	4.4–6.3	Blew to colourless
2 – Hydroxycinnamic acid	7.2–9.0	Colourless to green

(d) **Determination of vitamin B_1 :** Vitamin B_1 (thiamine) is non-fluorescent whereas its oxidation product, thiochrome fluoresces with blue colour. This property is used for the determination of vitamin B_1 in the food samples like meat, cereal, etc.

The food sample is treated with phosphatase which brings about hydrolysis of the phosphate esters of thiamine present in the sample. The solution on filtration removes phosphatase and other insoluble matter. Then, the filtrate is diluted to a known volume. From the filtrate, two equal aliquots are taken, one for analysis and other for a blank.

To both aliquots, equal quantities of sodium hydroxide and isobutyl alcohol are added. To the first oxidising agent like potassium ferricyanide is added. After shaking, the alcoholic solution is separated from the aqueous solution. Then, the alcoholic solution is examined in the fluorimeter. The whole procedure, including a blank, is repeat with a standard thiamine solution.

(e) **Determination of Vitamin B_2 (Riboflavin) :** Determination of vitamin B_2 is done by a fluorescence method because the fluorescent power depends upon the reaction conditions and upon the nature and amount of impurities.

Generally, the method of standard increment is used to become sure about the fact that impurities have the same effect upon the standard and unknown. In the method of standard increment, one measures fluorescence of a portion of the standard in the same solution with the unknown.

The procedure also considers the simple fact that riboflavin on oxidation yields a non-fluorescent substance.

An acid solution of the sample (a food stuff) is treated with different reagents to precipitate various interfering ions. This solution is then oxidised with dilute permanganate, Then, the residual fluorescence is measured as a blank. Now, a slight excess of solid sodium dithionite ($NagSaO_4$) is added.

Fluorescence is again determined. Then, a known volume of standard is added and fluorescence is again measured. Finally, the results are computed by the following scheme.

Table 5.3

Fluorescence of solution	Designation
10 ml oxidised sample + 1 ml water	F_A
Same + dithionite	F_B
Same + 1 ml standard	F_C

From the above results, the concentration of fluorescing material can be calculated by using the following formula.

$$\frac{F_B - F_A}{F_C - F_A} = \frac{m_x}{m_x + m_s}$$

In the above equation, m_x and m_s are the masses of riboflavin from sample and standard in the cuvette respectively.

(f) **Food-Stuffs :** The phenomenon of fluorescence is being utilised in examining conditions of food-stuffs.

When ultraviolet light is incident on newly laid eggs, they fluoresce with rosy colour, while bad eggs appear blue. Similarly butter, lard and different kinds of honey can be readily distinguished.

(g) **Police Work :** The difference in the fluorescence caused by ultraviolet rays in different types of inks enables police to detect forged documents.

(h) **Medicine :** Ringworm may be detected by the fluorescence caused by ultraviolet radiations.

When fluorescent liquids are injected into an animal body, the internal organs will fluoresce and can be observed by a microscope of special type called fluorescent microscope. In this way the internal organs of human or animal body may be diagnosed by doctors without any difficulty.

When, in fluoroscope. X-rays are allowed to fall on a screen of barium platinocyanide or other materials, there will be characteristic fluorescence which may be used for diagnosis by doctors. This is the principle of fluoroscope used in X-ray diagnosis.

(i) **Analysis :** The characteristic fluorescence of various substances when exposed to ultraviolet light offers a good method of quantitative analysis. This has been used in the analysis for the following.

(a) Drugs and dyes

(b) Textile and paper industry.

(c) Medicine and bacteriology,

(d) Fuels and chemicals.

(j) **Lamps :** Fluorescent lamps which are now widely used for lighting depend upon the fluorescence caused by ultraviolet light on phosphorus coated inside the fluorescent tubes.

(k) **Science :** Use of the phenomenon may be made in seeing the invisible radiations like ultraviolet rays.

(l) **Organic Analysis :** Fluorescence has been used to carry out qualitative as well as quantitative analysis for a great many aromatic compounds present in cigarette smoke, air-pollutant concentrates and automobile exhausts. A specific example is the determination of benzopyrene in the nanogram range.

APPLICATIONS OF PHOSPHORESCENCE

Similar to fluorescence, phosphorescence finds applications in biology and medicine. But these applications are not accepted for routine analysis. Some interesting applications are as follows.

(a) One can carry out the determination of aspirin (acetylsalicylic acid) in blood serum with high sensitivity by phosphorimetry at liquid nitrogen temperatures. By this method, 0.02 – 1.00 mg aspirin per ml of serum can be analysed. The metabolic product of aspirin, *i.e.,* salicylic acid, does not exhibit appreciable phosphorescence but is quite strongly fluorescent whereas aspirin is exhibiting strong phosphorescence but not fluorescence. Therefore, it becomes possible to carry out a simple chloroform extraction of the serum and then analysis is done for both the drug (aspirin) and its metabolite (salicylic acid) in the presence of one another by using these two complementary techniques (fluorimetry and phosphorimetry).

(b) Low concentrations of procaine, cocaine, phenobarbital and chloropramazine in blood serum have been determined by phosphorimetry in combination with extraction procedures.

(c) Cocaine and atropine in urine have been determined by employing phosphorimetry in combination with extraction procedures.

(d) Phosphorimetry has been employed in combination with thin layer or paper chromatography. An interesting example is given by Winefordner and Moye. They have separated three tobacco alkaloids (nicotine, nornicotine and anabasine) from crude tobacco by thin layer chromatography on alumina. The R_f values of three alkaloids are 0.80, 0.26 and 0.48 respectively. The separated drugs were scrapped from the plate and determined by phosphorimetry. This method is quite rapid.

COMPARISON OF FLUORESCENCE AND PHOSPHORESCENCE

As phosphorescence is more complicated experimentally than fluorescence, far fewer applications have been realised for phosphorimetry than for fluorimetry. But the potential of phosphorimetry is great and it is hopped that it will find valid applications in chemistry in the near future. However, it is clear that the two techniques are complementary.

Whenever we analyse complex samples having much interfering substances, we prefer phosphorimetry because it is more selective than fluorimetry.

One more advantage of phosphorimetry is that it is more sensitive than fluorimetry due to the following reasons.

(i) Scattering problems do not exist in phosphorimetry whereas these become severe in fluorimetry.

(ii) Quantum efficiencies are much more for phosphorescence than for fluorescence. The reason for this is that phosphorimetric determinations are carried out at – 196°C whereas fluorimetric determinations are carried out at room temperature. One more advantage for the low temperatures in phosphorimetry is that the quenching does not pose a serious problem, than in fluorimetry.

From the above discussion it follows that whenever fluorimetry and phosphorimetry are of equal sensitivity, fluorimetry is preferred because it is experimentally less complicated. However, phosphorimetry is preferred when fluorimetry is not sensitive or selective.

Antistoke's Behaviour : According to G.G. Stoke, the emitted radiation in fluorescence has greater wave-length than the absorbed radiations. There are however, cases where Stoke's law is violated. Two exceptions are.

(a) **Resonance Fluorescence** : *In certain cases, the fluorescent light has the same frequency as that of incident light. This fluorescent light is called resonance radiation and the phenomenon is known* as resonance fluorescence. This phenomenon is analogous to resonance of sound. For example, when mercury vapour with atoms in normal state ($^1S_0$) is exposed to ultraviolet radiations, it was found to absorb the radiation of wave-length 2537 Å. It then converts to $^3P_1$. If excited electrons return to the ground-state, they emit the radiations of the same wavelength, *i.e.,* 2537 Å.

$$Hg\ (^1S_0) + h\nu \rightarrow Hg^*\ (^3P_1)$$

$$Hg^*\ (3P_1) \rightarrow Hg\ (^1S_0) + h\nu$$

(b) **Sensitised Fluorescence** : *A substance which is normally non-fluorescent may be made fluorescent in the presence of other fluorescent substances. The phenomenon is known as* Sensitised fluorescence. If the vapour of thallium is added to mercury vapour and then exposed to radiations of wave-length, 2537 Å, the vapour of thallium fluoresces. Thus the thallium which is a non-fluorescent substance can be made fluorescent by mercury vapour. This can be explained by saying that in sensitised fluorescence, mercury vapour gets energy from incident light and thus its atoms get excited to higher energy levels.

$$Hg\ (^1S_0) + h\nu \rightarrow Hg^*(^3P_1).$$

Then, thallium atoms undergo collision with excited mercury atoms and thus a part of energy of excited mercury atoms is transferred to thallium atoms which are raised to higher energy levels.

$$Hg^* + T_l \rightarrow T_l^* + Hg$$

These excited atoms of thallium emit their own characteristic fluorescence on reverting to the ground state

$$T_l^* \rightarrow T_l + h\nu'$$

Quenching of Fluorescence : *When a photochemically excited atom has a chance to undergo collision with another atom or a molecule before it fluoresces, the intensity of the fluorescent radiation may be diminished or stopped. This phenomenon is known as quenching of fluorescence.* Some facts about quenching of fluorescence are as follows.

(i) Quenching of fluorescence depends greatly on the concentration of the fluorescent atoms and of the quenching substance.

(ii) At low pressures, very little quenching will occur. At high pressures, appreciable quenching takes place.

(iii) Quenching of fluorescence takes place appreciably in a liquid medium because collisions are very frequent.

Explanation. The quenching of fluorescence is due to transfer of energy from the photochemically excited atom to the molecule or atom with which it undergoes collisions. As a result of transfer of energy, the following changes may be possible.

(i) A photochemically excited atom may activate the atom with which it undergoes collision.

$$Hg^* + T_l \rightarrow Hg + T_l^*.$$

This results in sensitized fluorescence.

(ii) The photochemically excited atom may activate a molecule by collision.

$$Cd^* + H_2 \rightarrow Cd + H_2^*$$

(iii) A photochemically excited atom may react chemically with a molecule to form the products. Example is

$$Hg^* + H_2 \rightarrow HgH + H$$

(iv) An excited atom may undergo collision with another molecule and, thus, resulting its dissociation. Example is

$$Hg^* + H_2 \rightarrow Hg + 2H.$$

This phenomenon is known as photosensitisation.

Importance. The photochemical importance of the quenching of fluorescence is that the excited molecules or atoms obtained in the quenching process may undergo further reaction so as to continue the photochemical changes such as (a) photosensitisation (b) sensitised fluorescence, (c) photochemical reactions and (d) dissociation of molecules to form atoms

PHOTOELECTRIC EFFECT

It is the phenomenon of ejection of electrons from a metal plate when light of a suitable wave-length falls on it. The electrons emitted are called photo-electrons to indicate their mode of production.

Laws of Photo-Electric Emission : There are two important laws regarding the photoelectric **Emission.**

Law I : *The number of photo electrons ejected is directly proportional to the intensity of the incident light.* Thus, if the intensity of the incident light is increased, the number of electrons emitted is increased, but their velocity remains constant.

Law II: *The velocity of the emitted electrons increases with the increase in frequency of the incident light.* This law means that there is a limiting frequency for a metal, below which no electrons are emitted. This is known as the *threshold frequency.*

Explanation : The laws of photo-electric emission were explained by Einstein in 1905 on the basis of Planck's quantum theory of radiation. According to the quantum theory, the radiation consists of packets of energy called 'quanta' or 'photons' of energy hν, where *h* is the Planck's constant and ν the frequency of radiation. When a photon of energy hν is incident upon a metal surface, its energy is used up in two ways shown in Fig. 5.17.

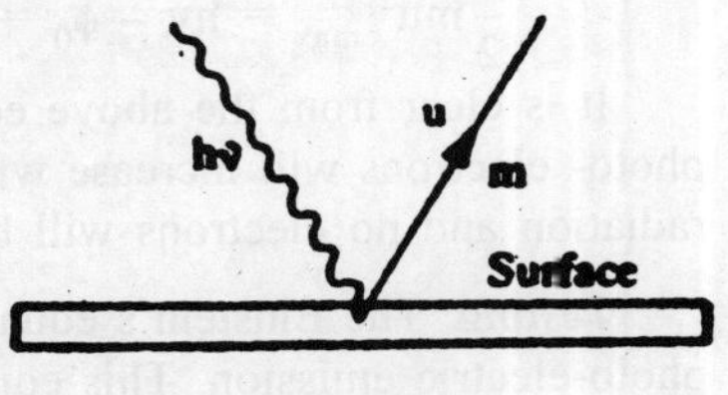

Fig. 5.17

(a) A part of its energy is used up in ejecting the electron just out of the surface. The energy depends upon the nature of the metal and is called work-function (ϕ_0).

(b) The rest part of the energy of the striking photon is used up in imparting kinetic energy $\frac{1}{2}$ mu² to the ejected electron.

Thus $$h\nu = \phi_0 + \frac{1}{2} mu^2 \qquad ...(1)$$

But $$\phi_0 = h\nu_0$$

where ν_0 is the threshold frequency. Therefore, equation (1) becomes as

$$hv - hv_0 + \frac{1}{2}\ mu^2 \text{ or } \frac{1}{2}\ mu^2 = h\ (v - v_0) \qquad ...(2)$$

The equation is called Einstein's photo-electric equation.

Explanation of IInd Law : From eq. (2), it follows that the velocity of a photo-electron increases with the increase in frequency. No electron will be emitted when the frequency of the incident radiation v is less than the threshold frequency v_0. This explains the II law of photoelectric emission.

Explanation of 1st Law : If the intensity of light is increased keeping the frequency same, more photons are incident on the metal surface but the energy of each photon will remain the same. Hence the number of electrons emitted per unit area per unit second will increase but their velocities remain the same. This is the I law of photoelectric emission. It is clear that the maximum value of velocity u_{max} for a given frequency v will occur for those electrons for which $\phi = \phi_0$. Hence, equation (2) becomes as

$$\frac{1}{2} mu^2{}_{max} = hv - \phi_0 = hv - hv_0 \qquad ...(3)$$

It is clear from the above equation that the maximum velocity of photo- electrons will increase with increase in frequency v of incident radiation and no electrons will be emitted when v is less than v_0.

Testing. The Einstein's equation explains the observed facts about photo-electric emission. This equation was tested in 1915 by Millikan who observed its results in agreement with experimental results. The successful explanation of photo electric effect by Einstein on the basis of quantum theory offers one of the evidences in favour of the quantum theory.

Importance. Photo-electric effect is used in the construction of photo electric cell which converts light energy into electric energy. The photo-electric cells are widely used in various fields.

PHOTOELECTRIC CELLS

It is a device for converting light energy into electric energy. There are *three* main types of photo-cells.

1. *Photo emission type.* It consists of an evacuated glass, or quartz tube which has its inner surface M coated with sodium, potassium or

caesium, and this coating is connected to – ve end of a battery. A window is left open through which light can enter the tube, and a wire loop P, or a cylindrical wire in the centre is connected to the +ve end of the battery and it collects the electrons, Fig. 5.18, 5.19 and 5.20. The two parts are insulated from each other and are brought out to two legs at the bottom of the photo cell. On admitting light, which is of a frequency above the threshold frequency, electrons from the photometal are emitted. They travel towards the loop wire and a current is produced in the circuit. Since a photo cell responds to the light which falls on it, as does the human eye, it is therefore, called an *electric eye.* There are two types of photo emission cells, (1) high vacuum type, and (2) gas filled type.

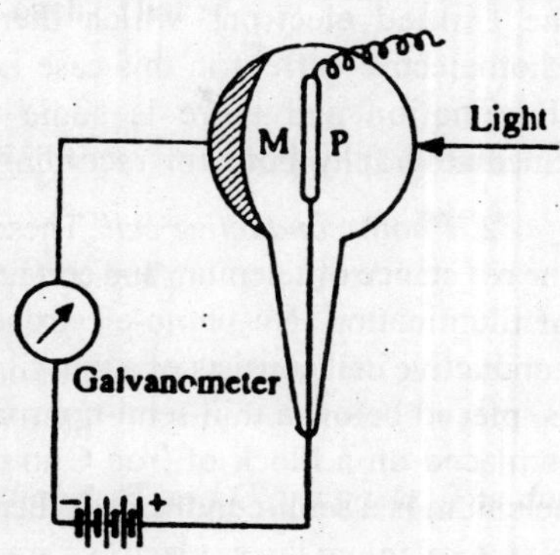

Fig. 5.18

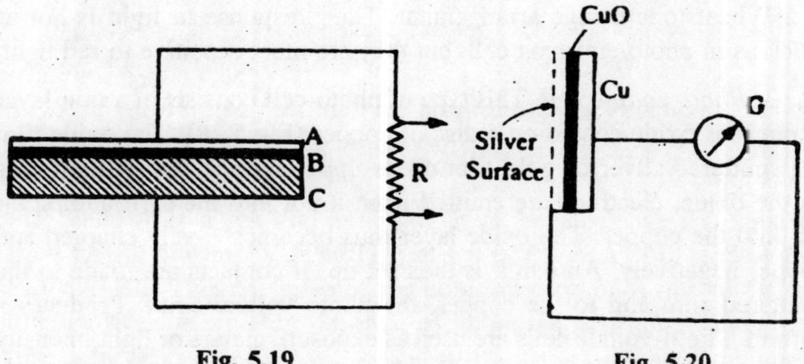

Fig. 5.19

Fig. 5.20

In vacuum cells, the photoelectric current is very small but these cells keep a strict proportionality between the current and intensity of light. The chief advantages of vacuum type are.

(i) The sensitivity of these cells remain unaltered for a very long time provided the cathode is properly selected.

(ii) There is no time lag between the incident light and the photoelectrons and the photoelectric current is proportional to the intensity of illumination. Thus, they are extremely accurate in response.

Due to the above mentioned advantages, these cells are being used in television and photometry.

In gas filled cells, the response to light is not so quick but since an inert gas is being used in the cell, ionisation of the gas is produced by the emitted electrons which therefore increase the photo current. Photoelectric current in this case is not proportional to the intensity of illumination and there is some time lag. Such cells are used in cinematography both for recording and reproduction of sound.

2. *Photo-conductive cell.* These cells are based on the property that the resistance of selenium and certain other metals decreases with increase of illumination. No photo-electrons are emitted in this case. A photo-conductive cell consists of a thin film of a semiconductor. *e.g.,* selenium, B, placed below a thin semi-transparent metal film A. The combination is placed on a block of iron C in contact with it (Fig. 5.19). Ordinary selenium is a semi-conductor. When light is incident on A, it is absorbed by the selenium layer, electrons are emitted and if a small external e.m.f, is applied, a current flows in the resistance and can operate a relay when the intensity of incident light is high. These cells are generally connected directly to a micrometer or a low resistance relay and are preferably used in a Wheatstone bridge arrangement. Their response to light is not as quick as in photo-emission cell, but they are more sensitive to red light.

3. *Photo voltaic cell.* This type of photo-cell consists of a thin-layer of cuprous oxide coated on a disc of copper (Fig 5.20). The oxide film has sputtered silver or gold film on its upper surface. When light falls on the oxide, electrons are emitted from it not into the surrounding air but into the copper. The oxide layer thus becomes + vely charged and copper negatively. An e.m.f. is thus set up. If contacts are made to the sputtered film and to the copper, the photo voltaic e.m.f. Produces a current. Photo-voltaic cells are used as exposure meters or light intensity meters.

Applications of Photoelectric cells. Main applications are.

(1) Reproduction of sound in films, (2) In televisions (3) In meteorology as day-light recorders, (4) For measuring the complexion of persons, (5) In automatic light switching for switching on and switching off the street light. (6) In determination of temperature of stars. (7) Inburglar alarms to detectthieves and in fire alarms to indicate the outbreaks of fire, (8) In traffic signals.

CHEMILUMINESCENCE

It is the phenomenon of emission of visible light as a result of a chemical change at a temperature at which a black body normally does not emit visible light.

From the above definition, it is evident that chemiluminescence is the reverse of photochemical reaction in which light absorption causes chemical action. Some examples are.

(i) A classical example of chemiluminescence is the phosphorus which produces a greenish-yellow glow. It is probably due to the oxidation of phosphorus vapours by atmospheric oxygen. Phosphorus trioxide also produces glow. This is probably due to the oxidation of P_2O_3 to P_2O_5.

(ii) When a solution of strontium chloride is added to dilute H_2SO_4 in dark, a feeble glow is produced along with the precipitate of $SrSO_4$.

$$SrCl_2 + H_2SO_4 \rightarrow SrSO_4 + 2HCl + h\nu.$$

(iii) It was observed by Evans that a solution of p-bromophenyl magnesium bromide (Grignard reagent) in ether produces greenish- blue glow consisting of a single broad band.

(iv) When alkali metal vapours react with halogens or with mercuric halides, luminescence occurs which consists of the spectrum of the alkali metal. The mechanism of this is based on the fact that sodium metal contains monatomic and diatomic molecules. The mechanism is.

$$Na + Cl_2 \rightarrow NaCl + Cl$$

$$Na + Cl \rightarrow NaCl$$

$$Cl + Na_2 \rightarrow Na + NaCl^*$$

$$NaCl^* + Na \rightarrow NaCl + Na^*$$

$$Na^* \rightarrow Na^* + h\nu \text{ (yellow spectrum)}$$

(v) When the surface of mercury is allowed to come in contact with a stream of atomic hydrogen, a blue fluorescence appears. The spectrum of this blue glow consists of the resonance line 2537 Å and a band of spectral lines from 4520 to 3250 Å due to mercuric hydride. The possible mechanism is due to

$$Hg + H \rightarrow HgH$$
$$H + H \rightarrow H_2 + \text{Energy}$$
$$HgH + \text{Energy} \rightarrow HgH^*$$
$$HgH^* + HgH \rightarrow Hg + Hg^* + H_2$$
$$Hg^* \rightarrow Hg + hv$$

Theoretical Explanation of Chemiluminescence : The phenomenon of chemiluminescence can be explained on the basis of quantum theory.- According to this theory, certain chemical reactions result in the formation of products which are in electronically excited states, When these excited products return to their ground levels, they do so by emitting the extra energy in the form of visible light radiations. Thus, by this process, the chemical energy is converted into light energy.

Sensitised Chemiluminescence : In 1925 Kautsky reported that oxidation of unsaturated silicon hydride by $KMnO_4$ is accompanied by a luminescent glow. If the same oxidation is carried out in the presence of contain dyestuffs such as rhodamine-B, a strong red fluorescence characteristics of the dye is observed. This type of change is known as *sensitised chemiluminescence.*

Explanation. During the oxidation of silicon hydride, the energy is given out and is subsequently transferred to the dyestuff, raising it to an excited state. When it returns to normal state, a fluorescence is observed.

PHOTOSENSITISATION

Certain reactions are known which are not sensitive to light. These reactions can be made sensitive by adding a small amount of foreign material which can absorb light and stimulate the reaction without itself taking part in the reaction. Such an added material is known as photo-senistiser and the phenomenon as photosensitisation.

Formerly, it was thought that photosensitisation might be akin to catalysis. It has now been confirmed that photosensitisation is different in nature from ordinary catalysis. Photosensitized reactions are spontaneous involving an increase in free energy of the system.

Role Played by a Photosensitiser : The formation of a photosensitiser is to absorb light, become excited and then pass on this energy to one of the reactants and thereby activate them for reaction, without itself taking part in the reaction. Thus, a photosensitiser acts as a carrier of energy. Some examples are.

REACTIONS SENSITISED BY MERCURY ATOMS

(i) When a mixture of mercury vapour and hydrogen gas is illuminated by light of wavelength 2537 Å, the dissociation of molecular hydrogen into atomic hydrogen takes place.

$$Hg + H_2 \rightarrow 2H + Hg$$

The *mechanism* which is now suggested is the excitation of mercury atom and then to transfer the energy to hydrogen molecule which will dissociated to form atoms.

$$Hg + h\nu \rightarrow Hg^*$$

$$Hg^* + H_2^* \rightarrow H_2^* + Hg$$

$$H_2^* \rightarrow 2H$$

(ii) When the reaction is carried out between hydrogen and oxygen in the presence of mercury vapour under the influence of light radiations, the reaction leads to the formation of H_2O_2 and H_2O.

$$3H_2 + 2O_2 \xrightarrow[\text{Light}]{\text{Hg}} 2H_2O + H_2O_2$$

The accepted *mechanism* of the above reaction is that the first stage is the formation of hydrogen atoms by collision between excited mercury atom and hydrogen molecule.

$$Hg + h\nu \rightarrow Hg^*$$

$$Hg^* + H_2 \rightarrow 2H + Hg$$

The above reaction may be followed by reactions such as

(1) $H + O_2 + X \rightarrow HO_2 + X$

(2) $HO_2 + HO_2 \rightarrow H_2O + O_2$

(3) $HO_2 + H_2 \rightarrow H_2O_2 + H$

Now H_2O_2 may either be isolated as such or further decomposed to form H_2O and O_2

$$H_2O_2 \rightarrow H_2O + \frac{1}{2}O_2$$

(iii) The reaction between hydrogen and carbon monoxide is photosensitised by mercury atoms. The main products are formaldehyde and glyoxal in similar amounts. The quantum yield for this reaction in nearly 2. The accepted *mechanism* is that the primary process is the formation of hydrogen atom which further carries out chain reactions as given below.

$$H + CO + X \rightarrow HCO + X \text{ (where X is the foreign body)}$$

$$\text{Either } 2HCO \rightarrow HCHO + CO \quad \text{or} \quad 2HCO \rightarrow (CHO)_2$$

(iv) Mercury photosensitises the decomposition of ammonia in the presence of light of wavelength 2537 Å. Quantum efficiency of this reaction is 7 and the rate of this photosensitised reaction is, about 200 times as great as the photolysis of ammonia (in the absence of mercury vapour), if both the reactions are carried out at the same wavelength.

The possible mechanism is

$$NH_3 + Hg^* \rightarrow NH_3^* + Hg$$

$$NH_3^* \rightarrow NH_2 + H$$

$$H + H \rightarrow H_2$$

$$NH_2 + NH_2 \rightarrow N_2H_4$$

$$N_2H_4 + H \rightarrow NH_3 + NH_2$$

$$NH_2 + NH_3 \rightarrow N_2 + 2H_2$$

Other example which are photosensitised by mercury atoms are. (a) Decomposition of phosphine, (b) Ozonolysis of oxygen, (c) Decomposition of acetone, (d) Decomposition of water vapour, (e) Decomposition of ethyl alcohol.

(b) Chlorine as A photosensitiser : In many reactions, chlorine acts as a photosensitiser. Let us consider the decomposition of ozone in the presence of chlorine (as a photosensitiser) under the influence of ultra-violet light.

$$2O_3 \xrightarrow[h\nu]{Cl_2} 2O_2 .$$

The rate of above reaction is independent of concentration, but is proportional to the intensity of absorbed light. Many mechanisms have been proposed from time to time. The most satisfactory mechanism is given below.

1. $Cl_2 + h\nu \rightarrow 2Cl$
2. $Cl + O_3 + X \rightarrow ClO_3^* + X.$

The excited ClO3* radical may be absorbed on the walls of the containing vessel to form a mixture of Cl_2O_6, Cl_2 and O_2.

3. $ClO_3^* + ClO_3^* + M \rightarrow Cl_2 + 3O_2 + M.$

In addition to the above reactions, $ClO_3^* + M \rightarrow Cl_2 + 3O_2 + M$

In addition to the above reactions, ClO_3^* in the gas phase may react to form ClO_3 and O_2.

4. $ClO_3^* + O_2 \rightarrow ClO_2 + 2O_2$

5. $ClO_2 + O_3 \rightarrow ClO_3 + O_2$

(b) *Bromine as a photosensitiser.* Bromine acts as photosensitiser in the conversion of maleic acid into fumaric acid.

$$\begin{array}{c}H—C—COOH\\ \|\\ H—C—COOH\\ \text{Maleic acid}\end{array} \xrightarrow[Br_2]{hv} \begin{array}{c}H—C—COOH\\ \|\\ HOOC—C—H\\ \text{Fumaric acid}\end{array}$$

(c) *Cadmium vapour as a photosensitiser.* Cadmium vapour acts as a photosensitiser for the polymerisation of ethylene, for the decomposition of ethane and propane to hydrogen, methane and higher hydrocarbons.

$$nC_2H_4 \xrightarrow[hv]{Cd} (C_2H_4)_n$$

Photosensitisation in Solid Phases : When silver halides in photographic plates are exposed to red (7500 Å) and yellow (5900 Å) light, they are not appreciably affected. But addition of certain photosensitisers like red dye to the emulsion makes the photographic plate to respond not only to red but even to far infrared radiations.

$$AgBr + hv \xrightarrow{\text{Red dye}} Ag + Br$$

Photosensitisation in Solution : When the decomposition of oxalic acid is carried out by light of shorter wavelength in presence of uranyl ions, the quantum yield for this reactions is 0.5 or more. The mechanism of this reaction is not clear.

$$UO_2^{2+}\ hv \rightarrow [UO_2^{+2}]$$

$$[UO^{2+}]^{2*} + \begin{array}{c}COOH\\ |\\ COOH\end{array} \longrightarrow CO_2 + CO + H_2O + UO_2^{+2}$$

Uranyl ions also act as a photosenitizer in the photolysis of formic acid.

Chlorophyll as a photosensitiser : Chlorophyll acts as a photosenitiser in the photosynthesis of carbohydrates from CO_2 and H_2O.

$$\text{Chlorophyll} + h\nu \rightarrow [\text{Chlorophyll}]^*$$

$$6CO_2 + 6H_2O + [\text{Chlorophyll}]^* \xrightarrow[\text{sunlight}]{} C_6H_{12}O_6 + 6O_2 + [\text{chlorophyll}]$$

PHOTOCHEMICAL INHIBITION

There are certain substances which are able to retard the rate of a photochemical reaction when present in trace amounts. Such substances are known as *inhibitors* and the phenomenon is known as *photochemical inhibition*. Some examples are.

(i) Traces of nitric oxide and propylene lower the quantum yield of the photochemical combination of hydrogen and chlorine.

(ii) Traces of impurities like NH_3 which when present in the hydrogen and chlorine reaction lower its quantum yield from 10^6 to 10^4. In 1905, Chapman showed that the inhibitor action of NH_3 is due to the side reaction of NCl_3 which may result due to the reaction of NH_3 with Cl_2.

$$NCl_3 + 3HCl \rightleftharpoons NH_3 + 3Cl_3$$

$$NCl_3 + 3H_2O \rightleftharpoons NH_3 + 3HClO$$

At equilibrium the concentration of NH_3 is maintained and the actual formation of HCl is reduced,

(iii) SO_2 and O_2 were also found to be inhibitors in various photochemical reactions.

Explanation. It is generally accepted that photo-inhibitors interrupt the chain reactions by removing chain carrier atoms or radicals.

PERIOD OF INDUCTION

Many reactions are characterised by an initial period during which the process appears to be silent. This time is known as period of induction. Some examples are.

(i) An induction period is observed in hydrogen and chlorine reaction when certain impurities are present. With pure hydrogen and chlorine, no induction period is observed.

(ii) An induction period is also observed in the absorption of As_2O_2 by mercuric oxide in the presence of light.

Period of induction is due to presence of impurities like ammonia. These substances act as inhibitors by breaking these chains. When these impurities are completely converted into nitrogen and ammonium chloride during the reaction, then only normal photochemical reaction can take place.

PHOTOSTATIONARY STATE OR PHOTOCHEMICAL EQUILIBRIUM

A state of photochemical equilibrium is said to exist in a reaction when the rates of two opposing reactions of which at least one is light sensitive, become equal under the influence of light radiation.

Types : A number of cases of the photostationary state have been studied. These cases fall into two categories.

(a) In the First Category, only one reaction is light sensitive

$$A + B \underset{\text{dark}}{\overset{\text{light}}{\rightleftharpoons}} C + D$$

Some examples of such type are.

(i) *Dissociation of Nitrogen Dioxide*. When the vapour of nitrogen dioxide is exposed to light radiation of wavelength less than 3700Å, it dissociates to form nitric oxide and oxygen but the combination of these products is a dark reaction. Hence the equilibrium is

$$2NO_2 \underset{\text{dark}}{\overset{\text{light}}{\rightleftharpoons}} 2NO + O_2$$

When the reaction just starts, the pressure of the gas rises due to the decomposition of the nitrogen dioxide but becomes constant as soon as the stationary state it reached.

Dimerisation of anthracene. Another example of a photo-stationary state is the dimerisation of anthracene in solution in which the forward process is light sensitive but the back ward reaction is a thermal reaction, and so the equilibrium is

$$2_{14}H_{10} \underset{\text{dark}}{\overset{\text{light}}{\rightleftharpoons}} C_{28}H_{20}$$

Let us calculate the equilibrium constant of the reaction of category (a), light $$A + B \underset{\text{dark}}{\overset{\text{light}}{\rightleftharpoons}} C + D$$

As the forward reaction is light sensitive, its rate will not depend upon the concentrations of A and B but upon the intensity of absorbed light. Hence.

Rate of the forward reaction = $k_1 I_{abs}$ where k_1 is constant for the forward reaction. As the backward reaction is a dark or thermal reaction, its rate will depend upon the concentrations of products C and D, *i.e.*,

Rate of backward reaction = k_2 [C] [D].

At photostationary state.

Rate of forward reaction = Rate of backward reaction

or $$\frac{k_1}{k_2} = \frac{[C][D]}{I_{abs}} \quad \text{or} \quad K = \frac{[C][D]}{I_{abs}}$$

where K is the equilibrium constant and is defined as the ratio of velocity constants of two opposing reactions.

The equilibrium constants for photochemical equilibria are constant only for a given light intensity, and vary as the latter is charged. Photochemical equilibrium constant is generally independent of temperature changes if the intensity of light is kept constant.

(b) In the Second Category, both the reactions are light-sensitive.

$$A + B \underset{\text{light}}{\overset{\text{light}}{\rightleftharpoons}} C + D$$

Few examples of this category are

(i) *Formation of sulphur trioxide.* It is an example of a photostationary state in which both the forward and back ward processes are light sensitive.

$$2SO_2 + O_2 \underset{\text{light}}{\overset{\text{light}}{\rightleftharpoons}} 2SO_3 .$$

The value of photochemical equilibrium constant of this reaction is almost independent of temperature from 500° to 800°C.

(ii) *Isomerisation of maleic acid into fumaric acid.* Another example is the isomerisation of maleic acid into fumaric acid in which both forward and backward reactions are light sensitive.

$$\begin{matrix} H-C-COOH \\ \| \\ H-C-COOH \end{matrix} \underset{\text{light}}{\overset{\text{light}}{\rightleftharpoons}} \begin{matrix} H-C-COOH \\ \| \\ HOOC-C-H \end{matrix}$$

Maleic acid   Fumaric acid

The value of equilibrium constant can be calculated in the similar manner as given in category (a).

Application. The concept of photostationary state has been used to explain the phenomenon of *vision.* When the light sensitive substance in eye called visual purple exposed to light, the latter gets bleached to form visual yellow.

Thus, an equilibrium state is established between visual purple and visual yellow. In dark the visual purple accumulates and eye becomes sensitive. When exposed to light further, it results in the phenomenon of *dazzling.*

BIOLUMINESCENCE

Certain living organisms emit light and show the phenomenon of chemiluminescence. It is known as bioluminescence. The phenomenon of bioluminescence was investigated by *E. N. Harvey* (1915). Some interesting examples are.

(i) The cold light produced by certain living organisms like fire-fly or glow worm is probably due to the oxidation of protein, luciferin, by the atmospheric oxygen in the presence of enzyme luciferase. The firefly emits light having a maximum intensity of wavelength of 5700Å. It can be easily judged by naked eye.

(ii) Certain marine animals also emit light. The protozoa Noctiluca of the sea produces phosphorescence. It has also been observed that many deep sea fishes have powerful organs which emit dazzling light.

PHOTOSYNTHESIS

The energy which supports the activities of most living organisms on the earth is derived directly or indirectly form the energy of sunlight through photosynthesis *Photo* = light, *synthesis* = to build up). *It is a process in which simple carbohydrates are synthesised from water and carbon dioxide in the chlorophyll containing tissues of plants in the presence of sunlight. Oxygen being a by-product is given out*

Kamen (1963) has defined photosynthesis as a *series of processes in which electromagnetic energy is converted into chemical free energy which can be used for biosynthesis.* By a single overall equation, it may be shown as below.

$$CO_2 + H_2O \xrightarrow[\text{chlorophyll}]{h\nu} 1/6\,(C_6H_{12}O_6) + O_2$$

In more general way, it is

$$CO_2 + H_2O \xrightarrow[\text{chlorophyll}]{h\nu} 1/n\,[CH_2O]_n + O_2$$

Scarce of Oxygen Liberated in Photosynthesis : The released oxygen (O_2) in photosynthesis comes from water [H_2O^{18}]. When green plants were supplied with water containing [H_2O^{18}]. The released oxygen was entirely of the O^{18} type, indicating that water is only source to release oxygen in photosynthesis. This can be represented by the following summary reaction.

$$6CO_2 + 12H_2O^{18} \rightarrow C_6H_{12}O_6 + 6H_2O + 6O_2{}^{18}.$$

Photosynthesis and Pigments. The photosynthetic products are energy-rich organic compounds. The potential chemical energy of these compounds comes from the light energy. The light' energy to be effective in photosynthesis must be absorbed by a suitable pigment. The vital role is performed by the green pigment chlorophyll, in plants.

Chlorophyll pigment. There are at least seven types of chlorophylls known : chlorophylls *a, b, c, d* and *e.* bacteriochlorophyll and bacterioviridin. All these chlorophyll molecules contain a *tetrapyrrole* skelton formed into a ring with an atom of magnesium in the centre of the ring. A so-called pyrrole molecule contains a skeleton of five atoms, four carbon and one nitrogen and five are arranged in a ring. Four such pyrroles arranged in a ring form the 'head' of a chlorophyll molecule. Attached to this *porhpyrin* ring at one point is an alcohol (phytol) "tail", a long chain of linked carbons. Relatively minor variations in the kinds and groupings of other atoms joined to this head and tail skeleton account for the differences among different kinds of chlorophylls.

Chlorophylls *a* and *b* are the two most abundant ones found in all the autotrophic plants except the pigment containing bacteria. Chlorophyll *b* is however, absent in blue-green, brown, and red algae. The other chlorophylls (c, d, e) are found only in algae in combination with chlorophyll a. Chlorophyll *a* possesses a $-$ CH_3, a methyl group which is replaced by a $-$ CHO, an aldehyde group, in chlorophyll *b.*

Path of Carbon Dioxide in Photosynthesis : The air contains 0.03% carbon dioxide. This gas diffuses into the air spaces between the cells of the mesophyll of the leaf through the open stomata. Here it combines with water to form carbonic acid. Transport of carbon dioxide in the form of carbonic acid takes place from the spongy cells to the palisade cells until the main site of photosynthesis is reached. It is now proved that in photosynthesis carbon dioxide is not reduced as such but it first forms a complex with cellular compounds and is then reduced by some 'activated' compounds produced as a result of photochemical reaction.

Since isolated choloroplasts are unable to reduce carbon dioxide in the light, it is concluded that either carbon dioxide uptake must take place outside the chloroplast or some mechanism has been destroyed during extraction of the chloroplasts. The first alternative is more probable. *Frenkel* (1941) prepared CO_2 from radioactive (C^{14}) carbon. He fed *Nitella* cells with this carbon dioxide ($C^{14}O_2$) in the dark and found that all the radioactive carbon was located in the cytoplasm. When exposed to light, 4/5 of radioactive carbon was found in the chloroplasts. The simplest interpretation of Frenkel's result is that carbon dioxide uptake occurs in the cytoplasm whereas reduction of carbon dioxide is associated with the chloroplasts.

Mechanism of Photosynthesis : Various theories have been propounded by various workers. But the earlier theories failed and still not theory is available which describes the mechanism satisfactorily. The uncertainty in the mechanism is die to.

(a) No products of the earlier stages of the process could be detected experimentally.

(b) When chlorophyll is extracted from plant, it fails to reproduce its photosynthetic property. So the photosynthesis cannot be studied under controlled experimental conditions.

(c) The complete structure of the chlorophyll was not known fully.

The mechanism of photosynthesis as currently understood is usually divided into three phases. (i) The absorption of the light energy by the pigment system, (2) conversion of light energy into chemical energy by photophosphorylation and (3) synthesis of organic compounds by the products of the photo-chemical reactions.

Enough evidence has accumulated in the recent past to show that photosynthesis consists of two stages, (a) light phase and a dark phase.

The reaction of light phase is light sensitive and 'therefore' called a photochemical reaction. The reactions of the dark phase do not require light and are temperature sensitive. Such reactions are after called *Blackman reactions*.

$$\xrightarrow[\text{Photochemical reaction}]{\text{Step I}} \xrightarrow[\text{chemical reaction}]{\text{Step II}} \text{Photosynthetic Product}$$

(The component reactions of photosynthesis).

Evidences for the Existence of Light and Dark Phases : There are following evidences.

1. *Experiments with intermittent light.* When Chlorella cells were exposed to continuous illumination, *Warbarg* found that the rate of photosynthesis was less when compared to those exposed alternately to light and darkness. The increase in the rate of photosynthesis was enhanced greatly if the span of the light period was reduced. This is because light period followed by darkness allows the cells to utilize all the products of the light reaction for the reduction of CO_2 in the dark reaction. When the light was continuous or interrupted; the products of light phase accumulate because the pace of light reaction is far greater than that of dark.

2. *Temperature Experiments.* When the photosynthesis is carried out in the excess of carbon dioxide but in the presence of low light intensity, the rate of photosynthesis does not increase with the increase in temperature. This indicates that some reaction requires light and is not affected by rise in temperature. When the concentration of carbon dioxide is low and the intensity of light is high, the rate of photosynthesis becomes double for every 10°C rise in temperature. This shows that there must be a reaction which is not affected by light. This is the dark reaction of photosynthesis.

   Let us discuss these two stages separately.

LIGHT PHASE

Evidence in support of light phase. Hill and *Scrisbrick* found that isolated chloroplasts when illuminated produced oxygen from water. If certain hydrogen acceptors (oxidants) were present in the medium, they were reduced by the chloroplasts. The oxidants are called *Hill reagents* and the reaction as *Hill reaction.* Hill attributed the production of oxygen and hydrogen from water by light. *Arnonin* 1954, working on isolated chloroplasts reported that in addition to carrying out the Hill reaction, the chloroplasts could also synthesis ATP in the light from ADP and inorganic phosphate (Pt). This *photosynthetic phosphorylation* may be represented as.

$$ADP + P_t \xrightarrow{\text{Light energy}} ATP.$$

Reaction occurring in Light phase.

When chlorophyll molecule absorbs one quantum of light, it gets excited. The excited chlorophyll molecule brings about the following changes.

(i) Excited chlorophyll molecule interacts chemically with water to liberate oxygen.

$$4H_2O + \text{chlorophyll}^* \rightarrow 4[H] + 4[OH] + \text{chlorophyll}$$

$$4[OH] \rightarrow 2H_2O + O_2.$$

The nascent hydrogen reacts with NADP to form $NADPH_2$ which acts as a reducing agent.

$$NADP + 2H \rightarrow NADPH_2$$

(ii) Excited chlorophyll also brings about the reaction between ADP and inorganic phosphate Pt to form an energetic compound ATP.

$$ADP + P_i + \text{chlorophyll}^* \rightarrow ATP + \text{chlorophyll}.$$

The products obtained from light reaction enter the dark reaction of photosynthesis.

Mechanism of light phase reactions. When sun light falls on the leaves, the chlorophyll present in the leaves absorbs light photon and becomes excited. This photo excitation results in the displacement of an electron from the normal orbit (of chlorophyll) into a anew orbit. The hole left by the displaced electron is soon filled up by the return of the same electron or another one.

If the same electron is returning the process is termed as *cyclic-transfer of electron.* However, if the different electron is returning, it is termed as non-cyclic transfer of electrons.

$$\text{Chlorophyll} + h\nu \rightarrow \text{Chlorophyll}^*$$

$$\text{Chlorophyll}^* \rightarrow (\text{chlorophyll})^* + e^-$$

Non-cyclic electron transfer. It involves the following steps.

(i) First of alkyl water dissociates into hydrogen and hydroxyl ions (Fig. 1).

$$4H_2O \rightarrow 4H^+ + 4OH^-$$

(ii) When a quantum of short wavelength of light is received by a pigment II of chlorophyll molecule, it loses an electron. The loss of electrons is immediately compensated by the electrons of 4 OH^- ions, *i.e.*, converted into water and oxygen.

$$4\,OH^- \rightarrow 2H_3O + O_2 + 4e^-$$

(iii) The electron released from above process is raised from + 0.8 eV to zero potential at which it is captured by an electron carrier called *Plastoquinone.*

(iv) After this the electron travels downhill and falls back to + 0.4 eV in a dark reaction through a series of carriers (cytochrome b_6, cyt. f, and plastocyanine) to pigment system I. The energy released in the dowahill passage of electron from cytochrome b_6 to cytochrome f is utilised to convert ADP and inorganic phosphate into ATP.

$$ADP + P_i \rightarrow ATP$$

(v) The electron is pumped from + 0.04 to – 0.4 eV with energy of another quantum absorbed by the system. I. According to *Tagawa* and *Arnon* (1962) the electron from system I is captured by any iron-containing protein, called *ferredoxin* with a potential of – 4.32 eV (Fig. 5.21).

(vi) According to *Aron* (1967) the electron is finally passed on at – 0.32 e.v. to NAD which together with two H+ released from water, becomes reduced to NHDPH + H^+ (expressed as $NADPH_2$ for convenience).

$$NADP^+ + 2e^+ + 2H^+ \rightarrow NADPH + H^+$$

Energy from ATP helps to more the electron from $NADPP_2$ to phosphoglyceric acid into carbon cycle.

Cyclic electron transfer. The expelled electron from the chlorophyll molecule is first accepted by *ferredoxin* or by an other electron acceptor of the chloroplast. The electron then traverses via cytochrome b_6 and cytochrome f, the two native cytochromes of chloroplasts, and ultimately reaches the original chlorophyll molecule from which it has been expelled. This scheme of electron transport is called cyclic photophosphorylation. In the cyclic photophosphorylation the electron, that is returned to the chlorophyll, is the same as the one that was expelled. In cyclic photophosphorylation energy is released during the transfer of electrons to cytochromes and the released energy is utilized to synthesize ATP.

II. Dark Phase : (Path of carbon in photosynthesis).

The present knowledge on the path of carbon in photosynthesis comes mainly from the work of *Melvin Calvin* and his associates,. They allowed the algae supplied with carbonlabelled CO_2 to photosynthesize for very brief periods of time and killed them instantaneously with boiling alcohol. Such studies using tracer technique revealed that products of CO_2 assimilation even for periods less than a minute were mainly sugars and amino acids shown in Fig. 5.21.

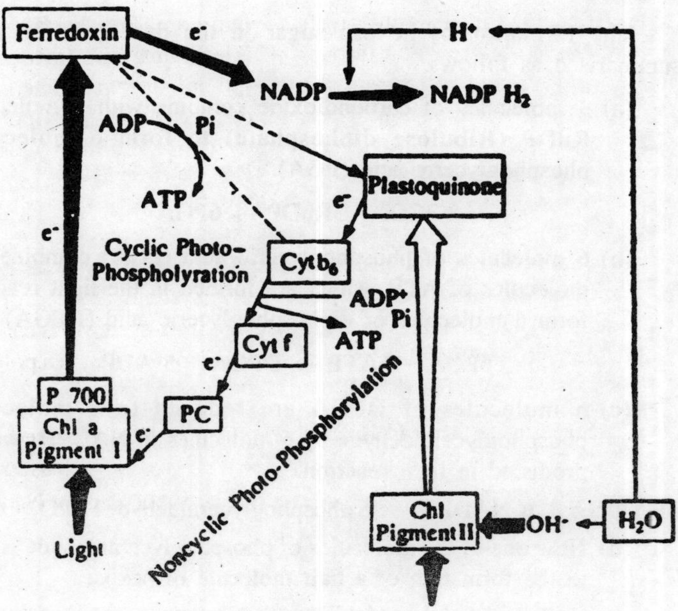

Fig. 5.21

Calvin's group was interested to know the first stable product of photosynthesis. For this they allowed the algae to photosynthesis for 5 second and extracted the compounds after killing the organisms. The first stable product of very brief period of photosynthesis was found to be 3 carbon compound phosphoglyceric acid (PGA). Since PGA is a 3 carbon compound, it was thought that CO_2 is accepted by a 2-carbon molecule to form 3 carbon PGA. But no such compound was detectable in plants. Later, *Benson* working in the same laboratory found that a 5-carbon sugar Ribulose diphosphate (RuDP) is the acceptor molecule of CO_2. On accepting CO_2 it was established that an unstable carbon intermediate is found and it cleaves into 2 molecules of PGA.

The end products of photosynthesis, namely sugars or starches, are synthesized from PGA by reversed sequence of EMP pathways of glycolysis. It was also established that some of the PGA is used to regenerate RuDP by accepting some more CO_2 molecules. Thus, there is a regular cycle of reactions that ultimately produce the end product of photosynthesis.

The formation of hexose sugar in the dark reaction may be summarised as follows.

(a) 3 molecules of carbondioxide combine with 3 molecules of RuDP (Ribulose diphosphate) to form 6 molecules of phosphoglyceric acid (PGA).

$$3CO_2 + 3RuDP \rightarrow 6PGA$$

(b) 6 molecules of phosphoglyceric acid (PGA) combine with 6 molecules of ATP, which are formed in the light reaction, to form 6 molecules of diphosphoglyceric acid (DPGA).

$$6PGA + 6ATP \rightarrow DPGA + 5\ ADP$$

(c) 6 molecules of DPGA are reduced to 6 molecules of phosphoglyceraldehyde by 6 molecules of $NADPH_2$ which are produced in light reaction.

$$6\ DPGA + 6\ NADPH_2 \rightarrow 6\ \text{phosphoglyceraldehyde}\ H_2PO_4 + 6\ NADP$$

(d) Now one of the molecules of phosphoglyceraldehyde is utilised in the formation of a half molecule of hexose.

$$1\ \text{Phosphoglyceraldehyde} \rightarrow 1/2\ (\text{hexose}).$$

The five other molecules of phosphoglyceraldehyde are utilised for the continued regeneration of ribulose diphosphate.

$$\text{5 Phosphoglyceraldehyde} \xrightarrow{\text{Changes}} \text{3 Ribulose Diphosphate.}$$

(i) The first molecule of phosphoglyceraldehyde is converted into its isomer, dihydroxy acetone phosphate. The latter molecule combines with second molecule of phosphoglyceraldehyde to form fructose 1, 6 diphosphate which in its turn gets dephosphorylated to form fructose monophosphate.

$$\text{Phosphoglyceraldehyde} \xrightarrow{\text{Isomerises}} \text{Dihydroxyacetone phosphate}$$

$$\text{Dihydroxy acetone phosphate} + \underset{\text{(II)}}{\text{phosphoglyceraldehyde}} \longrightarrow$$

$$\text{Fructose, 6 monophosphate} \xrightarrow{\text{ADP} \to \text{ATP}} \text{Fructose 1, 6 diphosphate.}$$

(ii) The fructose 6 monophosphate now combine with the third molecule of phosphoglyceraldehyde to give rise erthrose monophosphate (C_4) and xyluose monophosphate (C_5).

$$\text{Fructose} - 6 - \text{monophosphate} + \underset{\text{(III)}}{\text{Phosphoglyceraldehyde}}$$

$$\longrightarrow \text{Erythrose monophosphate } (C_4) + \text{Xylulose monophosphate } (C_5)$$

(iii) The erythrose monophosphate now combines with the fourth molecule of phosphoglyceraldehyde to form sedoheptulose diphosphate which loses one phosphate to become sedoheptulose monophosphate. It then combines with the fifth molecule of phosphoglyceraldehyde to give rise to one molecule each of xylulose monophosphate and ribose monophosphate.

$$\text{Erythrose monophosphate} + \text{Phosphoglyceraldehyde.} \longrightarrow \text{sedoheptulose diphosphate}$$

$$\downarrow \text{ADP} - \text{ATP}$$

$$\underset{+ \text{ Ribose monophosphate}}{\text{Xylulose monophosphate}} \xleftarrow[\text{(V)}]{\text{Phosphoglyceraldehyde}} \underset{\text{monophosphate}}{\text{Sedoheptulose}}$$

The ribose monophosphate gives rise to one molecule of ribulose monophosphate. The two molecule of xylulose monophosphate, one from this step and another form step (ii), are reversibly converted into its isomer ribulose monophosphate.

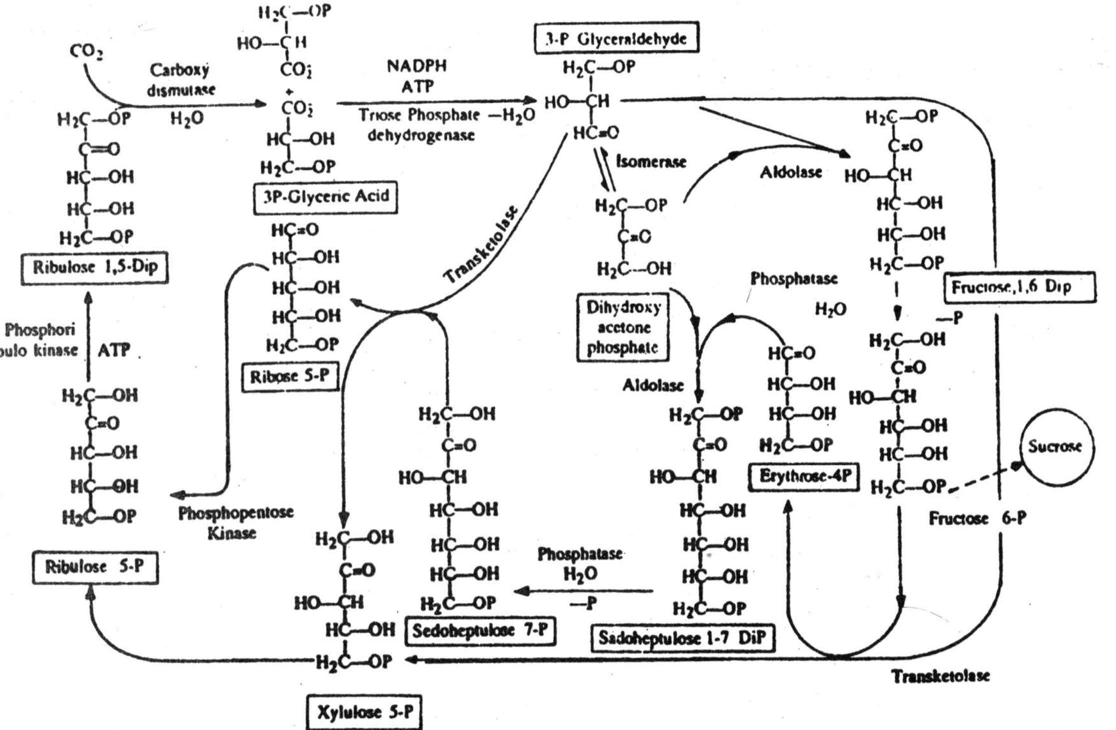

Fig. 5.21a : The path of carbon in the 'dark' phase of photosynthesis (Ater M. Calvin).

In this way five molecules of phosphoglyceraldehyde are converted into three molecules of ribulose monophosphate which in turn gets phosphorylated to form ribulose diphosphate. Now by doubling the number of CO_2 molecules of six, the path of carbon in dark phase of photosynthesis Fig. 5.21a.

LATENT IMAGE

The usefulness of present-day photographic materials is a result of the behaviour of the tiny individual particles of silver halide which are contained in photographic emulsion. In many cases these particles are extremely sensitive to light and store up the effect to an exceedingly small amount of light. The effect can, in turn, be multiplied many fold by the action of a developer. *This stored up effect of light is known as the latent image.* It is so small that it cannot be developed itself. Consequently, the nature of the latent image and the nature of the development process are closely related.

Def. *When a photographic plate is exposed to light radiation for a certain time, an invisible image is formed on it. This invisible image is known as latent image.*

FACTS

(i) In the formation of latent image, no change in the emulsion has been detected.

(ii) The formation of latent image is very sensitive to traces of substances such as Ag, Ag_2S etc.

(iii) When the photographic plate carrying latent image is exposed to mild reducing agent like pyrogallol, the latent image is developed to a negative plate.

Studies of the behaviour of the latent image under different conditions of development and exposure and its reactions with certain other chemicals have led to a fairly clear conception of its nature and the mechanism of its formation. An understanding of these processes requires knowledge of the structure of the silver halide grains themselves.

Theories of Latent Image Formation : When a film is exposed to light radiation, the absorbed energy ejects an electron from the bromide ion.

$$Br^- + h\nu \rightarrow Br + 1e^-.$$

The electron thus set free is captured by a silver ion to form a silver atom.

Thus, the overall reaction is the photochemical dissociation of silver bromide into silver and bromine by absorbing one quanta of light, *i.e.*,

$$AgBr + h\nu \rightarrow Ag + \dot{B}r.$$

Thus during the formation of latent image, silver is formed and bromine atom being removed by gelatin.

Objection. The main problem is that light is absorbed all over the surface of the photographic film but silver atoms make their appearance only at a few isolated points within each grain to form a latent image. How does it happen ? How is the latent image formed ? Various theories regarding the formation of latent image are.

I. Paul's View : According to Paul there are several empty holes in the crystal. When a positive ion is missing at one point a negative ion will be missing at some other point. Thus, there exists some holes in the crystal.

When light is incident on a photographic plate, the electrons are given out and finally fall inside the holes. Thus,

$$X^- + h\nu \rightarrow \quad X + e^-$$

the latent image was formed at these holes. When the exposed photographic plate is developed, negative image is formed at these holes. This theory does not explain the formation of latent image in complete manner.

+	–	+	–	+	–
–	+	–	+	–	+
+	–	hole	–	+	–
–	+	–	+	–	+
+	–	+	–	+	–
–	+	–	hole	–	+
+	–	+	–	+	–
–	+	–	+	–	+

Fig. 5.22 : Formation of latent image.

II. Webb's Theory : Recent advances in quantum mechanics and the applications of these principles to the structure and behaviour of crystals have contributed a great deal to knowledge of the problem of latent image formation. Webb (1936) first applied the principles of quantum mechanics to an explanation of the formation of the latent image.

Webb described the latent image as a concentration of electrons at a speck, or trap. The electrons are supposed to be transferred to a higher energy level by the absorption of light and in this energy level they could wander through the crystal at will as in the photo-conductance effect. The sensitivity specks were supposed to contain energy levels slightly lower than this conductances level, so that when an electron travels to such a specks, it gives up a bit of energy in going to this new level and is trapped. This gave an excellent explanation of the speck concentration of the effect of absorbed light but did not explain the formation of metallic silver as latent image substance.

III. Gurney and Mott's Theory : Gurney and Mott presented a theory for the formation of latent image. His main postulates are.

1. The energy structure of a crystal of silver halide can be described as in Fig. 5.23. The Ag and Br represent the silver and bromide ions which are present alternately in a crystal. The crosshatches labelled as P represent the energy level of the electrons connected with the bromide ions. This energy band is filled and electrons cannot move around in it. The clear energy band labelled S represents the energy level corresponding to the conductance level in metallic silver. In a crystal of silver halide all the valence electrons are associated with the bromide ions and none with silver so that this band is empty.

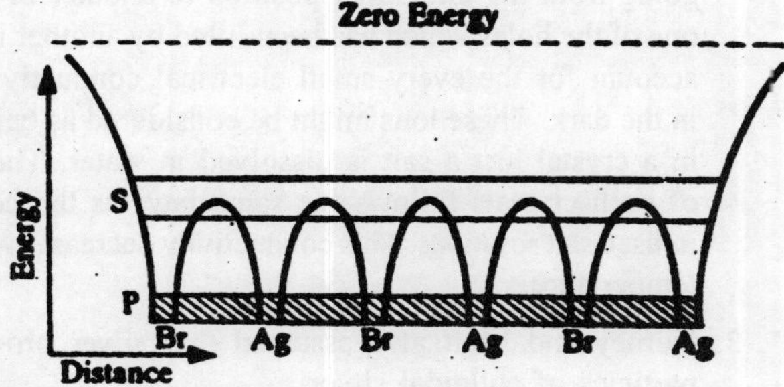

Fig. 5.23 : Quantum mechanical description of the energy level occurring in a crystal of silver bromide.

Electrons cannot have energy represented by the areas not included by these two bands. The photo-conductance phenomenon can be explained by this diagram–when light is absorbed, its energy is transferred to one of the electrons in the P and the electron is lifted to the S band where it is free of wander around just as the conductance electrons flow around in metallic silver.

2. Another concept which was applied to theory of photolysis by Gurney and Mott was the presence of interstitial ions in a crystal lattice. A perfect crystal would contain some ions which are not in their proper places but in the interstitial positions. Presumably these are caused by thermal motion of the ions in the crystals. This condition is shown i: Fig. 5.24.

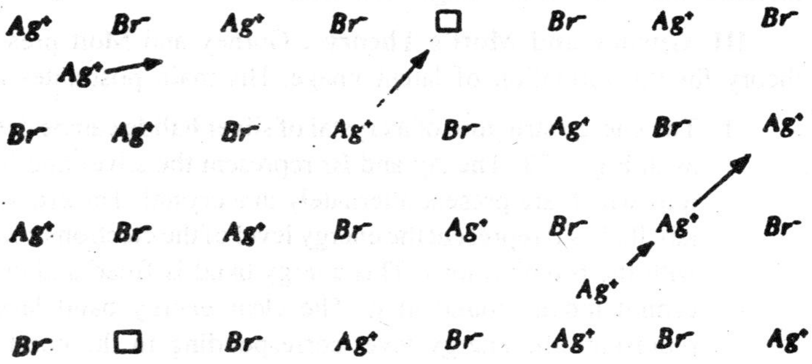

Fig. 5.24 : Interstitial silver ions in the silver bromide lattice.

These ions and their movement through the crystal, either by going from the interstitial position to another or by moving to one of the holes which has been lifted by another interstitial ion, account for the every small electrical conductivity of crystals in the dark. These ions might be considered as being in solution in a crystal just a salt is dissolved in water. The conductivity of such crystals follows the same laws as the conductivity of ionised salt solutions. This conductivity decreases with decreasing temperature.

3. Gurney and Mott also assumed that silver bromide contains particles of colloidal silver.

4. In order to apply his theory to the formation of latent image as well as to the photolytic formation of metallic silver he postulated

that the sensitivity specks, silver sulphide or whatever their complete composition might be, can serve the same function as the colloidal silver specks, *i.e.*, they can act as traps for electrons in the conductance level of the crystal. The theory involves the following points.

(i) When a film is exposed to light radiation, the primary absorption of light occurs in Br– ions of silver bromide crystal.

$$Br^- + h\nu \rightarrow Br^- + le$$

producing an excited electron.

(ii) The bromine atom is termed as a positive hole and is transferred from its original position in the crystal lattice to the surface of the crystal by the following process.

$$Br + Br^- \rightarrow Br^- \; Br$$

On the surface, the positive hole is trapped by coming in contact with an impurity centre line Ag_2S, probably by absorbing an electron from the sensitizer.

(iii) The excited electrons [from (i)] are free to move on through the crystal lattice. The moving electrons come in contact with a specks of silver and the trapped. This charges the speck negatively and this negative charge acts at some of interstitial silver ions which migrate to speck.

The positive charges on the speck due to silver ions are neutralised by the excited electrons.

$$Ag^+ + le \rightarrow Ag.$$

Thus, the above process is repeated and a large number of silver atoms collect as an invisible speck at one point in a crystal. The same process is repeated in other silver bromide crystals on the film and ultimately leads to the formation of latent image.

Objections : These are.

(i) This theory had not taken sufficient account of the possibility of the recombination of photo released electrons with the positive holes which are formed at the same time.

$$\underset{\text{positive hole}}{Br} + le \rightarrow Br^-$$

(ii) *Gurney* assumed the existence of interstitial silver ions on a photographic film. *Mitchell* showed that a very small number of interstitial silver ions is found to exist on a photographic plate. This number is not sufficient to explain Gurney-Mott's theory.

IV. Mitchell's Theory (1957) : It involves the following steps.

(i) Upon the absorption of a quantum of light, a positive hole and photo-electron are formed.

$$Br^- \; h\nu \rightarrow \underset{\text{Positivee hole}}{Br} + \underset{\text{Photo electron}}{1e}$$

(ii) The positive hole is trapped at a sensitivity speck in the crystal and this liberates an interstitial silver ion.

(iii) The liberated interstitial silver ions absorbed at some distortion or kink in the crystal lattice. Then, the absorbed positively charged silver ion absorbs the photo-electron.

$$Ag^+ + 1e \rightarrow Ag.$$

This single silver atom is called the latent pre-image.

(iv) Another interstitial silver ion released by the trapping of a positive hole is absorbed this latent pre-image.

$$Ag^+ + Ag \rightarrow \overset{\displaystyle Ag^+}{\underset{}{\overset{|}{Ag}}}$$

Again, the silver ion traps a photo-electron to give a two-atom specks. This is called latent sub-image.

$$\underset{\displaystyle Ag}{\underset{|}{Ag^+}} + 1e \rightarrow \underset{\displaystyle Ag}{\underset{|}{Ag}}$$

(v) A third interstitial silver ion released by trapping of positive hole is absorbed on the latent sub-image and again, the absorbed silver ion traps a photo-electron, leading to the formation of latent image.

General Views : The series of experiments carried out in recent years have shown that the above theories are not satisfactory and they require further investigations.

PHOTOCHEMICAL KINETICS

The rate laws which photochemical reactions follows are generally more complex than those for thermal reactions because more variables are involved.

Like the kinetics of reactions as discussed in chemical kinetics we will apply the steady state treatment. We will now apply steady state-treatment to reactions which do not involve chains. Examples are

(i) **Dissociation of HI :** The dissociation of hydriodic acid was investigated by *Warburg* (1911) in ultraviolet light of wavelength 2070, 2530 and 2820 Å.

$$2HI \xrightarrow{hv} H_2 + I_2$$

The experimental quantum efficiency of this reaction was 2.0. In order to account for this,. Warburg postulated the following mechanism.

(a) $HI + hv$ ¾→ $H + I$ Rate = I_{abs}

(b) $H + HI \xrightarrow{k_2} H_2 + I$ Rate = k_2 [H] [HI]

(c) $I + \xrightarrow{k_3} I_2$ Rate = k_3 $[I]^2$

In the above mechanism, hydrogen iodide is consumed insteps (a) and (b). Thus, its rate of dissociation is given by

$$\frac{-d[HI]}{dt} = I_{abs} + k_2 [H] [HI] \quad ...(1)$$

Also, the net rate of formation of hydrogen atoms is given by

$$\frac{d[H]}{dt} = I_{abs} - k_3 [H] [HI] \quad ...(2)$$

As [H] atom is short lived, the steady treatment, therefore, can be applied to it. Thus, equation (2) becomes

$$\frac{d[H]}{dt} = 0 \; I_{abs} - k_2 [H] [HI] \text{ or } k_2 [HI [HI] = Iabs$$

Substituting this value in equation (1), we get

$$\frac{-d[HI]}{dt} = 2I_{abs} \quad ...(3)$$

Quantum yield. By definition, the quantum yield is given by

$$\phi = \frac{\frac{-d[HI]}{dt}}{I_{abs}} = \frac{2I_{abs}}{I_{abs}} = 2 \qquad \text{[From (3)]}$$

This value of quantum efficiency is quite in agreement with experimental value Table 5.3.

Table 5.3 : Photochemical Decomposition of HI

Wavelength	Moles per K cal	Quantum yield
2070 Å	1.44×10^{-2}	1.98
2530 Å	1.85×10^{-2}	2.08
2830 Å	2.09×10^{-2}	2.10

(ii) Dissociation of HBr. The quantum yield for the dissociation of HBr is also two. Spectroscopic studies revealed that the mechanism followed by this reaction is the same as represented above for the dissociation of hydrogen iodide.

We will now apply steady-state to following reaction which involves chain mechanism.

1. Hydrogen and chlorine reaction and
2. Hydrogen and bromine reaction.

For details of chain reactions place see the chapter on 'Chemical Kinetics''

(i) Hydrogen and Chlorine Reaction : It is a classical example of photochemistry which involves chains. The reaction is

$$H_2 + Cl_2 \xrightarrow{hv} 2HCl$$

This reaction was first of all studied by *W. Cruickshank* and later by *J. W. Draper, R Bunsen* and so many other workers.

Facts : Special facts about this reaction are.

(i) The reaction is provoked by the light absorbed by chlorine in the region of 4785 Å.

(ii) An induction period is often observed. This is a period after the start of illumination during which the reaction is either very slow or does not take place at all.

The induction period is due to the presence of impurities like ammonia, NO_2, etc. That is why with pure hydrogen and chlorine no induction period is observed.

(iii) It was found by Draper (1841) that hydrogen-chlorine combination was followed by an increase in volume at constant pressure. This is known as Draper's effect.

(iv) The effect of oxygen is to influence the reaction after it has started. It has been found that the rate of reaction is inversely proportional to the concentration of oxygen.

(v) The quantum yield varies from 10^4 to 10^6 in the absence of oxygen. The value of quantum yield depend upon the impurities and oxygen concentration in the reaction vessel.

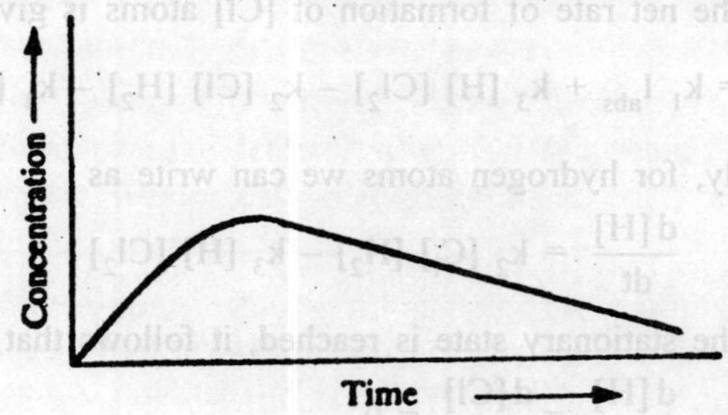

Fig. 5.25 : Combination of H_2 and Cl_2.

On the suggestion of Bodenstein, Nernst (1918) explained the reaction by chain mechanism which is universally accepted.

We will now consider the chain mechanism in two ways to explain.
(a) Combination of hydrogen and chlorine in the absence of oxygen and
(b) Combination of hydrogen and chlorine in the presence of oxygen.

In the Absence of Oxygen : (a) When a mixture of hydrogen and chlorine is exposed to light in the region of chlorine spectrum, chlorine molecule dissociates into atoms.

Initiation.

(i) $Cl_2 + hv \xrightarrow{k_1} 2Cl$] Rate = $k_1 I_{abs}$

The initiation reaction is followed by the following reactions. *Chain Propagation.*

(ii) $Cl + H_2 \xrightarrow{k_2} HCl + H$ Rate = k_2 [Cl] [H_2]

(iii) $H + Cl_2 \xrightarrow{k_3} HCl + Cl$ Rate = k_3 [H] [Cl_2]

Chain Termination.

(iv) $Cl + \text{wall} \xrightarrow{k_4} \frac{1}{2}Cl_2$ Rate = k_4 [Cl]

Derivation of Rate Law : The total rate of formation of hydrogen chloride is given by reactions (ii) and (iii). Therefore, its rate is given by

$$\frac{d[HCl]}{dt} = k_2 [Cl] [H_2] + k_3 [H] \{Cl_2] \qquad ...(4)$$

The rate of formation of chlorine atoms is given by step (i) and (ii) and the rate of removal of chlorine atoms is given by steps (ii) and (iv). Therefore, the net rate of formation of [Cl] atoms is given by

$$\frac{d[Cl]}{dt} = k_1 I_{abs} + k_3 [H] [Cl_2] - k_2 [Cl] [H_2] - k_4 [Cl] \qquad ...(5)$$

Similarly, for hydrogen atoms we can write as

$$\frac{d[H]}{dt} = k_2 [C_l] [H_2] - k_3 [H] [Cl_2] \qquad ...(6)$$

When the stationary state is reached, it follows that

$$\frac{d[H]}{dt} = \frac{d[Cl]}{dt} = 0.$$

$$k_1 I_{abs} + k_3 [H] [Cl_2] - k_2 [Cl] [H_2] - k_4 [Cl] = 0 \qquad ...(7)$$

and $k_2 [Cl] [H_2] - k_3 [H] [Cl_2] = 0$ ...(8)

By adding Eqs. (7) and (8), we get

$$k_1 \text{ Iabs} - k_4 [Cl] = 0 \text{ or } [Cl] = \frac{k_1}{k_4} I_{abs}. \qquad ...(8A)$$

Also, Eq. (8) gives as

$$k_2 [Cl] [H_2] = k_3 [H] [Cl_2] \qquad ...(9)$$

Substituting Eqs. (9) in (4) we get

$$\frac{d[HCl]}{dt} = 2k_2 [H_2] \frac{k_1}{k_4} I_{abs}$$

$$= 2\left(\frac{k_1 k_2}{k_4}\right) I_{abs} [H_2] \qquad ...(11)$$

Expression (11) is in perfect agreement with experimental data. In the presence of high chlorine content, step (iv) is replaced by following complicated steps.

(v) $Cl + Cl_2 \rightarrow Cl_3$

(vi) $2Cl_3 \rightarrow 3Cl_2$

If we calculate the rate expression by using steps (i), (ii), (iii), (v) and (vi), we get an expression which is very complex.

(b) **In the Presence of Oxygen :** When the reaction is carried out in the presence of oxygen, the following mechanism was proposed in 1921 by Gohring.

(i) $Cl_2 + h\nu \xrightarrow{k_1} 2Cl$ Rate $= k_1 I_{abs}$

(ii) $Cl + H_3 \xrightarrow{k_2} HCl + H$ Rate $= k_2 [Cl] [H_2]$

(iii) $H + Cl_2 \xrightarrow{k_3} HCl + Cl$ Rate $= k_3 [H] [Cl_2]$

(iv) $H + O_2 \xrightarrow{k_4} HO$ Rate $= k_4 [H] [O_2]$

(v) $Cl + O_2 \xrightarrow{k_5} ClO_2$ Rate $= k_5 [Cl] [O_2]$

(vi) $Cl + X \xrightarrow{k_6} ClX$ Rate $= k_6 [Cl] [X]$

In step (iv), X is any substance which removes chlorine atom. Applying the steady-state treatment to chlorine atoms, we get

$$\frac{d[Cl]}{dt} = k_1 I_{abs} - k_2 [Cl] [H_2] + k_3 [H] [Cl_2]$$

$$- k_5 [Cl] O_2] - k_6 [Cl] [X] = 0$$

This equation simplifies to

$$[Cl = \frac{k_1 I_{abs} + k_3 [H][Cl_2]}{k_2 [H_2] + k_5 [O_2] + k_6 [X]} \quad ...(12)$$

The rate of formation of hydrogen atoms is given by

$$\frac{d[H]}{dt} = k_2 [Cl [H_2] - k_3 [H] [Cl_2]$$

$$- k_4 [H] [O_2] = 0 \quad ...(13)$$

This equation on simplification yields.

$$[Cl] = \frac{k_3 [H][Cl_2] + k_4 [H][O_2]}{k_2 [H_2]} \qquad ...(14)$$

By equating Eqs. (12) and (14), we get

$$k_2 [H_2] (k_1 I_{abs} + k_3 [H [Cl_2]) = (k_2 [H_2] + k_5 [O_2] + k_6 [X]) + (k_3 [H] [Cl_2] + k_4 [H] [O_2])$$

By omitting small term $k_4 k_5 [H] [O_2]^2$, the above equation is simplified to give the value of [H], *i.e.*,

$$[H] = \frac{k_1 I_{abs} k_2 [H_2]}{k_3 k_6 [Cl_2][X] + [O_2](k_1 k_4 (H_2] + k_3 k_5 [Cl_2] + k_4 k_6 [X])} \qquad ...(15)$$

The rate of formation of hydrogen chloride is given by steps (ii) and (iii). Thus,

$$\frac{d[HCl]}{dt} = k_2 [Cl] [H_2] + k_3 [H] [Cl_2] \qquad ...(16)$$

Reaction (iii) is much faster than (ii), consequently the rate of formation is given by step (iii) only. Therefore, Eq. (16) becomes as

$$\frac{d[HCl]}{dt} = k_3 [H] [Cl_2] \qquad ....(17)$$

Introduction of expression (15) for [H] into this equation, we obtain:

$$\frac{d[HCl]}{dt} = \frac{k_1 k_2 k_3 [H_2][Cl_2] I_{abs}}{k_3 k_6 [Cl_2][X] + [O_2](k_2 k_4 [H_2] + k_3 k_5 [Cl_2] + k_4 k_5 [X])} \qquad ...(18)$$

$$\text{or} \quad \frac{d[HCl]}{dt} = \frac{\left(\frac{k_1 k_3}{k_4}\right) I_{abs} [H_2][Cl_2]}{\frac{k_3 k_6}{k_2 k_4} [Cl_2][X] + [O_2]\left([H_2] + \frac{k_3 k_5}{k_2 k_4} [Cl_2] + \frac{k_6}{k_2} [X]\right)}$$

$$= \frac{k I_{abs} [H_2][Cl_2]}{m [Cl_2] + [O_2]\left([H_2] + \frac{[Cl_2]}{10}\right)} \qquad ...(19)$$

where $k = \frac{k_1 k_3}{k_4}$, $m = \frac{[X][k_3 k_6]}{k_2 k_4}$

and $$\frac{1}{10} = \frac{k_3k_5}{k_2k_4} \quad ...(20)$$

Eq. (19) is in full agreement with the experimental rate equation as given by Bodenstein and Unger.

(2) Reaction Between Hydrogen and Bromine : In 1924, Bodenstein and Lurke-Meyer found that the photochemical reaction proceeds according to the empirical equation

$$\frac{d[HBr]}{dt} = \frac{k'[H_2]I_{abs}{}^{1/2}}{1+\frac{[HBr]}{m'[Br_2]}} \quad ...(21)$$

where k' and m' are constants, and I_{abs}, is the intensity of the light absorbed.

Mechanism : The photochemical combination between hydrogen and bromine to form hydrogen bromide is an example of a chain reaction. Its mechanism is similar to the thermal reaction with the difference that the initiation is brought about by the absorption of a photon by a bromine molecule.

Chain initiation

(i) $Br_2 \xrightarrow[k_1]{hv} 2Br$ Rat = $k_1\ I_{abs}$

The rest part of the mechanism is similar to the thermal reaction

Chain Propagation. Rate = k_2 [Br] [H_2]

(ii) $Br + H_2 \xrightarrow{k_2} HBr + H$ Rate = k_2 [Br] [H_2]

(iii) $H + Br_2 \xrightarrow{k_3} HBr + Br$ Rate = k_3 [H] [Br_2]

Chain inhibition :

(iv) $H + HBr \xrightarrow{k_4} H_2 + Br$ Rate = k_4 [H] [HBr]

Chain breaking : Rate = $k_5\ [Br]^2$

(v) $Br + Br \xrightarrow{k_5} Br_2$

Here k_1, k_2, k_3, k_4, and k_5 represent the specific rates of all the five reactions respectively.

Derivation of Rate Law : Since HBr is produced by reactions (ii) and (iii) and removed by the reaction (iv), the net rate of formation of HBr is given by

$$\frac{d[HBr]}{dt} = k_2 \, Br] \, [H_2] + k_2 \, [H] \, [Br_2] - k_4 \, [H] \, [HBr] \qquad ...(22)$$

As. Eq. (22) involves concentrations of hydrogen and bromine atoms which are too small quantities to be measured directly, it is required that their concentration must be expressed in measurable quantities. This is done by writing down and solving steady-state equations for [H] and [Br].

The Br atoms are formed by (i), (ii) and (iv) and are removed by (ii) and (v), so the net rate of formation is given by

$$\frac{d[Br]}{dt} = k_1 \, I_{abs} + k_3 \, [H] \, [Br_2] + k_4 \, [H] \, [HBr] - k_2 \, [H_2] \, [Br] - k_5 \, [Br]^2 \qquad ...(23)$$

As [H] atoms are formed by reaction (ii) and removed by (iii) and (iv), (v), so the net rate of formation is given by

$$\frac{d[H]}{dt} = k_2 \, [H_2] \, [Br] - k_3 \, [H] \, [Br_2] - k_4 \, [H] \, [HBr] \qquad ...(24)$$

When the stationary state is reached, it follows that

$$\frac{d[H]}{dt} = 0 \text{ and } \frac{d[Br]}{dt} = 0$$

Therefore, Eqs. (23) and (24) become as

$$k_1 \, I_{abs} + k_3 \, [H] \, [Br_2] + k_4 \, [H] \, [HBr] - k_2 \, [H_2] \, [Br] - k_5 \, [Br]^2 = 0 \qquad ...(25)$$

$$k_2 \, [H_2] \, [Br] - k_3 \, [H] \, [Br_2] - k_4 \, [H] \, [HBr] = 0 \qquad ...(26)$$

By adding Eqs. (25) and (26), we get

$$k_1 I_{abs} - k_5 \, [r]^2 = 0 \quad \text{or} \quad [Br] = \sqrt{\left(\frac{k_1}{k_5} I_{abs}\right)} \qquad ...(27)$$

Putting the value of Br in Eq. (26). we get

$$k_2 \, [H_2] \sqrt{\left(\frac{k_1 \, I_{abs}}{k_5}\right)} - k_3 \, [H] \, [Br_2] - k_4 \, [H] \, [HBr] = 0$$

$$\text{or} \quad k_3 \, [H] \, [Br_2] + k_4 \, [H] \, [HBr] = k_2 \, [H_2] \sqrt{\left(\frac{k_1 \, I_{abs}}{k_5}\right)}$$

$$\text{or}\quad [H]\,(k_3\,[Br_2] + k_4\,[HBr]) = k_2\,[H_2]) = k_2\,[H_2]\,\sqrt{\left(\frac{k_1\, I_{abs}}{k_5}\right)}$$

$$[H] = \frac{k_2\,[H_2]\sqrt{\left(\frac{k_1\, I_{abs}}{k_5}\right)}}{k_3\,[Br_2] + k_4\,[HBr]} \quad ...(28)$$

On substituting the values of [Br] and [H] from Eqs. (27) and (28) in Eq. (22), we obtain

$$\frac{d\,[HBr]}{dt} = k_2\,[H_2]\sqrt{\left(\frac{k_1\, I_{abs}}{k_5}\right)} + \frac{k_3\,[Br_2]\,k_2\,[H_2]\sqrt{\left(\frac{k_1\, I_{abs}}{k_5}\right)}}{k_3\,[Br_2] + k_4\,[HBr]}$$

$$- \frac{k_4\,[HBr]\,k_2\,[H_2]\sqrt{\left(\frac{k_1\, I_{abs}}{k_5}\right)}}{k_3\,[Br_2] + k_4\,[HBr]}$$

$$\frac{d\,[HBr]}{dt} = k_2\,[H_2]\sqrt{\left(\frac{k_1\, I_{abs}}{k_5}\right)}$$

$$\left(1 + \frac{k_3\,[Br_2]}{k_3\,[Br_2] + k_4\,[HBr]} - \frac{k_4\,[HBr]}{k_3\,[Br_2] + k_4\,[HBr]}\right)$$

$$\text{or}\quad \frac{d\,[HBr]}{dt} = \frac{k_2\,[H_2]\sqrt{\left(\frac{k_1\, I_{abs}}{k_5}\right)}}{k_3\,[Br_2] + k_4\,[HBr]}$$

$$(k_3\,[Br_2] + k_4\,[HBr] + k_3\,[Br_2] - k_4\,[HBr])$$

$$= \frac{2k_2\,[H_2]\,k_3\,[Br_2]\sqrt{\left(\frac{k_1\, I_{abs}}{k_5}\right)}}{k_3\,[Br_2] + k_4\,[HBr]} \quad ...(29)$$

Dividing the numerator and denominator by $k_3\,[Br_2]$, we get

$$\frac{d\,[HBr]}{dt} = \frac{2k_2\,[H_2]\sqrt{\left(\frac{k_1\, I_{abs}}{k_5}\right)}}{1 + \frac{k_4\,[HBr]}{k_3\,[Br_2]}}$$

$$= \frac{k'[H_2]\sqrt{\left(\frac{k_1 I_{abs}}{k_5}\right)}}{1+\frac{HBr}{m'[Br_2]}} \quad ...(30)$$

where $k' = 2k_2$ and $m' = k_3/k_4$ are two constants. Equating (30) in similar to Eq. (2) obtained by Bodenstein on the basis of this experimental data. Inspite of the chain mechanism similar to hydrogen and chlorine reaction, the quantum yield of this photochemical reaction is very low, being about 0.01 at ordinary temperatures. This is due to following reasons.

1. Reaction (ii) is highly endothermic and therefore, takes place so slowly at ordinary temperatures that most of the bromine atoms recombine to form bromine molecules. Therefore, the reactions (iii), (iv) and (v) which are due to reaction (i) cannot occur. Hence, the quantum yield is extremely slow.
2. With increasing time, the reversal reaction becomes predominant and hence, the rate of formation of HBr decreases.

3. Photolysis of Acetaldehyde : The photolysis of acetaldehyde has attracted much attention, in recent years. While discussing this reaction, the following facts must be kept in mind.

(a) The quantum yield of carbon monoxide formation increases with decreasing wave-length.

(b) When the temperature is raised, the quantum yield increases.

(c) The quantum yield of carbon monoxide formation at room temperature increases with decreasing pressure at 3130 A°.

(d) The photolysis of acetaldehyde in the light of 2500=3100 A° wavelength yields methane, carbon monoxide, ethane and a number of other products.

$$CH_3CHO + h\nu \rightarrow CH_4 + C_2H_6 + CO + \text{other products}$$

To account for the above facts, the following mechanism has been postulated.

(a) $CH_3CHO + h\nu \longrightarrow CH_3 + CHO$ Rate = Iabs

(b) $CH_3CHO + CH_3 \xrightarrow{k_2} CH_4 + CH_3CO$

Rate = k_2 [CH_3CHO] [CH_3]

(c) $CH_3CO \xrightarrow{k_3} CO + CH_3$ Rate = k_3 [CH3Co]

(d) $CH_3CH_3 \xrightarrow{k_4} C_2H_6$ Rate = $k_4 [CH_3]^2$

In the above mechanism, carbon monoxide is formed only in step (c), hence the rate of formation of this substance must be

$$\frac{d[CO]}{dt} = k_3 [CH_3CO] \quad \text{.. (31)}$$

The CH_3CO radical is formed by step (b) and is used up in the step (c). Therefore, the net rate of its formation is given by

$$\frac{d[CH_3CO]}{dt} = k_2 [CH_3CHO] [CH_3] - k_3 [CH_3CO] \quad \text{...(32)}$$

Again, $[CH_3]$ radical is formed by steps (a) and (c) and is consumed in steps (b) and (d). Hence its net rate of formation is given by

$$\frac{d[CH_3]}{dt} = I_{abs} + k_3 [CH_3CO] - k_2 [CH_3CHO] [CH_3] - k_4 [CH_3]^3 \quad \text{...(33)}$$

Applying the steady-state treatment to $[CH_3CO]$ radical, we obtain

$$\frac{d[CH_3CO]}{dt} = k_2 [CH_3CHO] [CH_3] - k_3 [CH_3CO] = 0$$

or $k_2 [CH_3CHO] [CH_3] - k_3 [CH_3CO] = 0$ ...(34)

Similarly, the steady-state treatment on applying to $[CH_3]$ radical yields

$$\frac{d[CH_3]}{dt} = I_{abs} - k_2 [CH_3CHO] [CH_3] + k_3 [CH_3CO] - k_4 [CH_3]^2 = 0$$

$$I_{abs} - k_2 [CH_3CHO] [CH_3] + k_3 [CH_3CO] - k_4 [CH_3]^2 = 0 \quad \text{...(35)}$$

By adding equations (34) and (35), we obtain

$$I_{abs} - k_4 [CH_3]^2 = 0$$

or $$k_4 [CH_3]^2 = I_{abs}$$

or $$[CH_3] = \left[\frac{I_{abs}}{k_4}\right]^{1/2}$$

On substituting the value of $[CH_3]$ in equation (34), we get

$$k_2\ [CH_3CHO]\left(\frac{I_{abs}}{k_4}\right)^{1/2} - k_3\ [CH_3CO] = 0$$

or $$k_3\ [CH_3CO] = k_2\ [CH_3CHO]\ \left(\frac{I_{abs}}{k_4}\right)^{1/2} \qquad ...(36)$$

Substituting this value of k_3 $[CH_3CO]$ in equation (36), we obtain

$$\frac{d\,[CO]}{dt} = k_2\ [CH_3CHO]\left(\frac{I_{abs}}{k_4}\right)^{1/2}$$

$$= \frac{k_2}{(k_4)^{1/2}}\ [CH_3CHO]\ I_{abs})^{1/2}$$

$$= I_{abs}^{\ 1/2}\ [CH_3CHO] \qquad ...(37)$$

where $$k = \frac{k_2}{(k_4)^{1/2}}$$

Quantum Yield : If the rate of formation of carbon monoxide is divided by the intensity of absorbed radiation, the value of quantum yield is obtained. Thus,

$$\text{Quantum yield} = \frac{d\,[CO]/dt}{I_{abs}}$$

$$= \frac{k\ I_{abs}^{\ 1/2}\ [CH_3CHO]}{I_{abs}} \qquad \text{[From eq. (37)]}$$

$$= \frac{k\,[CH_3CHO]}{I_{abs}} \qquad ...(38)$$

GENERAL DISCUSSION

1. From equation (38), it follows that the quantum yield of carbon monoxide formation increases as the intensity of light is decreased.
2. The activation energy of chain carrying process (b) has an appreciable energy, *e.g.*, about 10 K cal.

(b) $CH_3CHO + CH_3 \rightarrow CH_4 + CH_3CO$

This high activation energy accounts for the fact that quantum yield decreases as the temperature is lowered.

3. The effect of pressure on quantum yield may be explained by assuming that primary process

(a) $CH_3CHO + h\nu \rightarrow CH_3 + CHO$

is the main one at 3130 Å. At this wavelength, the CHO radical does not possess enough energy to decompose spontaneously. At high pressure, the CHO radicals may combine to form glyoxal. At low pressure, the CHO radicals may diffuse to the walls to form carbon monoxide.

LASER TECHNIQUES

When a large percentage of an ensemble of molecules can be brought into an excited state, a way frequently can be found by which large numbers of these excited molecules can be triggered into an almost simultaneous joint transition back to the inactive state, with the resulting emission of a beam of intense coherent radiation. Such a device is now known as laser, an acronym for light amplification by stimulated emission of radiation.

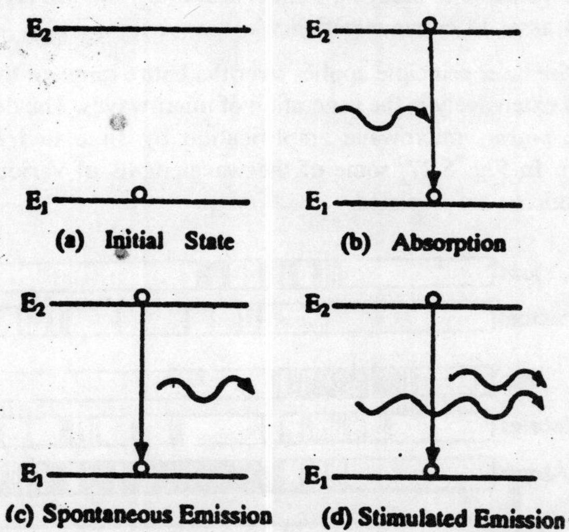

Fig. 5.26 : Four stage os laser action.

The four steps associated with such a device are illustrated in Fig. 5.26. In (a) the molecule is inactive, in (b) the molecule gets activated by a photon, in (c) the molecule returns spontaneously to its initial state, emitting radiation, and in (d) the transition to the ground state is triggered by an incoming beam, which is then amplified by the emitted radiation. For effective laser operation the amount of emission form (c) must be small compared with (d).

In a typical solid laser like a ruby crystal, the active atoms are those of chromium, which are held in a lattice of aluminium oxide. A flash of intense light raises a large percentage of the chromium atoms to the excited state.

Then radiation can trigger the return transition, resulting in the emission of light which is coherent both in space and in time because the stimulating photon on entering the atom causes it to emit a photon which is precisely "in step" with it.

In this way, monochromatic beams os great intensity can be produced which travel great distances without spreading. There are many important applications possible for such beams both in science and technology. They are valuable in studying Raman scattering and are certainly going to be an asset in communication.

As the laser principle applies over the entire range of the spectrum, it is used extensively in the generation of microwaves. The device is then called a *maser*, microwave amplification by simulated emission of radiation. In Fig. 5.27, some of the wavelengths of various laser and maser sources are depicted.

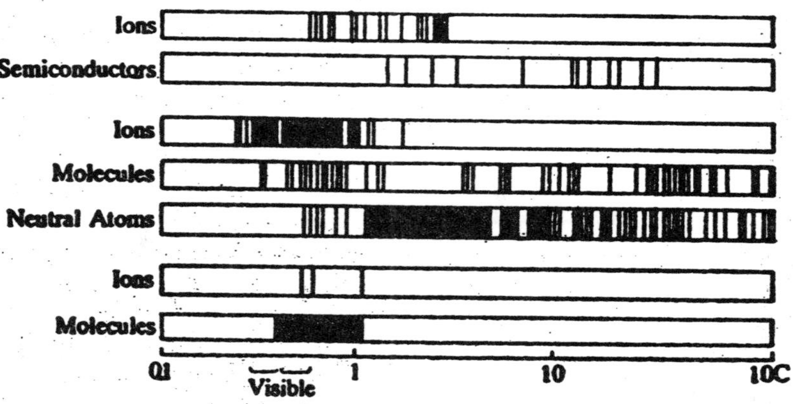

Fig. 5.27 : Laser and maser radiation sources available.

SOLVED EXAMPLES

Example 1:

Calculate the energy in calories per mole or per Einstein for radiations of wavelength 100 Å..

Solution:

Energy per quantum = hv

But $h = 6.62 \times 10^{-27}$ **erg-sec.**

$$\therefore \quad V = \frac{c}{\lambda} = \frac{3\times10^{10}}{1000\,\text{Å}} = \frac{3\times10^{10}\,\text{cm.sec}^{-1}}{1000\times10^{-8}\,\text{cm}} = 3\times10^{15}\ \text{sec}^{-1}$$

Energy per quantum = hv

$$= (6.62\times10^{-27}\ \text{erg. sec})\ (3\times10^{15}\ \text{sec}^{-1})$$

$$= 19.86\times10^{-12}\ \text{ergs.}$$

But $N = 6.02 \times 10^{28}$ **molecules/mole**

and $hv = 19.86 \times 10^{-12}$ **erg.**

Therefore, the energy per mole or per Einstein

$$= N\,hv$$

$$= (6.02\times10^{23}\ \text{molecules mole}^{-1})\ 19.86\times10^{-12}\ \text{ergs/mole}$$

$$= 11.94\times10^{12}\ \text{ergs/mole}$$

$$= \frac{11.94\times10^{12}}{10^{7}}\ \text{Jule/mole} \qquad [\because 1\ \text{Joule} = 10^{7}\ \text{ergs}]$$

$$= \frac{11.94\times10^{12}}{4.184\times10^{7}}\ \text{cal/mole} \qquad [\because 1\ \text{cal} = 4.184\ \text{Joules}]$$

$$= \frac{2.8590\times10^{5}}{10^{3}}\ \text{K cal/mole} \ [\because 1\ \text{K cal} = 10^{3}\ \text{cal}]$$

$$= 285.90\ \text{K cal/mole}$$

$$= \frac{285.90}{23.06}\ \text{electron volt} \ [\because 1\ \text{eV} = 23.06\ \text{K cal/mole}]$$

$$= 12.390\ \text{electron-volts.}$$

Example 2:

$$B \rightarrow C$$

1.0 × 10^{-5} mole of B was formed on absorption of 6.62 × 10^7 ergs at 3600 Å. Calculate the quantum yield or efficiency.

Solution:

No. of moles reacting

$$= 1.0 \times 10^{-5} \times 6.02 \times 10^{23} \text{ molecules}$$

$$= 6.02 \times 10^{18} \text{ molecules}$$

No. of quanta absorbed

$$= \frac{\text{Total energy absorbed}}{\text{Energy of one quantum}} = \frac{6.62 \times 10^7 \text{ ergs}}{h\nu}$$

$$= \frac{6.62 \times 10^7}{hc/\lambda}$$

$$= \frac{6.62 \times 10^7 \times \lambda}{hc} \qquad \left[\because \nu = \frac{c}{\lambda}\right]$$

But $\quad l = 3600 \text{ Å} = 3600 \times 10^{-8}$ cm.

$c = 3 \times 1010$ cm/sec, $h = 6.62 \times 10^{-27}$ erg/sec.

$$\text{No. of quanta absorbed} = \frac{6.62 \times 10^7 \times 3600 \times 10^{-8}}{6.62 \times 10^{-27} \times 3 \times 10^{10}} = 1.2 \times 10^{19}$$

$$\text{Therefore, equation yield} = \frac{\text{No. of molecules reacting}}{\text{NO. of quanta absorbed}}$$

$$= \frac{6.02 \times 10^{18}}{1.2 \times 10^{19}} = \mathbf{0.506.}$$

Examples 3:

Radiation of wave-length 2540 A° was passed through a cell containing 10 ml of a solution of 0.0495 molar oxalic acid and a 0.01 molar uranyl sulphate. After the absorption of 8.81 × 10^3 ergs of radiation the concentration of oxalic acid was reduced to 0.0383 molar. Calculate the quantum yield for the photo-chemical decomposition of oxalic acid at given wave-length.

Solution:

10 ml. of 0.0495 molar oxalic acid

$$= \frac{10 \times 0.0495}{1000} = 0.000495 \text{ mole}$$

10 ml of 0.0383 molar oxalic acid

$$= \frac{10 \times 0.0383}{1000} = 0.000383 \text{ mole}$$

$\therefore$ Amount of oxalic acid decomposed

$$= (0.000495 - 0.000383 = 0.000112 \text{ mole}$$

No. of molecules of oxalic acid decomposed by light.

$$= 0.000112 \times 6.023 \times 10^{23} \text{ molecules.}$$

Also, the No. of quanta absorbed.

$$= \frac{\text{Energy absorbed}}{\text{Energy of one quantum}} = \frac{8.81 \times 10^{8} \text{ ergs}}{h\nu}$$

$$= \frac{8.81 \times 10^{8} \text{ ergs}}{hc/\lambda} \qquad \left[\because \nu = \frac{c}{\lambda}\right]$$

$$= \frac{8.81 \times 10^{8} \lambda}{h\nu} = \frac{8.81 \times 10^{8} \times 2540 \times 10^{-8}}{6.625 \times 10^{-27} \times 3 \times 10^{10}}$$

$$= \frac{8.81 \times 2540}{6.625 \times 3 \times 10^{-17}}$$

$$\because \quad \text{Quantum yield} = \frac{\text{No. of molecules reacted}}{\text{No. of quanta absorbed}}$$

$$\text{Quantum yield} = \frac{0.000112 \times 6.023 \times 10^{23}}{\dfrac{8.81 \times 2540}{6.625 \times 3 \times 10^{-17}}} = \mathbf{0.5937.}$$

Example 4:

The bond energy in a molecule is 142.95 cal/mole. What is the longest wave-length of light capable of dissociating this molecule ?

Solution:

Energy per mole = N hν

where N= 6.02×10^{23} molecules, h = 6.62×10^{-27} erg-sec

$$\nu = \text{frequency} = \frac{c}{\lambda} = \frac{3.0 \times 10^{10}}{\lambda}$$

$$\therefore \text{ Energy per mole} = \frac{6.02 \times 10^{23} \times 6.62 \times 10^{-27} \times 3 \times 10^{10}}{\lambda} \quad ...(A)$$

The energy per mole = 142.95 cal/mole

$= 142.95 \times 10^3$ cal/mole

$= 142.95 \times 10^3 \times 4.184 \times 10^7$ ergs/mole ...(B)

Equating eqs. (A) and (B), we get

$$\frac{6.02 \times 10^{23} \times 6.62 \times 10^{-27} \times 3 \times 10^{10}}{142.95 \times 10^3 \times 4.184 \times 10^7}$$

$= 142.95 \times 103 \times 4.184 \times 107$

or $$\lambda = \frac{6.02 \times 10^{23} \times 6.62 \times 10^{-27} \times 3 \times 10^{10}}{142.95 \times 10^3 \times 4.184 \times 10^7} \text{ cm}$$

$= 2000 \times 10^{-8}$ cm = **2000 Å.**

Example 5:

The dissociation energy of hydrogen is 102900 cal/mole. If H_2 is dissociated by illumination with radiation of wave-length 2537 Å. What fraction of the radiant energy will be converted in to kinetic energy ?

Solution:

Dissociation energy= 102900 cal/mole

$$= \frac{102900}{6.023 \times 10^{23}} = 1.708 \times 10^{-19} \text{ cal/molecule.}$$

Also, one quantum of light is able to dissociate one molecule of hydrogen. Therefore, Energy of one quantum $= h\nu = \frac{hc}{\lambda}$.

But $h = 6.625 \times 10^{-27}$ erg sec,

$c = 3 \times 10^{10}$ cm/sec

$\lambda = 2537$ Å $= 2537 \times 10^{-8}$ cm

$$\therefore \text{ Energy of one quantum} = \frac{6.625 \times 10^{-27} \times 3 \times 10^{10}}{2537 \times 10^{-8}} \text{ ergs}$$

$$= \frac{3 \times 6.625 \times 10^{-9}}{2537} \text{ ergs}$$

$$= \frac{3 \times 6.625 \times 10^{-9}}{2537 \times 4.18 \times 10^{7}} \text{ cal}$$

$[\because$ 1 cal 4.18 × 10^7 ergs]

But the dissociation energy per mole = 1.708 × 10^{-19} cal

$\therefore$ Energy converted into kinetic energy

$= (1.874 \times 10^{-19} - 1.708 \times 10^{-19})$ cal

$= 0.166 \times 10^{-19}$ cal

Hence, fraction converted into K.E. $= \frac{0.166 \times 10^{-19}}{1.874 \times 10^{-19}}$.

Example 6:

A substance when dissolved in water at 10^{-3} M concentration absorbs 15 per cent of an incident radiation in a path of 1 cm length. What should be the concentration of the solution in order to absorb 90 per cent of the same incident radiation ?

Solution:

First case. Concentration, c = 10–3 M, absorption = 10% = 0.1

or $I/I_0 = 1 - 0.1 = 0.9$, t = 1 cm

According to Beer's law, $I/I_0 = 10^{-act}$

or $0.90 = 10^{-a \times 10^{-3} \times 1}$ or a = 45.8

Second case. Concentration c = ?. Absorption = 90% = 0.90

$\therefore$ $I/I_0 = 1 - 0.90 = 0.10$, t = 1 cm

$\therefore$ $I/I_0 = 10$ act or $0.10 = 10^{-45.8 \times c \times 1}$

or c = 0.0218 M

6

KINETICS OF HOMOGENEOUS REACTIONS AND CATALYSIS

THEORY OF HOMOGENEOUS CATALYSIS

A chemical reaction to occur it is essential that the reacting substances must possess sufficient minimum energy called *activation energy*. In many cases it is not found to be so.

According to theory of homogeneous catalysis, the function of the catalyst is to bring about reaction between such; molecules which do not possess enough energy to enter into chemical combination by providing and alternative path in which lesser energy of activation is required.

We can thus compare the function of a catalyst to that of tunnel in crossing mountain. In order to cross a mountain at its full height through a tunnel at some lower height, the catalyst provides, thus, an alternative path to the reaction, requiring lower energy of activation. Thus, the catalyst can provide an alternative path, for a reaction to occur at lower activation energy (Fig. 6.1). The various postulates of the theory of homogeneous catalysis are :

(i) The catalyst first forms an intermediate compound with ;the reactant. This reactant with which catalyst combines is often termed as "substrate": when A is the substrate, X is the catalyst, AX is the intermediate compound and k_1 and k_2 are the velocity constants for the forward and backward reactions.

(ii) The intermediate compound then reacts with other reactant molecule (B) to form the product and catalyst.

$$AX + B \xrightarrow{k_3} AB + X$$

This reaction is slow and is the rate determining step. Thus

$$\text{Rate of reaction} = k_3\,[AX]\,[B] \qquad ...(1)$$

where k_3 is the velocity constant.

(iii) The catalyst X which is regenerated in last step may further under go steps (i) and (ii) to form more and more of the products. Thus, the rate of homogeneous catalytic reactions depends upon the concentration of catalyst X. It means that the rate of the reaction will increase if the concentration of catalyst is increased. This can be proved as follows :

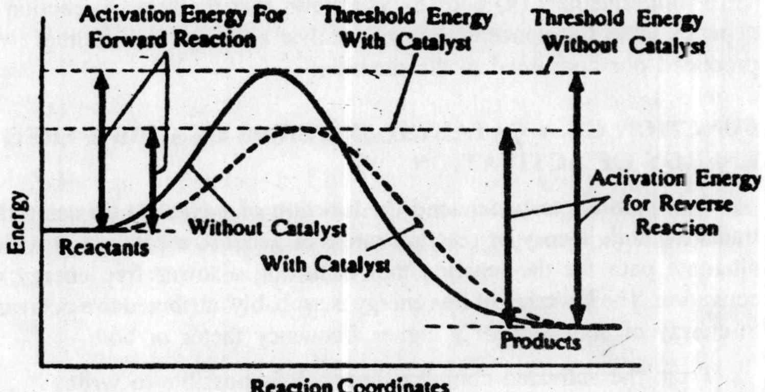

Fig. 6.1

The concentration of intermediate compound [AX], by using postulate of steady state concept, is given by

$$\frac{d[AX]}{dt} = k_1[A][X] - k_2[AX] - k_3[AX][B] = 0$$

or $$k_1\,[A]\,[X] = k_2\,[AX] + k_3\,[AX]\,[B]$$

or $$[AX] = \frac{k_1[A][X]}{k_2 + k_3[B]} \qquad ...(2)$$

Substituting this value of [AX] in equation (1), we get

$$\text{Rate of reaction} = \frac{k_1k_3[A][X][B]}{k_2 + k_3[B]} \quad ...(3)$$

From equation (3), two case may arise :

(i) If $k_2 << k_3$ [B] then Eq. (3) becomes as :

$$\text{Rate of reaction} = \frac{k_1k_3[A][X][B]}{k_3[B]} = k_1[A][X]$$

$$= k1\ [A]\ [X] \quad ...(4)$$

(ii) If $k_2 << k_3$ [B], then Eq. (3) becomes as

$$\text{Rate of reaction} = \frac{k_1k_3[A][X][B]}{k_2} \quad ...(5)$$

From equations (4) and (5), it follows that the rate of reaction depends upon the concentration of catalyst although it is neither produced nor consumed in the reaction.

FUNCTION OF A CATALYST IN TERMS OF GIBB'S FREE ENERGY OF ACTIVATION

It is possible to understand the function of a catalyst by using the transition state theory or reaction rates. In general, a catalyst provides alternate path for the reaction that is having a lower free energy of activation. The lowering of free energy is probably attributed to a decrease in energy of activation or a higher frequency factor or both.

From the activated complex theory, it is possible to write

$$k_f = \frac{k_BT}{h}\exp\left(-G_f{}^{o}/RT\right) \quad ...(i)$$

where k_f is the rate constant for the forward reaction, k_B is the Boltzmann constant and ΔG_f^o is the standard Gibb's free energy of activation.

In the presence of a catalyst, it is possible to write

$$k_f{}' = \frac{k_BT}{h}\exp\left(-\Delta G_f{}'^{o}/RT\right) \quad ...(ii)$$

where (') denotes a catalysed reaction.

From Fig. 6.1, it is evident that the free energy of activation gets lowered for both the forward and backward reactions without altering the overall free energy change of the reaction. This means that a catalyst is able to change the rates of both the forward and backward reaction. But ΔG^o does not get changed. Therefore, the equilibrium constant K will not get changed in the presence of the catalyst because ΔG^o is related to the equilibrium constant by the relation $-\Delta G^o = RT \text{ in } k$. It further reveals that a catalyst helps in attaining the equilibrium position rapidly but does not help in changing the relative proportion of reactants and products at equilibrium.

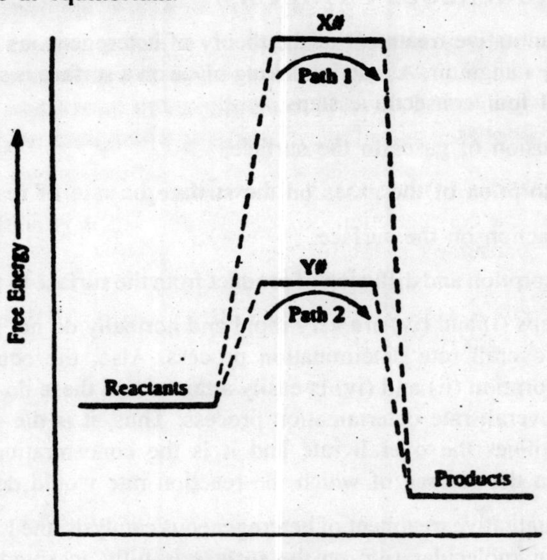

Fig. 6.2 : Lowering of Gibbs free energy of activation of a reaction by a catalyst. Path 1 is without catalyst; path 2 is with catalyst.

Mathematically it is possible to write the rate constants for the forward and backward reactions as follows:

$$k_f = (k_B T/h) \exp\left(- \Delta G_f'/RT\right) \quad \text{...(iii)}$$

$$k_b = (k_B T/h) \exp\left(- \Delta G_b'/RT\right) \quad \text{...(iv)}$$

On dividing Eq (iii) by Eq. (iv) we get

$$K_{eq} = \frac{k_f}{k_b} = \frac{(k_B T/h)\exp\left(-\Delta G_f'^{o}/RT\right)}{(k_B T/h)\exp\left(-\Delta G_b'^{o}/RT\right)}$$

$$= \exp[-(\Delta G_f'^{o})/RT - (-\Delta G_b'^{o})/RT]$$

$= \exp(-\Delta G^0/RT)$

where $\Delta G^o = (\Delta G_f'^o - \Delta G_b'^o)$

Since ΔG^o has the same value in the presence or absence of a catalyst, K_{eq} remains the same.

QUANTITATIVE TREATMENT OF ADSORPTION (THEORY OF HETEROGENEOUS CATALYSIS)

The quantitative treatment of the theory of heterogeneous catalysis was given by Langmuir. A reaction taking place on a surface is supposed to consist of four consecutive steps.

(i) Diffusion of gases to the surface,

(ii) Adsorption of the gases on the surface,

(iii) Reaction on the surface,

(iv) Desorption and diffusion of product from the surface to the bulk.

Here steps (i) and (iv) are very rapid and normally do not play any role in the overall rate determination process. Also, the equilibrium between adsorption (ii) and (iv) is easily attained and these do not take part in the overall rate determination process. Thus, it is the step (iii) which determines the overall rate and it is the concentration of the molecules on the surface of which the reaction rate would depend.

In the qualitative treatment of heterogeneous catalysis, the Langmuir concept of unimolecular film on the surface is fully accepted.

Irving and Langmuir's concept of unimolecular film on the surface satisfactorily explains the observed kinetics. Langmuir postulated that :

1. The gases adsorbed by a solid surface are not able to form a layer more than a single molecule in depth, *i.e.*, adsorbed gas is unimolecular in thickness.
2. There exists a dynamic equilibrium between the adsorbed gas and the gas in the bulk phase, *i.e.*, rate of condensation and rate of adsorption of molecules will be equal.

It is possible to obtain relationship between the fraction of the surface covered and the pressure of the gas at constant temperature mathematically using these ideas. Suppose

θ = fraction of the surface area covered by gas at any instant.

$(1 - \theta)$ = fraction of the bare surface available for adsorption.

P = pressure of the gas.

Now, when gas molecules colloide with unit area of the surface, the rate of condensation of molecules would be expected to be proportional to the pressure, P, and fraction of uncovered surface $(1 - \theta)$. Thus,

$$\text{Rate of adsorption} = k_1 (1 - \theta) P \quad ...(1)$$

$$\text{Rate of evaporation} = k_2\theta \quad ...(2)$$

where k1 and k_2 refer to the constants for a given system. At equilibrium, these two rates are equal. Then

$$k_1 (1 - \theta) P = k_2\theta$$

or

$$\theta = \frac{k_1 P}{k_2 + k_1 P} \quad ...(3)$$

$$= \frac{bP}{1 + bP} \quad ...(4)$$

where $b = k_1/k_2$ and is termed as the Langmuir adsorption isotherm. Graphical representation of Eq. (3) is shown in Fig. 6.3.

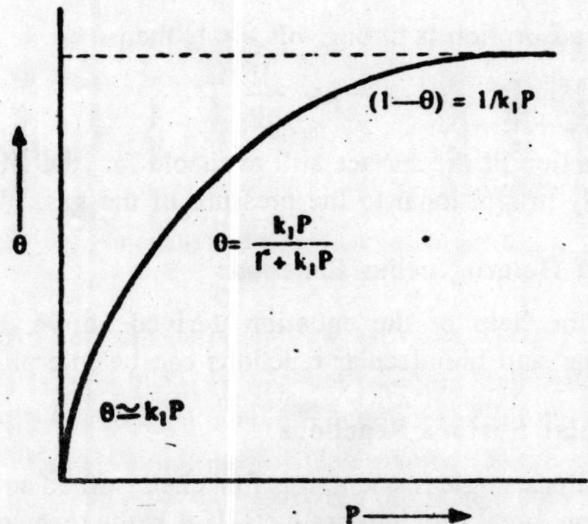

Fig. 6.3 : Langmuir adsorption isotherm.

Limiting cases of Eq. (4)

(a) If the gas is slightly adsorbed, *i.e.*, when adsorption is slight either due to very low pressure or due to the low-adsorption capacity of the surface, b is small, and bP may be neglected compared to unity. Hence, Eq. (4) gets reduced

$$\theta = bP$$

i.e., extent of adsorption is directly proportional to the pressure and the reaction will behave as one of the first order.

(b) If a gas is strongly adsorbed, the surface gets covered by a monomolecular thick layer of the gas, *i.e.*, when b or P or both are large, then bP >> 1, and Eq. (4) is transformed into

$$\theta = \frac{bP}{bP} = 1$$

Under these conditions, the reaction rate is constant or independent of pressure and the reaction is considered to be of kinetically zero order.

(c) Another deduction can be made from Eq. (4)

$$1 - \theta = 1 - \left(\frac{bP}{1 + bP}\right)$$

$$= \frac{1}{1 + bP}$$

If the adsorption is strong, bP >> 1, then

$$1 - \theta = \frac{1}{bP}$$

i.e., the fraction of the surface still available for the adsorption would be inversely proportional to the pressure of the gas.

Kinetics of Heterogeneous Reactions

With the help of the equation derived above, the kinetics of unimolecular and bimolecular reactions can be interpreted.

Unimolecular Surface Reactions

If there is a single reactant, it is first chemisorbed and subsequently, on activation, breaks up into products. If A is the reactant molecule and S the surface atom of the solid, the elementary processes may be depicted as follows :

$$A + S \underset{k_{-1}}{\overset{k_1}{\rightleftharpoons}} AS$$

$$AS \xrightarrow{k_2} \text{Products}$$

where AS refers to the adsorbed molecule.

Suppose, θ be the fraction of the surface covered by A at any instant t and pressure P. According to Langmuir-Hinshelwood the reaction rate should be :

$$\text{Rate} = r = k_2\theta \quad \text{...(5)}$$

If we assume a steady-state approximation for [AS], we get

$$\frac{d[AS]}{dt} = k_1[A][S] - k_2[AS] - k_2[AS] = 0$$

or
$$[AS] = \frac{k_1[A][S]}{k_{-1} + k_2} \quad \text{...(6)}$$

Now let concentration of vacant sites, $[S] = C_s(1 - \theta)$

and concentration of occupied sites, $[AS], = C_s$

where C_s refers to the total concentration of the surface sites of the catalyst. On substituting the values of [S] and [AS] in Eq. (6), we get

$$C_s\theta = \frac{k_1[A]C_s(1-\theta)}{k_{-1} + k_2}$$

or
$$\theta = \frac{k_1[A]}{k_1[A] + k_{-1} + k_2} \quad \text{...(7)}$$

and
$$r = \frac{k_1 k_2[A]}{k_1[A] + k_{-1} + k_2} \quad \text{...(8)}$$

On inverting expression (8), we get

$$\frac{1}{r} = \frac{1}{k_2} + \frac{k_{-1} + k_2}{k_2 k_1[A]} \quad \text{...(9)}$$

For the gaseous reactions, concentration term [A] can be replaced by partial pressures and Eq. (9) can be modified as :

$$\frac{1}{r} = \frac{1}{k_2} + \frac{k_{-1} + k_2}{k_2 k_1} \cdot \frac{1}{P_A} \quad \text{...(10)}$$

A plot of $1/r$ against $1/P_A$ would give a straight line having $1/k_2$ as the intercept and $\frac{k_{-1} + k_2}{k_2 k_1}$ as the slope.

Limiting Cases of Eq. (10)

Case I : In the rate expression

$$r = \frac{k_1 k_2 P_A}{k_1 P_A + k_{-1} + k_2}$$

when $k >> (k_1 P_A + k_{-1})$,

$$r = k_1 P_A \qquad ...(11)$$

On integration, Eq. (11) gives :

$$k_1 = \frac{2.303}{t} \log_{10}\left(\frac{P_i}{P}\right)$$

where, P_i refers to the initial pressure of A and P its pressure, at any time t. This equation is of first order with respect to concentration of A.'

Case II : If $k_2 << (k_1 P_A + k_{-1})$, the rate equation (10) becomes as follows:

$$r = \frac{k_1 k_2 P_A}{k_1 P_A + k_{-1}} = \frac{(k_1 / k_{-1}) k_2 P_A}{(k_1 / k_{-1}) P_A + 1} = \frac{k k_2 P_A}{k P_A + 1} \qquad ...(12)$$

where, K refers to adsorption equilibrium constant. Expression (12) which is identical to Langmuir adsorption isotherm can be analysed further, as follows :

(a) At low pressure, *i.e.*, $KP_A << 1$, thus

$$r = K.k_2 P_A \qquad ...(13)$$

and the reaction would be of first order with respect to A.

(b) At high pressure, *i.e.*, $KP_A >> 1$

$$r = k_2 \qquad ...(14)$$

and the reaction rate would be independent of pressure and the reaction is of zero order with respect to A.

Eqs. (13) and (14) may be explained by variation of rate of reaction with pressure as shown in Fig. 6.4

Bimolecular Surface Reaction

Now, consider a reaction in which two molecules A and B react on a surface and get adsorbed on neighbouring sites. The process may take place in two ways :

$$A + S \underset{k_{-1}}{\overset{k_1}{\rightleftharpoons}} AS$$

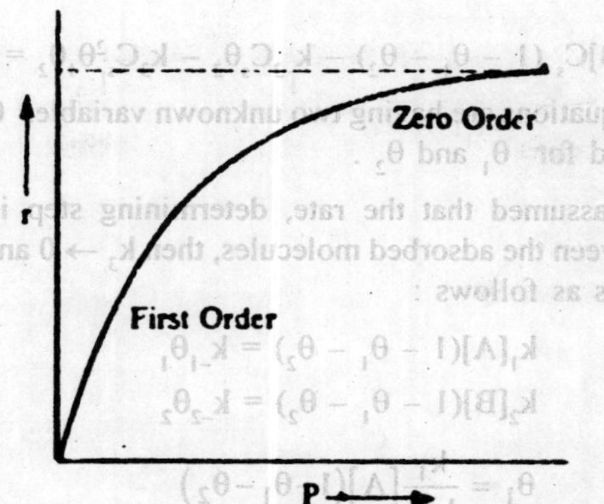

Fig. 6.4 : Variation of rate with pressure.

$$B + S \underset{k_{-2}}{\overset{k_2}{\rightleftharpoons}} BS$$

$$AS + BS \rightarrow \text{Products} + 2S$$

Suppose θ_1 and θ_2 be the fractions of the surface covered by adsorption of A and B, respectively. The fraction of the surface, which is vacant, is $(1 - \theta_1 - \theta_2)$.

The rate of formation of products is given as follows :

$$r = k_3\theta_1\theta_2 \qquad ...(15)$$

On applying steady-state approximation to [AS] and [BS], we obtain

$$\frac{d[AS]}{dt} = 0 = k_1[A][S] - k_{-1}[AS] - k_3[AS][BS] \qquad ...(16)$$

$$\frac{d[BS]}{dt} = 0 = k_2[B][S] - k_{-2}[BS] - k_3[AS][BS] \qquad ...(17)$$

If C_s refers to the total concentration of the surface sites, then we have

$$[AS] = C_s\theta_1$$

$$[BS] = C_s\theta_2$$

$$[S] = C_s(1 - \theta_1 - \theta_2)$$

On inserting [AS], [BS] and [S] in Eqs. (16) and (17), we get

$$k_1[A]C_s(1 - \theta_1 - \theta_2) - k_{-1}C_s\theta_1 - k_3C_s^2\theta_1\theta_2 = 0 \qquad ...(18)$$

$$k_2[B]C_s(1-\theta_1-\theta_2) - k_{-2}C_s\theta_2 - k_3C_s^2\theta_1\theta_2 = 0 \qquad ...(19)$$

These equations are having two unknown variables θ_1 and θ_2 hence, can be solved for θ_1 and θ_2 .

If it is assumed that the rate, determining step is the chemical reaction between the adsorbed molecules, then $k_3 \rightarrow 0$ and Eqs. (18) and (19) becomes as follows :

$$k_1[A](1-\theta_1-\theta_2) = k_{-1}\theta_1$$

$$k_2[B](1-\theta_1-\theta_2) = k_{-2}\theta_2$$

$$\theta_1 = \frac{k_1}{k_{-1}}[A](1-\theta_1-\theta_2)$$

$$= k_1[A](1-\theta_1-\theta_2) \qquad ...(20)$$

and $$\theta_2 = \frac{k_2}{k_{-2}}[B](1-\theta_1-\theta_2)$$

$$= k_2[B](1-\theta_1-\theta_2) \qquad ...(21)$$

where $$k_1 = \frac{k_1}{k_{-1}} \text{ and } k_2 = \frac{k_2}{k_{-2}}$$

On solving for θ_1 and θ_2 we obtain

$$\theta_1 = \frac{k_1[A]}{1+k_1[A]+k_2[B]}$$

and $$\theta_2 = \frac{k_2[B]}{1+k_1[A]+k_2[B]}$$

On inserting the values of θ_1 and θ_2 in Eq. (15), we obtain

$$r = k_3 = \frac{k_1k_2[A][B]}{\{1+k_1[A]+k_2[B]\}^2} \qquad ...(22)$$

When P_A and P_B denote the partial pressures of A and B for the gaseous reaction, Eq. (22) changes to

$$r = k_3 = \frac{k_1k_2P_AP_B}{1+k_1P_A+k_2P_B} \qquad ...(23)$$

On examining the expression (23), three special cases of important significance may arise.

Case I : If each gas (A and B) gets adsorbed very slightly. In such an even, $K_1P_A << 1$ and also $K_2P_B << 1$, so

$$r = k_3 k_1 k_2 P_A P_B \qquad ...(24)$$

This implies that the reaction will be of second order and first each with respect to A and B.

Examples of this type include the hydrogenation of ethylene on Cu or reaction between NO and O_2 on glass.

Case II : If one reactant, A, is relatively more strongly adsorbed than B.

Here, $K_1 P_A >> K_2 P_B$

Hence, $$r = \frac{k_3 k_1 k_2 P_A P_B}{(1 + k_1 P_A)^2}$$

i.e., the rate would be of first order with respect to B. But as the partial pressure of A increases, the rate increases to a maximum and then decreases.

Such complicated kinetics has been followed in the reaction between CO_2 and H_2 on platinum.

Case III : When one reactant, A, is very strongly adsorbed. In such cases, we get $K_1 P_A >> K_2 P_B$ and $k_1 P_A >> 1$.

Therefore, the reaction may be put as follows :

$$\frac{k_3 k_2 P_A P_B}{k_1 P_A{}^2} = \frac{k_3 k_2}{k_1} \cdot \frac{P_B}{P_A}$$

Thus, the rate is dependent strongly on the concentration of the strongly absorbed component.

The reaction between CO and O_2 on platinum follows such kinetics.

Retarded Reactions

Surface reactions sometimes become complicated by the adsorption of not only the reactants but also a product of the reaction. In fact, the product may get adsorbed more strongly than the reactant, thus decreasing the effective surface area available for the adsorption of reactant. This results in a retardation of the reaction rate. Suppose there is a reaction

$$A \rightarrow B + C$$

where, A refers to the reactant which gets weakly adsorbed and the product B gets strongly adsorbed. Then the fraction of the surface covered (θ_1) by A can be put as follows :

$$\theta_1 = \frac{K_A P_A}{1 + K_A P_A + K_B P_B}$$

The rate of reaction A will then be proportional to θ_1 and the pressure P_A.

Hence $$r = \frac{k_2 K_A P_A}{1 + K_A P_A + K_B P_B}$$

But when the product B gets very strongly adsorbed, then we have

$$r = \frac{k_2 K_A P_A}{K_B P_B}$$

i.e., rate would be directly proportional to the pressure of the reactant and inversely to the pressure of the product responsible for retardation.

Hinshelwood and Burk found that ammonia decomposes on platinum at 1138°C according to the reaction

$$2NH_3(g) \Leftrightarrow N_2(g) + 3H_2(g)$$

Nitrogen is having no effect on the reaction but the hydrogen strongly adsorbed and retards the reaction rate.

Effect to Temperature on Heterogeneous Reactions

Arrhenius equation, also applicable to heterogeneous reactions, may be given as follows :

$$\frac{d \ln k}{d T} = \frac{E_a}{RT^2}$$

The activation energy (E_a) for a heterogeneous reaction, evaluated from the plot of log k against l/T, is called apparent energy of activation. As the adsorption is temperature dependent, the evaluated E_a would be equal to the algebraic sum of true activation energy (E_t) and the heat of adsorption of reactions (λ_R) and products (λ_P) such that

$$E_a = E_t - \lambda_R + \lambda_P$$

The exact relation is dependent on the nature of kinetics.

Unimolecular Surface Reactions

Case I : If the reactant is very strongly adsorbed (Zero order surface reaction). Here $\theta \rightarrow 1$, and Eq. (5) becomes as follows :

$$\text{rate} = k_2$$

In this case the fraction of surface covered has been independent of the temperature and therefore,

$$E_a = E_t .$$

Case II : If the reactant is slightly adsorbed. Under the condition, *i.e.*, $KP_A << 1$

$$\text{Rate} = K.\ k_2 .\ P_A$$

From Arrhenius equation, we have

$$\frac{d \ln(K.k_2)}{d T} = \frac{E_a}{RT^2}$$

or
$$\frac{d \ln K}{d T} + \frac{d \ln k_2}{d T} = \frac{E_a}{RT^2}$$

But
$$\frac{d \ln K}{d T} = \frac{\Delta H_{ads}}{RT^2}$$

and
$$\frac{d \ln k_2}{d T} = \frac{\Delta E_t}{RT^2}$$

Therefore
$$\frac{E_a}{RT^2} = \frac{E_t}{RT^2} + \frac{\Delta E_{ads}}{RT^2}$$

As ΔH_{ads} is generally negative,

$$E_a < E_t$$

i.e., activation energy of this type of reaction gets lowered.

Bimolecular Surface Reactions

Case I : If both reactants are weakly adsorbed. Under the conditions, *i.e.*, $K_1\ P_A << 1$ and also $K_2P_B << 1$.

$$\text{Rate} = k_3\ K_1\ K_2\ P_A\ P_B$$

From Arrhenius equation, we have

$$\frac{d \ln(K_1 K_2 K_3)}{d T} = \frac{E_a}{RT^2}$$

or
$$\frac{d \ln K_1}{d T} + \frac{d \ln K_2}{d T} + \frac{d \ln K_3}{d T} = \frac{E_a}{RT^2}$$

or
$$\frac{(\Delta H_{ads})_1}{RT^2} + \frac{(\Delta H_{ads})_3}{RT^2} + \frac{E_t}{RT^2} = \frac{E_a}{RT^2}$$

$$E_a = E_t + (\Delta H_{ads})_1 + (\Delta H_{ads})_2$$

or $\quad E_a < E_t$, since ΔH_{ads} is generally negative.

Case II : If one of the reactants gets very strongly adsorbed. If, out of A and B, B is more strongly adsorbed, then we have

$$\text{rate} = \left(k_3 \frac{K_2}{K_1}\right) \frac{P_B}{P_a}$$

From Arrhenius equation, we have

$$\frac{d \ln\left(k_3 \frac{K_2}{K_1}\right)}{d T} = \frac{E_a}{RT^2}$$

or
$$\frac{d \ln K_2}{d T} - \frac{d \ln K_1}{d T} + \frac{d \ln K_3}{d T} = \frac{E_a}{RT^2}$$

or
$$\frac{(\Delta H_{ads})_2}{RT^2} - \frac{(\Delta H_{ads})_1}{RT^2} + \frac{E_t}{RT^2} = \frac{E_a}{RT^2}$$

or
$$E_a = E_t + (\Delta H_{ads})_2 + (\Delta H_{ads})_1$$

It means that the strongly adsorbed product increases the activation energy, thereby inhibiting the overall reactions.

ABSOLUTE RATE THEORY IN HETEROGENEOUS GAS REACTIONS

The absolute rate theory offers a very satisfactory mean to calculate the rates of heterogenous reactions involving the gases. Let us consider a bimolecular reaction between A and B involving a reaction centre, S, *e.g.*, an atom, on the surface. Then

$$A + B + S \Leftrightarrow [A - B - S] \neq \text{Products.} \qquad \text{...(1)}$$

In this type of reactions there exists an equilibrium between, the reactants (A and B) and the activated complex. If C_A, C_B and C_S denote the concentration of A, B and S respectively, the value of equilibrium constant $K \neq$ of the activated state is obtained by applying the law of mass action to Eq. (1).

$$K \neq = \frac{C \neq}{C_A C_B C_S} \qquad \text{...(2)}$$

where, C is the concentration of the activated complex. If we consider the partition functions, Eq. (2) becomes as

$$K^{\neq} = \frac{\theta^{\neq}}{\theta_A \theta_B \theta_S} e^{-E/RT} \quad ...(3)$$

where, E denotes the energy of activation for the heterogeneous reacting. On combining Eqs. (2) and (3), we get

$$\frac{C^{\neq}}{C_A C_B C_S} = \frac{\theta^{\neq}}{\theta_A \theta_B \theta_S} e^{-E/RT}$$

or $$C^{\neq} = \frac{1}{C_A C_B C_S} \frac{\theta^{\neq}}{\theta_A \theta_B \theta_S} e^{-E/RT} \quad ...(4)$$

The rate of the reaction dx/dt for bimolecular reaction is

$$\frac{dx}{dt} = C^{\neq} \frac{kT}{h} \quad ...(5)$$

Substituting Eq (4) in Eq. (5), we get

$$\frac{dx}{dt} = \frac{kT}{h} \cdot \frac{1}{C_A C_B C_S} \frac{\theta^{\neq}}{\theta_A \theta_B \theta_S} e^{-E/RT} \quad ...(6)$$

The rate of the reaction may also be given by

$$dx/dt = k_r C_A C_B \quad ...(7)$$

where, k_r is the specific rate constant. On comparing Eqs. (6) and (7), we get

$$k_r = C_s \frac{kT}{h} \frac{\theta^{\neq}}{\theta_A \theta_B \theta_S} e^{-E/RT} \quad ...(8)$$

The S part of the surface possesses no translational, no rotational but its only vibrational contribution is of the order of unity. It means that the partition function contains only the vibrational factor. Thus $\theta^{\neq}/\theta_S$ may be taken as unity. Hence Eq. (8) becomes as

$$k_r = C_s \frac{kT}{h} \frac{1}{\theta_A \theta_B} e^{-E/RT} \quad ...(9)$$

The partition functions θ_A and θ_B are determined by the known properties of the reactants. The value of C_S may be evaluated for a sparsely covered surface where it is equal to the number of atoms.

INTRODUCTION TO CATALYSIS

'*Catalysis*' was first used by Berzelius in 1836 to describe a number of experimental observations which included the discovery by *Thenard* that ammonia was decomposed by metals and by *Dobereiner* (1825) that manganese dioxide affected the rate of decomposition of potassium chlorate. Berzelius defined the catalysis as : "It is the phenomenon in which the presence of a foreign substance could accelerate its rate without being used up in that reaction."

He called the foreign substance as catalyst Later on, it was reported that catalyst could also retard the rate of reaction. Thus, the definition of Berzelius was generalised. The new definition of catalyst and catalysis are: "Catalyst is any substance which can change the speed of the reaction without being used up in that reaction and phenomenon is known as catalysis."

In many reactions, one of the products itself acts as a catalyst. An example of such reaction is the oxidation of oxalic acid by acidified potassium permanganate which may be represented as follows :

$$2KMnO_4 + 3H_2SO_4 + 5\begin{matrix}COOH\\COOH\end{matrix} \rightarrow$$

$$K_2SO_4 + 2MnSO_4 + 8H_2O + 10CO_2$$

The speed of reaction increases as the reaction progresses. This acceleration is due to the presence of Mn^{2+} ions which get formed in the reaction. This type of phenomenon in which one of the products itself acts as a catalyst is known as auto-catalysis.

TYPES OF CATALYSIS

These can be divided into two classes :

(a) Homogeneous Catalysis

In homogeneous catalysis, the catalyst and reactants are in this same phase. Some examples from gaseous and liquid phases are give below :

(i) *Example from Gaseous Phase :* In the lead chamber process for the manufacture of sulphuric acid, nitric oxide gas catalyses the reaction between SO_2 and O_2.

$$2SO_2 + O_2 \xrightarrow{NO} 2SO_3$$

In this example, the reactants (SO_2 and O_2) and the catalyst (NO) are in the gaseous phase.

(ii) *Example from Liquid Phase :* The inversion of cane sugar is catalysed by a mineral acid.

$$\underset{\text{cane sugar}}{C_{12}H_{22}O_{11}} + H_2O \xrightarrow{H+} \underset{\text{glucouse}}{C_6H_{12}O_6} + \underset{\text{fructose}}{C_6H_{12}O_6}.$$

Here the catalyst and reactants are in the liquid phase.

The hydrolysis of an ester like methyl acetate gets catalysed by hydrogen ions :

$$CH_3COOCH_3 + H_2O \xrightarrow{H+} CH_3COOH + CH_3OH$$

The conversion of acetone into diacetone alcohol is catalysed by hydroxyl ions.

$$\begin{matrix} CH_3 \\ CH_3 \end{matrix}> C=O + HCH_2COCH_3 \xrightarrow{OH^-}$$

$$\begin{matrix} CH_3 \\ CH_3 \end{matrix}> \underset{OH}{C}\; CH_2COCH_3$$

Diacetone alcohol

The decomposition of hydrogen peroxide is catalyzed in the presence of chloride ions.

$$2H_2O_2 \xrightarrow{Cl^-} 2H_2O + O_2$$

(b) Heterogeneous Catalysis

In heterogeneous catalysis, the catalyst is present in a different phase; than that of reactants. Examples are :

(i) *Heterogeneous Catalysis Involving Solid Reactants* : The example for this type is the decomposition of postassium chlorate in the presence of solid MnO_2 which acts as a catalyst.

$$2KClO_3 \xrightarrow{MnO_2} 2KCl + 3O_2$$

Thus, the catalyst (MnO_2) is present as a separate phase from the solid reactant ($KClO_3$).

(ii) *Heterogeneous Catalysis Involving Liquid Reactants* : The decomposition of hydrogen peroxide is catalysed by colloidal solution of gold and platinum.

$$2H_2O_3\ (l) \xrightarrow[\text{Au}]{\text{Pt or}} 2H_2O + O_2$$

Here H_2O_2 is a liquid reactant.

(iii) *Heterogeneous Catalysis Involving Gaseous Reactants* : Combination of SO_2 and O_2 in the presence of finely divided platinum is an example of this type :

$$SO_2(g) + O_2(g) \xrightarrow{Pt(s)} 2SO_3$$

Criteria of Characteristics of Catalyss : A catalysed reaction is found to possess the following characteristics which can also serve as criteria of catalysis.

Effect of the Equilibrium of a Reversible Reaction

In case of reversible reactions, the catalyst does not influence the composition of reaction mixture at equilibrium. Thus, it helps to attain the equilibrium quickly. In other words it affects the forward and backward reactions to the same extent and the value of the equilibrium constant remains unchanged.

This was proved by Bodenstein who showed that the use of a catalyst hastened the approach of equilibrium in the decomposition of hydrogen iodide,

$$2HI \rightleftharpoons H_2 + I_2$$

but did not change the concentrations of the reactants or products. For example, the use of platinised asbestos as a catalyst in the combination of sulphur dioxide and oxygen causes an appreciable increase in the rate of reaction but it is not able to increase the yield of sulphur dioxide under give conditions to temperature and pressure.

$$2SO_2 + O_2 \rightleftharpoons 2SO_3$$

Inability to Start a Reaction

According to Ostwald, a catalyst cannot start a reaction, but can only decrease its rate. This point since long has been a matter of controversy as there are some reactions in which it appears as if catalyst actually starts the reaction, *e.g.*, perfectly dry H_2 and O_2 do not combine to form water even if they are left in contact for years, but in presence of a little water (catalyst) the reaction proceeds quite rapidly. The reaction in the presence of a catalyst occurs through some alternative path which needs much lower energy of activation. Hence it is speeded up.

Specific Nature of a Catalyst

Action of a catalyst is highly specific in nature. That is, a particular substance can act as a catalyst only in a particular reaction and not in all. For example MnO_2 may catalyse the decomposition of $KClO_3$ but not of KCl_4. Highly specific action of a catalyst can be compared to the specific use of a key which can open a particular lock and not every lock (Emil Fischer). Enzymes have also specific catalytic actions. Transition metals such as iron, cobalt, nickel, platinum and palladium are able to catalyse reactions of various types.

Nature of the Products is Unaltered by the Presence of the Catalyst

H_2 and N_2 always combine to give NH_3 in the absence or presence of the catalyst. Similarly, SO_2 and O_2 will always combine to form SO_3 whether a catalyst is present or not. But this is not always true. Some exceptions are given below :

(i) CO and H_2 combine to give three different products when three different catalysts are used.

$$CO + 3H_2 \xrightarrow{Ni} CH_4 + H_2O$$

$$CO + H_2 \xrightarrow{Cu} HCHO$$

$$CO + 2H_2 \xrightarrow{ZnO + Cr_2O_3} CH_3OH$$

(ii) Similarly, chlorination of toluene in the presence of a halogen carrier such as iron or iodine and in absence of sunlight, takes place in the benzene ring. But chlorine is substituted in the side chain in the absence of a catalyst but in the presence of sunlight or at higher temperature.

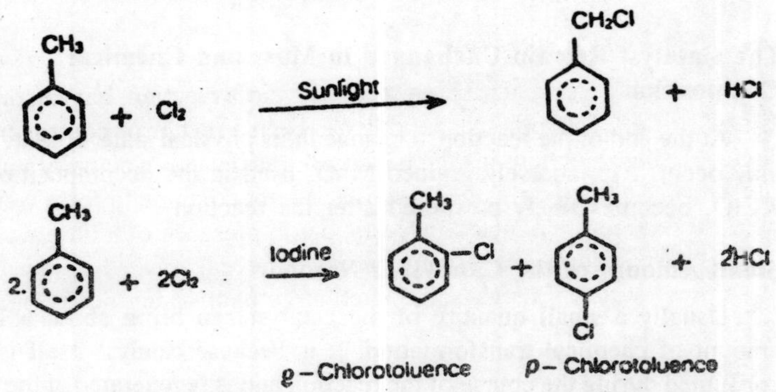

$$C_6H_5CH_3 + Cl_2 \xrightarrow{Sunlight} C_6H_5CH_2Cl + HCl$$

$$2.\ C_6H_5CH_3 + 2Cl_2 \xrightarrow{Iodine} o\text{-}ClC_6H_4CH_3 + p\text{-}ClC_6H_4CH_3 + 2HCl$$

g – Chlorotoluence p – Chlorotoluence

g– Chlorotoluence p – Chlorotoluence

Fig. 6.5

Optimum Temperature

There is a particular temperature at which the efficiency of a catalyst is most marked. The temperature is known as the optimum temperature. The activity of an enzyme which acts as a catalyst increases exponentially with temperature. If the temperature is raised sufficiently the enzymes are coagulated and lose their activity. It has been reported that activity of enzymes is maximum between 35° to 37° C.

Action of Promoters

The addition of small amounts of foreign substances which are not themselves catalytically active, sometimes increases the activity of the catalyst. Such substances are called promoters. In the manufacture of NH_3 by Haber's process, finely divided iron, acts as catalyst, while molybdenum acts as a promoter.

A Catalyst is Poisoned by Certain Substances

The activity of a catalyst is inhibited or completely destroyed by the presence of even minute traces of certain substances called catalytic poisons or anticatalysts. For example, in the manufacture of H_2SO_4 by the contact process, a trace of As_2O_3 destroys the catalytic efficiency of spongy platinum. However, vanadium pentoxide catalyst is preferred because it has been found to be less sensitive to poisoning. Similarly, traces of mercury are able to reduce the catalytic of copper for the combination of ethylene and hydrogen to form ethane.

$$\begin{matrix} CH_2 \\ CH_2 \end{matrix} + H_2 \rightarrow \begin{matrix} CH_3 \\ CH_3 \end{matrix}$$

The Catalyst Remain Unchanged in Mass and Chemical Composition

At the end of the reaction, a change in its physical state, colour, etc., may occur, *e.g.*, coarsely grained MnO_2 used in the decomposition of $KClO_3$ becomes finely powdered after the reaction.

Small Amount of the Catalyst is Needed

Usually a small quantity of the catalyst can bring about a large amount of chemical transformation. It is because catalyst itself is not consumed during the course of the reaction and is regenerated at the end. For example, 1 gm. of copper in 10^6 litres catalyses the oxidation of

$NaHSO_3$ by air. Another example is that a low concentration as one gm atom of colloidal platinum in 10^8 litres can catalyse decomposition of hydrogen peroxide. This is only true in the case of heterogeneous catalysis. However, in case of homogeneous reactions, the rate is proportional to the concentration of the catalyst, *e.g.*, in the inversion of cane sugar, dilute acids act as catalysts and the rate of inversion is proportional to the $[H^+]$ ion concentration.

In many heterogeneous reactions, the rate increases with increase in the area of a catalytic surface. This is able to explain why the efficiency of a solid catalyst gets increased when it is present in a finely divided state or deposited on some active material like asbestos (cf. platinised asbestos).

CLASSIFICATION OF CATALYSIS

Autocatalysis

The process in which one of the products of reaction acts as the catalyst is known as autocatalysis. A few examples of autocatalysis are :

(i) Hydrolysis of an ester is autocatalysed by the acid which is a product of the reaction.

$$RCOOR' + H_2O \Leftrightarrow RCOOH + R'OH$$

(ii) During titration of warm solution of oxalic acid by $KMnO_4$ solutions, the first few drops takes appreciable time before they are decolourised Since the reaction is initially very slow, but after some time the decolourisation goes fairly rapid as the Mn^{+2} ions, formed in the course of reaction, catalyse the reaction.

$$5C_2O_4^{2-} + 2MnO_4^- + 16h^+ \rightarrow 2mN^{2+} + 10CO_2 + 8H_2O$$

(iii) In the action of HNO_3 on copper, NO_2^- ions produced during the reaction act as catalyst.

$$3Cu + 8HNO_3 \rightarrow 3Cu(NO_3)_2 + H_2O + 2NO$$

Nitric oxide formed in the above reaction dissolves in water in the presence of air to provide NO_2^- ions.

Negative Catalysis or Inhibition

Examples are known in which the presence of a catalyst decreases or retards the rate of the chemical reaction. Such substances are called

Negative Catalysts or inhibitors and the phenomenon is called Negative Catalysis or Inhibition. Some examples are :

1. Presence of 1% alcohol retards the oxidation of chloroform into phosgene.

$$4CHCl_3 + 3O_2 \xrightarrow{C_2H_5OH} 4COCl_2 + 2Cl_2 + 2H_2O$$

2. The decomposition of H_2O_2 is retarded by the presence of acetanilide or H_2SO_4 or H_3PO_4.
3. Presence of lead tetra ethyl, nickel carbonyl 1 serves as antiknock material in internal combustion engines.

Explanation : Two possible explanations are available to describe the action of an inhibitor :

(i) Inhibitor deactivates or destroys some other foreign substances present in the reacting system as a positive catalyst, and thus does not involve directly in the reaction.

(ii) In some cases an inhibitor operates by dislocating a step in a multistep reaction. For example, addition of phosphate to a solution of H_2O_2 inhibits its decomposition by combining with existing traces of ferric ions which catalyse the decomposition of H_2O_2.

Catalytic Poisons or Anticatalysts

Certain substances when present in the reactants decrease the efficiency of the catalyst. These substances are called catalytic poisons or anticatalysts. Such an effect is observed with a solid catalyst and is found to be of two types :

(a) *Temporary Poisoning* : If the catalyst-regains its activity when the substance responsible for its poisoning is eliminated from the reactants, the poisoning is regarded as temporary. Presence of water vapour or oxygen acts as a temporary poison for iron, the catalyst used in manufacture of ammonia :

(b) *Permanent Poisoning* : If the poisoned catalyst fails to regain its activity even when the catalytic poisons are subsequently removed, the poisoning of the catalyst is said to be permanent. Examples of permanent poisons are given below in the tabular form :

Table 6.1

Name of Poison	*Name of the catalyst*	*Process*
CO	*Iron*	*Haber's process*
AS_2O_3	Platinised asbestos	Contact process
HCN	Colloidal platinum	Decomposition of H_2O_2
Bromine vapour	Nickel	Hydrogenation of o ils

Action of the poisons : It is probably due to the preferential absorption of poison on the active spots of the catalyst and thus reducing the number of free active spots available for the absorption of the reacting molecules. For example, the poisoning of platinum catalyst by CO takes place as shown in Fig. 6.9.

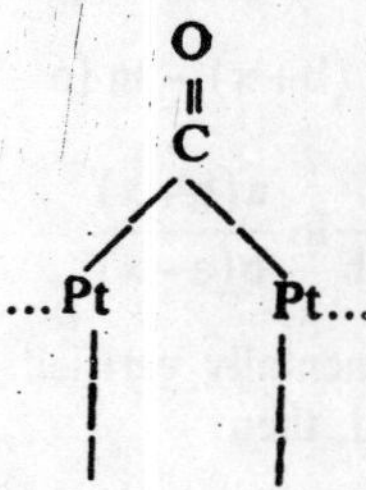

Fig. 6.6

Mathematical Expression of Autocatalytic Reactions

Let us deduce this expression by considering the hydrolysis of methyl acetate in which the product acetic acid formed catalyses the reaction.

$$\underset{\substack{a \\ a-x}}{CH_3COOC_2H_5} + H_2O \Leftrightarrow \underset{\substack{b \\ b+x}}{CH_3COOH} + C_2H_5OH \quad \begin{matrix}\text{Initial} \\ \text{After time t}\end{matrix}$$

Suppose, a and b are the initial amount of ester and acetic acid respectively. Suppose, x the extent of reaction, according to the law of mass action, is given by

$$\frac{dx}{dt} - kb(a-x) + kx(a-x) = k(a-x)(b+x)$$

or
$$\frac{dx}{(a-x)(b+x)} = k\,dt$$

On separating the variables, we get

$$\frac{1}{a+b}\left[\frac{1}{a-x}+\frac{1}{b+x}\right]dx = k\ dt$$

On integration, we get

$$\frac{1}{a+b}\left[\ln(b+x) - \ln(a-x)\right] = kt + C \qquad ...(2)$$

In order to get the value of C,

put x = 0 when t = 0, then

$$C = \frac{1}{a+b}\ln\frac{b}{a} \qquad ...(3)$$

On substituting eq. (3) in (2), we get

$$\frac{1}{a+b}\left[\ln(b+x) - \ln(a-x)\right] = kt + \frac{1}{a+b}\ln\frac{b}{a}$$

or

$$k = \frac{1}{a+b}\ln\frac{a(b+x)}{b(a-x)} \qquad ...(4)$$

This has been experimentally verified. If the initial catalyst be an acid other than acetic acid, then

$$dx/dt = k_1 (a - x)\, b + k_2 (a - x)\, x$$

$$= (k_1 b + k_2 x)(a - x)$$

This on integration gives,

$$k_1 b + k_2 a = \frac{1}{t}\ln\frac{a(k_1 b + k_2 x)}{k_1 b(b-x)} \qquad ...(5)$$

Equation (5) is also verified experimentally.

Catalytic Promoters or Activators

There is a group of substances which are not catalysts by themselves but when present along with some catalyst greatly enhance their action. They are called catalytic promoters or activators. Some examples of promoters are given below :

Action of Promoters : It is explained by assuming that a loose compound is formed between the catalyst and the promoter, which possesses an increased absorption capacity than that of pure catalyst only.

Table 6.2

Name of promoters	*Name of the Catalyst*	*Process*
Molybdenum	Iron	Manufacture of NH_3 by Haber's process
Copper	Iron (finely divided)	Bosch process for H_2 from water gas.
HCl or H_2O	$AlCl_3$	Isomerisation of hydrocarbons.

Enzyme Catalysis.

Enzymes are complex organic compounds which are produced by living plants and animals. They disperse in water forming colloidal solutions and are very specific catalysts. As enzymes are having the dimensions in the colloidal range (1000 – 10000 A°) and their kinetic behaviour is similar to that of heterogeneous process, enzyme catalysis has been referred to as "Microheterogeneous catalysis". Each enzyme can catalyze a specific reaction. For instance, the enzyme diastase produced in the germinated barley seeds is able to convert starch into maltose sugar.

$$\underset{\text{Starch}}{2(C_6H_{10}O_5)_N} + nH_2O \xrightarrow{\text{Diastase}} \underset{\text{Maltose}}{nC_{12}H_{22}O_{11}}$$

Another enzyme known as maltase is able to convert maltose into glucose.

$$\underset{\text{Maltose}}{C_{12}H_{22}O_{11}} + H_2O \xrightarrow{\text{Maltase}} \underset{\text{Glucose}}{2C_6H_{12}O_6}$$

The enzyme known as zymase, produced by living yeast calls, is able to convert glucose into ethyl alcohol.

$$\underset{\text{Glucose}}{C_6H_{12}O_6} \xrightarrow{\text{Zymase}} \underset{\text{Ethyl alcohol}}{2C_2H_5OH} + 2CO_2$$

The enzyme urease present in soyabeans is able to cause quantitative hydrolytic decomposition of urea into ammonia and carbon dioxide.

$$\underset{\text{Urea}}{NH_2CONH_2} + H_2O \xrightarrow{\text{Urease}} 2NH_3 + CO_2$$

Point of differences between enzyme catalysis and general heterogeneous catalysis. The main points are :

(i) *Nature of Catalysis* : Enzymes are non-lying complex nitrogeneous compounds produced by living organisms. They cannot be synthesised by artificial means in the laboratory. On the other hand, the general heterogeneous catalyst may be anything occurring naturally or produced artificially.

(ii) *Enzyme is Consumed During the Reaction* : Enzymes are consumed during the reaction which they catalyse. On the other hand, the general inorganic catalysis remains unchanged chemically during the reaction process.

(iii) *Presence of Certain Inorganic Substances* : Certain inorganic substances such as ammonium salts are needed as food for the enzyme producing. No substances are required for the general type of catalysts.

(iv) *Presence of Co-enzymes* : Co-enzymes are substances which are similar in nature to the enzymes. The presence of coenzyme is essential for the action of the enzyme. In the ordinary heterogeneous catalysis the presence of a promoter is not always essential. A few enzymes, their source and the reactions catalysed by them are shown below :

Table 6.3

Enzymes	*Sources*	*Reaction Catalysed*
Zymase	Yeast	Conversion of glucose into alcohol
Lactic bacilli	Curd	Conversion of milk into curd.
Lipase	Castor oil seed	Conversion of cane-sugar into glycerol.

Characteristics of Enzyme Catalysis

1. All enzymes are proteins and share common properties, They form colloidal solutions and are of high molecular weight.
2. Enzymes are associated with every chemical reaction that occurs in the living system.
3. Enzymes generally accelerate biochemical reactions by reducing the energy requirement (activation energy).
4. Enzymes do not alter the amount or nature of the product.

5. Enzymes do not affect the amount of energy released or absorbed during the reaction.
6. Enzymes retain identity at the end of the reaction, as in the beginning. In other words, they are not destroyed by the reactions they catalyze and so can be used again. However, a given molecule of an enzyme cannot be used indefinitely because it is readily inactivated by heat or action of acid, and so no. Conversely, inorganic catalysis are highly stable and can be used over and over again.
7. Extremely small amounts of enzymes are able to bring about measurable changes.
8. Enzymes catalyze biochemical reactions at significantly lower temperatures. For example, to hydrolyze 5 ml of 1 % starch solution, it is essential to add five drops of concentrated sulphuric acid and boil the mixture for over fifteen minutes. On the other hand, even 100 times diluted saliva containing salivary amylase would require just 8-10 minutes to hydrolyze the same quantity of starch at only 37°C.
9. Enzymes are specific in their action, *i.e.*, they are capable of acting on a predetermined substrate. An enzyme that can hydrolyze starch is unable to hydrolyze cellulose. For this, another enzyme, cellulose is needed. The degree of specificity, however, varies. Most intracellular enzymes are specific, whereas certain digestive enzymes work on a comparatively wide range of related compounds.
10. Most enzymes can work in either direction, *i.e.*, they are capable of operating reversibly. In other words, one and the same enzyme can catalyze the breakdown as well as the synthesis of the substance. However, in many instances, the reversibility of reaction has not yet been demonstrated.

Factors Governing the Rate of an Enzyme Reaction

1. *Effect of Temperature on Enzyme Catalysis :* Similar to chemical catalysts, the enzyme catalyst is able to decrease the activation energy of a reaction at a given temperature. In reality, the decrease of activation energy by an enzyme catalyst is far greater than that by a non-enzyme catalyst. It is known that the rate of a reaction generally increases with

an increase in temperature. However, this condition is highly unfavourable for a living call. Enzymes are, in fact, very sensitive to high temperatures. Due to the proteinous nature of an enzyme, increase in temperature brings about denaturation of the enzyme protein which gives rise to a decrease in effective concentration of the enzyme and hence a decrease in reaction rate. Upto about 45°C, the reaction rates of enzyme-catalysed reactions get increased with temperature. At temperatures greater than 45°C, thermal denaturation of enzyme becomes increasingly significant. At about 55°C, rapid denaturation completely destroys the catalytic function of the enzyme protein. The effect of temperature on the rates of enzyme-catalyzed reactions has been shown in (Fig. 6.7, 6.8).

Effect of pH. Enzymes are sensitive to change in pH (hydrogen ion concentration) in the reaction medium (Fig. 6.8). In fact, every enzyme works best at a specific pH called the optimal pH. Most enzymes display maximum activity at or around neutral pH. Excessive acidity or alkalinity renders them inactive. However, a few digestive enzymes operate either in distinctly acidic or alkaline medium. For instance, the stomach enzyme, pepsin, functions best at a pH of 2.0, where as the pancreatic enzyme, trypsin, has an optimal pH of 8.0.

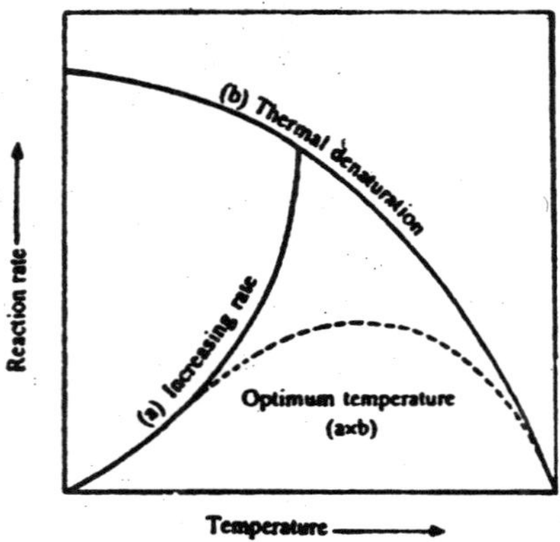

Fig. 6.7 : Temperature-dependence of the rate of an enzyme catalyzed reaction. (a) represents the increase in rate with increase in temperature. (b) represents the decrease in the rate as a function of thermal denaturation of the enzyme; the dashed line curve shows the combination of (a × b) .

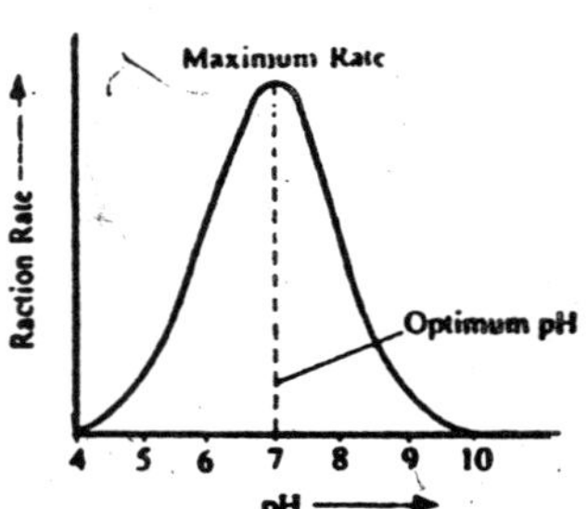

Fig. 6.8 : Effect of pH on the reactions rate of enzyme.

Effect of activator. Many enzymes cannot act on their own and require the help of some substances for their activation and for speeding up the rate of reaction (Fig. 6.9). Such substances are called activators. This activation is due to the effect of certain ions such as Na^+, K^+, Ca^{++}, Mg^{++}, Mn^{++}. You are now with the stomach enzyme pepsin. This is produced in the inactive form as pepsinogen by one type of cells in the gastric glands. Pepsinogen is converted into the active form, pepsin by HCl secreted by certain other cells of the glands. Similarly, when pancreatic juice containing the inactive trypsinogen is poured into the duodenum, it comes in contact with another enzyme, enterokinase, secreted by the mucosal layer of the duodenum, which in turn, converts trypsinogen into the active form, trypsin.

Effect of inhibitors. Certain substances called inhibitors slow down the rate of enzymatic reaction (Fig. 6.10). For example, mercury ;and arsenic ions inhibit the activity of a wide range of enzymes. Also, some antibiotics are effective inhibitors for bacterial growth since they hamper their protein synthesis.

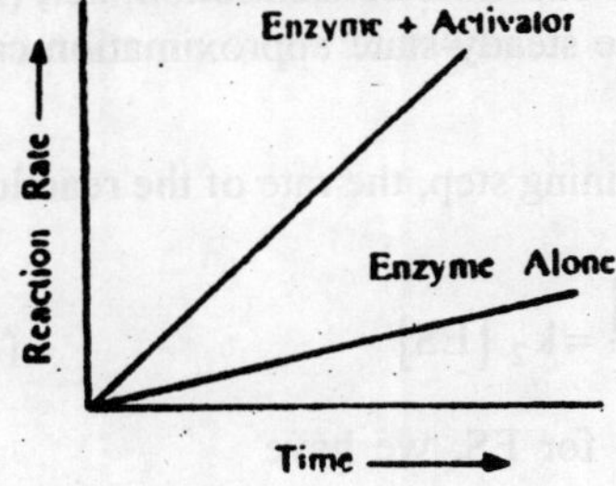

Fig. 6.9 : Effect of activators on the reaction rate of enzyme.

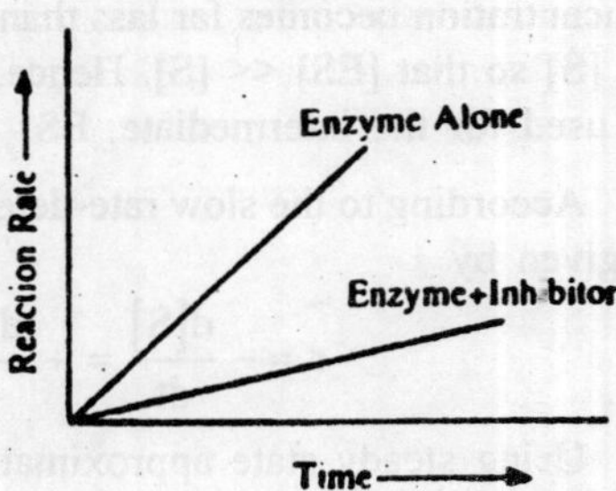

Fig. 6.10 : Effect of inhibitor on the reactions rate of enzymes.

Concentration : The rate of reaction is approximately proportional to the cocentration of enzyme.

Mechanism and Kinetics of Enzyme-Catalyzed Reactions

Biochemists L. Michaelis and M. Menten proposed, in 1913, a mechanism for the kinetics of enzyme-catalyzed reactions which envisages the following steps :

(i) *Step (1) :* The enzyme combines with the reactant (substrate) to form an enzyme substrate complex which remains in equilibrium with the enzyme and substrate.

$$E + S \underset{K_{-1}}{\overset{k_1}{\Leftrightarrow}} ES$$

where, E is the enzyme, S is the substrate, and ES is an enzyme substrate complex.

(ii) *Step (2)* : The enzyme-substrate, complex, ES can decompose to form products with simultaneous regeneration of the enzyme.

$$ES \xrightarrow{k_2} E + P$$

where, P is the product. In the overall reaction S → P the enzyme is consumed in the step (1) and regenerated in step (2).

The problem can be tackled using either the equilibrium approximation or the steady-state approximation. Experiment shows, however, that true equilibrium would not be achieved in the fast step because the subsequent slow reaction would be constantly removing the intermediate enzyme-substrate complex. ES. Generally, the enzyme concentration becomes far lass than the substrate concentration, *i.e.*, [E] << [S] so that [ES] << [S]. Hence, the steady-state approximation can be used for the intermediate, ES.

According to the slow rate-determining step, the rate of the reaction is given by

$$r = -\frac{d[S]}{dt} = +\frac{d[P]}{dt} = k_2\,[ES] \qquad ...(1)$$

Using steady state approximation for ES, we have

$$d[ES]/dt = k_1\,[E][S] - k_{-1}\,[ES] - k_2\,[ES] = 0 \qquad ...(2)$$

Now, [E] cannot be experimentally measured. The equilibrium between the free and bound enzyme is given by the enzyme conservation equation, *viz.*,

$$[E]_0 = [E] + [ES] \qquad ...(3)$$

where, $[E]_0$ refers to the total enzyme concentration (which can be measured); [E] refers to the free enzyme concentration and [ES] refers to the bound (or reacted) enzyme concentration. Hence

$$[E] = [E]_0 - [ES] \qquad ...(4)$$

On substituting for [E] in Eq. (2), we get

$$d[ES]/dt = k_1\,\{[E]_0 - [ES]\}[S] - k_{-1}\,[ES] - k_2\,[ES] = 0 \qquad ...(5)$$

Collecting terms and simplifying, we get

$$k_1 [E]_0 [S] = \{k_{-1} + k_2 + k_1 [S]\} [ES] \quad ...(6)$$

$$[ES] = \frac{k_1 [E]_0 [S]}{k_{-1} + k_2 + k_1 [S]} \quad ..(7)$$

On substituting for [ES] in Eq. (1), we get

$$r = \frac{k_1 k_2 [E]_0 [S]}{k_{-1} + k_2 + k_1 [S]} \quad ...(8)$$

On dividing the numerator and the denominator by k_1, we get

$$r = \frac{k_2 [E]_0 [S]}{(k_{-1} + k_2)/ k_1 + [S]} = \frac{k_2 [E]_0 [S]}{k_m + [S]} \quad ...(9)$$

where, the new constant K_m, called the Michaelis constant, is given as follows :

$$k_m = (k_{-1} + k_2)/k_1 \quad ..(10)$$

It is to be noted that k_m is not an equilibrium constant. Eq. (9) is known as Michaelis-Menten equation.

Further simplifications of Eq (9) can be made. If it is assumed that all the enzyme has reacted with the substrate at high concentrations, the reaction will be going at maximum rate. No free enzyme will remain so that $[E]_0 = [ES]$. Hence, from Eq. (1) we get

$$r_{maximum} \equiv V_{max} = k_2 [E]_0 \quad ...(11)$$

where V_{max} refers to the maximum rate, using the notation of enzymology.

The Michaelis-Menten equation can now be written as follows.

$$r = V_{max} [S]/(k_m + [S] \quad ...(12)$$

Two cases may arise :

(a) If $k_m >> [S]$, [S] can be neglected in the denominator of Eq. (12), giving

$$r = V_{max} [S]/k_m = k' [S]$$

(first-order reaction) ...(13)

(b) If $[S] >> k_m$, K_m can be neglected in the denominator, giving

$$r = V_{max} = \text{constant}$$

(zero-order reaction) ...(14)

These two cases may be shown diagrammatically in Fig. (6.5)

Again, if, $k_m = [S]$, $r = 1/2\ V_{max}$

Hence, Michaelis constant may be defined as equal to that concentration of S at which the rate of formation of product becomes half the maximum rate obtained at high concentration of S.

The constant k_2 in Eq. (11) is termed as the turnover number of the enzyme. The turnover number may be defined as the number of molecules converted in unit time by one molecule of enzyme. Typical values of k_2 are 100 – 1,000 per second though may be as large as 10^5 to 10^6 per second.

Now one must know the physical reason why the reaction rate of an enzyme-catalyzed reaction gets changed from first-order to zero-order as the substrate concentration gets increased. The answer to that question is that each enzyme molecule is having one or more 'active sites' at which the substrate must be bound in order that the catalytic action may take place. At low substrate concentrations, most of these active sites remain unoccupied at any time. As the substrate concentration gets increased, the number of active sites which get occupied increases and hence the reaction rate also get increased. However, at very high substrate concentrations, virtually all the active sites get occupied at any time so that further increase in substrate concentration will not further increase the formation of enzyme-substrate complex.

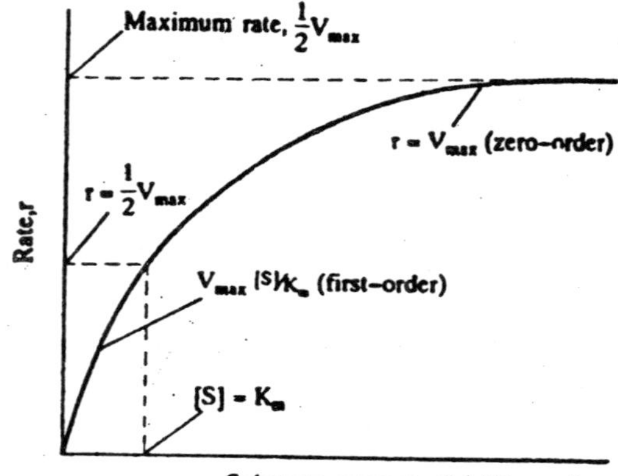

Fig . 6.11 : Kinetics of an enzyme catalysed reaction.

It rather becomes difficult to determine V_{max} (and hence k_m) directly from the plot of r against [S]. However it becomes possible to rearrange Eq. (12) so as to permit some alternative plots for easy determination of V_{max}. Two of the best known methods which make use of the re-arranged equations are described as follows.

1. *The Lineweaver-Burk Method* : This method employs the rearranged equation

$$\frac{1}{r} = \frac{K_m}{[S]\,V_{max}} + \frac{1}{V_{max}} \qquad ...(16)$$

A plot of 1/r against 1/[S] will give a straight line whose intercepts on the x and y-axes are $(-1/k_m)$ and $1/V_{max}$, respectively and whose slope is (k_m/V_{max}).

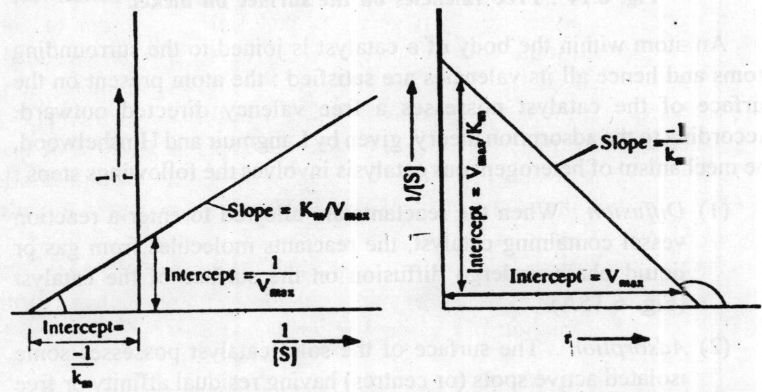

Fig. 6.12 : The Lineweaver-Burk method of plotting enzyme kinetic data.

Fig. 6.13 : Eadie-Hofstee method of plotting enzyme kinetic data.

2. *The Eadie-Hofstee Method* : The method employs the rearranged equation

$$r/[S] = V_{max}/K_m - r/k_m \qquad ...(17)$$

A plot of r/[S] against r would give a straight line with slope equal to $-1/k_m$ and an intercept on the y-axis equal to V_{max}/k_m. From the graph both k_m and V_{max} can be calculated.

THEORY OF HETEROGENEOUS CATALYSIS

The theory of heterogeneous catalysis is based upon the phenomenon of adsorption. The adsorption or contact theory was postulated by Faraday

(1833) and later revived by many others. It explains the action of *heterogeneous catalysis*. The action of a heterogeneous catalyst is due to the presence of *free valencies* on its surface. These free valencies offer an opportunity to the reactant molecules to undergo chemical reaction on the surface of the catalyst. The situation may be readily understood by the diagram shown in Fig. 6.14.

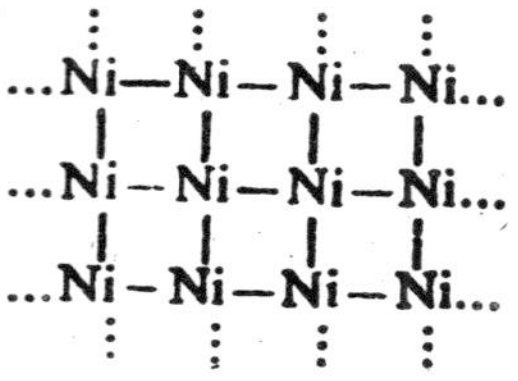

```
  :    :    :    :
...Ni—Ni—Ni—Ni...
   |    |    |    |
...Ni—Ni—Ni—Ni...
   |    |    |    |
...Ni—Ni—Ni—Ni...
  :    :    :    :
```

Fig. 6.14 : Free valencies on the surface on nickel.

An atom within the body of a catalyst is joined to the surrounding atoms and hence all its valencies are satisfied : the atom present on the surface of the catalyst possesses a free valency directed outward. According to the adsorption theory, given by Langmuir and Hinshelwood, the mechanism of heterogeneous catalysis involves the followings steps :

(1) *Diffusion* : When the reactants are allowed to enter a reaction vessel containing catalyst, the reactants molecules from gas or liquid phase undergo diffusion on the surface of the catalyst (Fig. 6.15A).

(2) *Adsorption* : The surface of the solid catalyst possesses some isolated active spots (or centres) having residual affinity or free unsatisfied valency forces. Due to these unsatisfied valency forces on the catalyst surface, the molecules of the gaseous reactants get adsorbed to form layers of unimolecular thickness (Fig. 6.15B).

Adsorption leads to higher concentrations of the adsorbed reactants on the surface of the catalyst and this, by the law of mass action, should give an enhanced rate of reaction. Best situation is when all the reactants are equally adsorbed. Adsorption being an exothermic process, the heat of adsorption decreases the need of energy of activation.

(3) *Formation of Activated Complex* : Due to the close proximity between adsorbed molecules on the catalyst surface, they undergo interaction to form an activated complex, (Fig. 6.15C). Whenever

an activated complex is formed, it is accompanied by an increase in its energy.

(4) *Decomposition of Activated Complex :* As soon as the activated complex is formed, it starts decomposing to give the products and the free surface of the catalyst is regenerated (see Fig. 6.15D).

(5) *Diffusion of the Products :* The molecules of reaction products diffuse away from the catalyst surface. The heat of adsorption helps in this connection. Now, the fresh molecules may come on the surface of catalyst and react to form the desired products.

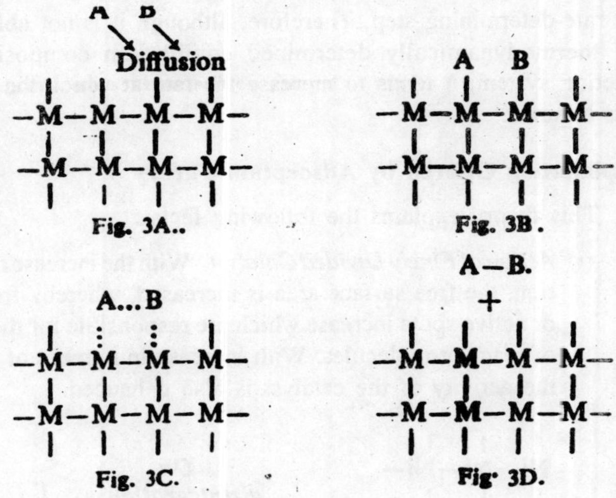

Fig. 6.15

The steps (1), (2) (3), (4), and (5) are consecutive. If any one of these steps is having a much slower rate constant than the others, it will become the rate determining step. Steps (1) and (5) are generally very fast. Further, steps (2) and (4) are usually faster than step 3 though they may sometimes be slower. It is generally accepted that the kinetics of surface reactions can be treated successfully on the basis of the assumptions given below :

1. The rate-determining step refers to the chemical reaction at the surface, *i.e.*, reaction of the adsorbed molecules on the surface, *i.e.*, step 3 given above.

2. Chemisorption plays a significant role in heterogeneous catalysis. In chemisorption chemical bonds get formed between the adsorbate and the surface resulting in a monolayer (Langmuir adsorption).

3. The reaction-rate per unit surface area has been proportional to q, the fraction of the surface covered. The value of q is given by the Langmuir adsorption isotherm.

It is to be remembered that the reason for heterogeneous catalytic activity remains the same as for homogeneous catalysis, *i.e.*, the catalyst increases the rate of the reaction by lowering the activation energy of the rate-determining step. Therefore, although it is not able to disturb the thermodynamically determined equilibrium composition of the reaction system, it tends to increase the rate at which the equilibrium gets attained.

Explanation Offered by Adsorption Theory

This theory explains the following facts :

(i) *Action of Finely Divided Catalyst :* With the increase of disintegration, the free surface area is increased, whereby free valencies or active spots increase which are responsible for the adsorption of reactant molecules. With increase in number of active spots the activity of the catalyst is also enhanced.

```
 |    |    |
—Ni—Ni—Ni—                On
 |    |    |          disintegration       [   |   ]
—Ni—Ni—Ni—          ——————————————→      9 [ —Ni— ]
 |    |    |                               [   |   ]
—Ni—Ni—Ni—
 |    |    |
```

Total Valencies = 12 **Tolal Valencies $9 \times 4 = 36$**

Fig. 6.16

(ii) *Action of Promoters :* Promoters themselves get adsorbed on the surface of the catalyst producing discontinuity or unevenness on the surface, thereby increasing the number of active centres. The increase in the number of active centres results in greater adsorption of the reactant molecules on the interface between the promoter and the catalyst where free valencies are crowded.

Increased adsorption means greater concentration, hence higher rate of the reaction.

(iii) *Action of Poisons :* It is probably due to the preferential adsorption of poisons on the active spots of the catalyst and thus reducing the number of free active spots available for the adsorption of reacting molecules.

(iv) *Specific Action of the Catalyst :* The specific action of the catalyst is due to the fact that the extent of adsorption of the reactant molecule son the catalyst surface and their subsequent conversion into products, depend upon the chemical affinity of the catalyst for the reactant molecules.